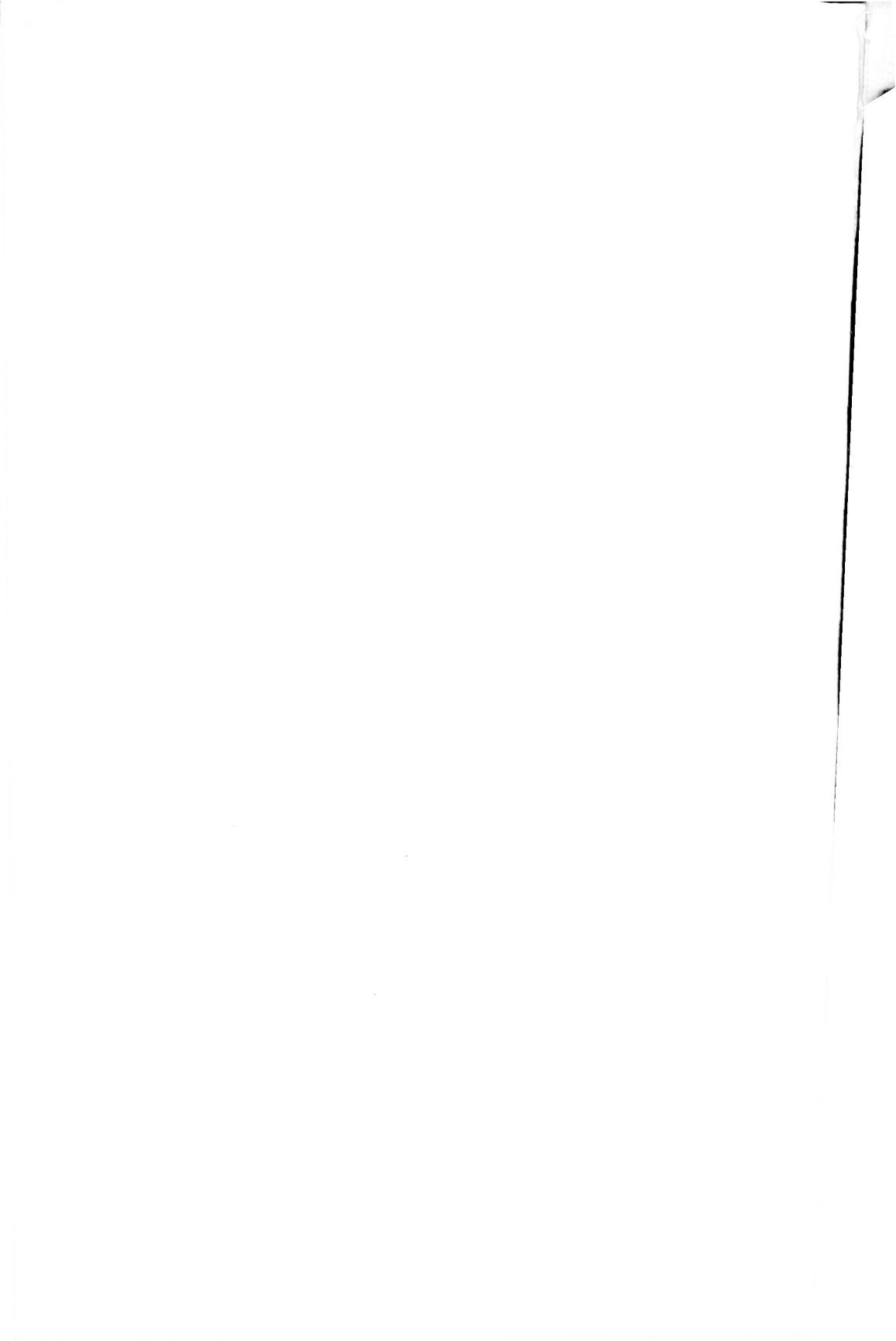

Handbook of Geodetic Science

Handbook of Geodetic Science

Edited by **Russell Sands**

R CALLISTO REFERENCE

New York

Published by Callisto Reference,
106 Park Avenue, Suite 200,
New York, NY 10016, USA
www.callistoreference.com

Handbook of Geodetic Science
Edited by Russell Sands

International Standard Book Number: 978-1-63239-399-9 (Hardback)

Printed in the United States of America.

Contents

Preface

Every book is a source of knowledge and this one is no exception. The idea that led to the conceptualization of this book was the fact that the world is advancing rapidly; which makes it crucial to document the progress in every field. I am aware that a lot of data is already available, yet, there is a lot more to learn. Hence, I accepted the responsibility of editing this book and contributing my knowledge to the community.

Geodetic science is described as a branch of earth sciences and applied mathematics. Space geodetic techniques like global navigation satellite systems (GNSS), very long baseline interferometry (VLBI), satellite gravimetry and altimetry, and GNSS reflectometry & radio occultation, can precisely measure even the minute changes in the Earth's shape, its rotation and gravitational pull; and are capable of calculating mass changes in earth's system to great accuracy as well. This book provides an insight into the latest experiments and evolution in space geodetic methods and theories, including GNSS, VLBI, gravimetry, geoid, geodetic atmosphere, geodetic geophysics and geodetic mass transport related with ocean, hydrology, cryosphere and solid-Earth. The book will act as a fine guide on geodetic methods to engineers, scientists and user community as well.

While editing this book, I had multiple visions for it. Then I finally narrowed down to make every chapter a sole standing text explaining a particular topic, so that they can be used independently. However, the umbrella subject sinews them into a common theme. This makes the book a unique platform of knowledge.

I would like to give the major credit of this book to the experts from every corner of the world, who took the time to share their expertise with us. Also, I owe the completion of this book to the never-ending support of my family, who supported me throughout the project.

Editor

Geodetic Techniques

VLBI Geodesy: Observations, Analysis and Results

Robert Heinkelmann

Additional information is available at the end of the chapter

1. Introduction

Besides the Global Navigation Satellite Systems (GNSS), Satellite and Lunar Laser Ranging (SLR, LLR) and the Doppler distance measurement technique DORIS (Doppler Orbitography and Radiopositioning Integrated by Satellite), Very Long Baseline Interferometry (VLBI) is one of the space-geodetic techniques. In addition to the aforementioned, satellite missions such as radar and laser altimetry and geodetically used components of gravity field satellites in particular CHAMP (CHAllenging Mini satellite Payload), GRACE (Gravity Recovery And Climate Experiment), and GOCE (Gravity field and steady-state Ocean Circulation Explorer) can be counted to the space-geodetic techniques. In radio astronomy VLBI is a technique for astrophysics and astrometry. The later has many things in common with geodetic VLBI; only the schedule varies among terrestrial and celestial motivated observing sessions in terms of the number of observed radio sources as well as the number and sequence of observations to radio sources. Together with geodynamics, oceanography, glaciology, meteorology, and climatology, geodesy provides the metric basis for interdisciplinary research within the geosciences.

In this chapter geodetic VLBI is introduced. Section 2 describes the fundamentals of the VLBI technique up to the provision of observables. Then sections 3 and 4 give an introduction to the analysis of the various observables and the derived operational and scientific results. The chapter finishes with remarks and conclusions on the current and future role of geodetic VLBI.

2. VLBI technique

The VLBI system can be described by the following components:

i.   the astronomical object, the radio source,

ii.   the propagation of radio waves, the media of propagation of the electromagnetic wave, in particular the Earth's atmosphere,

iii.   the antenna- and receiver system, the mechanical and electronic instrumentation,

iv.   the Earth as being the carrier of the interferometer baselines formed by antenna pairs,

v.   the correlator, and

vi.   the analysis of VLBI observations, i.e. the application of physically motivated mathematical models through the software based on the objective and subjective decisions of the operator(s).

In spite of the large parts in common for the analyses, the scientific aims of astrometric and geodetic and those of astrophysical VLBI significantly differ. While radio astronomy aims to investigate a large variety of astronomical objects and their astrophysical characteristics, geodetic and astrometric VLBI focusses on precise point positioning and derivates, such as the accurate determination of very long distances on Earth, plate motion, or Earth orientation. In the upcoming sections I will describe the various system components in some more detail, necessary for the understanding of the scientific results obtained by geodetic and astrometric VLBI.

2.1. Space segment: radio sources

While astronomical VLBI deals with a large variety of objects, such as supernovae, pulsars, blazars, flare-stars, areas of star formation like globules, OH- and H_2O-maser sources, close and distant galaxies, gravitational lenses, starburst-galaxies, and active galactic nuclei (AGN), geodetic and astrometric VLBI prefers extra-galactic, radio-loud, and compact objects like quasars (quasi stellar radio source), radio galaxies (see Figure 1.), and objects of type BL Lac(ertae). The radio emission of quasars is caused by the accretion of mater into a black hole in the center of the so called host galaxy, where the spectrum is generally dominated by optical, ultra-violet or X-ray emission. During their fall into the black hole matter and electrons are relativistically accelerated. Thus, besides a smaller amount of thermic radiation the natural radiation of radio sources is due to the synchrotron effect. This radiation has a number of beneficial characteristics, such as high intensity and continuity, i.e. not distinct individual spectral lines are emitted but a noise over a broad bandwidth. Quasars contain extreme radio-loud AGNs dominating the emission of their host galaxies. Radio galaxies contain or at least contained an AGN as well, for without the existence of such a center the formation of the observed mater outflow (jets) and radio bubbles (lobes) could not be explained. Around the gravitating center of these objects usually there is a dust torus. The difference between a quasar and a radio galaxy is due to the geometry of observation. Quasars are radio galaxies where the edge of the dust torus obscurs the AGN in the line of sight of the observer (Haas & Meisenheimer, 2003). Objects of type BL Lac belong also to radio galaxies (Tateyama et al., 1998). For this type of object the angle between the line of sight and the direction of the jet are very small, i.e. one is looking into the jet. While the physical characteristics of the astronomical objects are still subject of scientific discussion, their strong emission of noise in a broad radio spectrum is a fact. For geodesy a radio source has to fulfill a number of criteria to be useful as a celestial reference point.

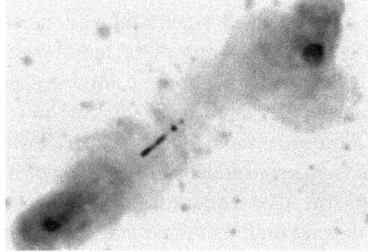

Figure 1. Negative black-white image of radio galaxy 3C219 (0917+458) from a superposition of radio and optical images (NRAO/AUI/NSF). The black dot in the middle shows the AGN.

The brightness, i.e. the intensity of the radiation has to be strong enough at the observed frequency bands. Depending on the antenna characteristics, the minimum lies around 0.01 Jy (Jansky; 1 Jy = 10^{-26} J m^{-2}). This condition has to be fulfilled to achieve an appropriate signal to noise ratio (SNR). The average intensity is 0.38 Jy in X-band and 0.47 Jy in S-band, the two so-called NASA (National Aeronautics and Space Administration, USA) frequencies of geodetic VLBI. The maximal intensity of a radio source may reach 20 Jy in both bands. So far about 4500 sources have been observed by geodetic VLBI. The number of compact bright (intensity > 0.06 Jy) radio sources is expected to be about 25,000 (Preuss, 1982) and thus the number of observable astrometric radio sources is by far not exploited. Besides, the intensity of the radio flux should not vary too much with time to enable a continuous observation.

The compactness, i.e. the spatial extension of the intensity maximum, specified through the angle diameter of the radio source core, has to be smaller than the intended coordinate, declination and right ascension, precision. Since only a very little number of radio sources is ideal compact in X- and in particular in S-bands, the sources show an intrinsic structure within the angle diameter and, consequently, structure corrections of the observations need to be considered (Charlot, 1990). Neglecting the structure would lead to an average error of about 8 ps (2.4 mm) on the group delay observation (Sovers et al., 2002).

The maximum of intensity within a radio source would ideally not shift among the observed frequencies. This effect is, however, in general not fulfilled and the observed position varies up to 700 μas within a source depending on the frequency. For geodesy/astrometry it does not matter whether the core frequency shift is caused by actually different locations of emission or by self-absorption of the object. A structure correction has to be evaluated depending on the observed frequencies; in our case a dual-frequency structure correction (Charlot, 2002). For comparison and for providing a link to catalogues in other, e.g. optical frequencies, the inclusion of radio optic counterparts is another criterion for choosing reference sources. Thus, some radio sources are observed for this purpose when the position of the optical counterpart is known even if other reference point characteristics are not optimal.

The stability, i.e. the temporal invariability of the position of the intensity maximum, has to be given to a certain amount and the radio source should not exhibit significant proper motions or parallaxes. Considering the very large distances of Earth-bound or near-Earth baselines to extra-galactic objects, the two later conditions are evidently fulfilled in a sufficient way. Nevertheless, several radio sources show significant variations of the topology of their intensity maximum. The emission areas of most of the radio sources are not ideal symmetric and centered at the core, but elongate with a bright component at the beginning of a diffuse tail: core-jet-structure. Fortunately, only a few radio sources show significant deformations of their topology up to 300 μas. Besides the radio images, which allow for astrophysical modeling of the structure variations, mathematical modeling of source coordinate time series is achieved through statistical methods such as Allan-variance or hypothesis testing. While the statements about the specific structure depend on the number and quality of the radio images and the astrophysical assumptions, the statistical methods rely on the number and quality of the delay observations and the de-correlation of source and other parameters. The two stability criteria have found to be conflicting in some cases (Moor et al., 2011).

For the realization of radio catalogues, a repeated observation of the radio sources and thus an appropriate visibility from Earth-bound baselines is necessary. This criterion competes with the desired geometrical distribution of the radio sources, which ideally aims for an evenly and consistently sky coverage.

2.2. Propagation media: Space-time, particles, and electrons

On its way through space-time the radio wave might by affected not only by space curvature, but also by charged and neutral particles. According to the distance from Earth, these effects can be divided into inter-stellar effects, which are to a large extend ionizing effects, gravitational effects through our solar system bodies, and ionizing and delaying effects through the atmosphere, ionosphere and neutrosphere of Earth.

Little is known about inter-stellar impact on VLBI group delays. Since VLBI is an interferometric technique, all effects, which are common to both interfered signals are absorbed by the clock model and consequently not visible in the observation. A large part of the inter-stellar propagation media is assumed to have large spatial extend with little variation on the spatial scales of Earth baselines of up to about 12.000 km. Nevertheless, if VLBI observations would be ionized through inter-stellar media, a comparison of VLBI-derived ionospheric delays with those obtained by other space-geodetic techniques, e.g. GNSS, would be appropriate for investigation. In our comparisons (Dettmering et al., 2011a) during two weeks and another separate day, we found no evidence for additional ionization besides the one caused by Earth's ionosphere. Of course the comparisons should be extended to all available data.

While the radio waves propagate through our solar system, the distances to massive objects can get considerably small and the space-curvature may significantly affect the two signals in a different way. For the history of VLBI observations and the precision of 1 ps of the current theoretical VLBI group delay model, called the consensus model, the effects of Sun, Earth and in some cases Jupiter have shown impact on the results. For Sun and Earth the effects of space-curvature on group delays, the so-called Shapiro-delay, are to be considered not only for a particular geometry of the observation. In the case of Earth, the gravitational delay becomes theoretically maximal, if one antenna observes in zenith and the other antenna at zero degree elevation. Due to the largest mass in the solar system the gravitational delay in the vicinity of Sun has to include another post-Newtonian term. At the limb of Sun the delay can reach 169 ns (as seen from a 6000 km baseline) and it is still about 17 ps almost at the opposite direction (175 degree away from the line of sight from an 6000 km baseline). The higher order term in the vicinity of Sun is about 307 ps at the limb and with 6 ps still significant in one degree distance to the heliocenter; but then it drops considerably fast. For Jupiter and the other solar system planets, corrections would be only necessary, if the ray path is almost in the direction of the object, i.e. grazing the limb of the object. The gravitational delays by the various massive bodies can be finally added by superposition and are usually considered in the theoretical delay model.

By far the largest contributions on the delay observable due to propagation effects are caused by the Earth's atmosphere. For electromagnetic waves the modeling of the effects can be conveniently separated into dispersive (frequency dependent) and non-dispersive (frequency independent) parts. The dispersive characteristics of ionosphere are the main reason for dual-frequency observations in geodetic VLBI. Radio observations are delayed by the ionosphere in the order of several tens of meters, while the neutrosphere's contribution is around 15 m. Both are primarily depending on the elevation angle of the observation, since the path lengths through a spherical shell is approximately proportional to the sine of the elevation angle. The difference between the atmospheres is, that the ionospheric effects can by reduced to mm precision by dual-frequency calibration, while only the hydrostatic part of the neutrosphere, about 90% of the neutrosphere delay, can be effectively reduced, if the surface air pressure at the location of the observatories is accurately known. The remaining non-hydrostatic part has to be estimated (Dettmering et al., 2010).

2.3. Ground segment: radio telescopes and further instrumentation

For radio telescopes one can primarily discern single dish antennae and multi dish antenna arrays or cluster, which are arranged in a certain configuration, e.g. along a straight line, in star formation, Y-or T-form. While the arrays, connected via phase-stable cables, are typically used for image reconstruction in astronomical VLBI, single dish antennae are widely distributed for geodetic and astrometric purposes. Nevertheless, the application of antenna arrays has been investigated for geodetic purposes as well (Saosao & Morimoto, 1991). Geodetic VLBI antennae (see Figure 2.) are usually full steerable constructions made of steel with a concrete foundation attached to a fixed point of geometric reference. In the 80s and 90s of the last century there were a number of mobile VLBI stations active: the U.S. American systems MV1, MV2, and MV3 (Clark et al., 1987), which were eccentrically installed above one of about 40 platforms (Ma et al., 1990). The German Transportable Integrated Geodetic Observing system (TIGO, Hase, 1999) was after a test phase at Wettzell, Germany, steadily installed at Concepción, Chile, for improving the terrestrial network of core IVS VLBI sites. With the advent of GNSS, the application of mobile VLBI has been ceased, since it became economically unviable.

Most VLBI antenna reflectors are of Cassegrain type. In addition to a parabolic main reflector, the Cassegrain antenna has a sub reflector at the focal point of the main reflector, which is of convex hyperbolical shape. One of the focal points of the sub reflector lies in the middle of the main reflector. After the radio signal has been focused in the opening of the main reflector, covered by a feed horn, it is not immediately received but undergoes several electronic conversions, so-called heterodyne reception. The situation of the sub reflector may lead to shadowing effects of the main reflector. The inflicted signal loss, however, is negligibly small (Rogers, 1991). For gaining a sufficient signal to noise ratio, the antenna diameter and directivity play a significant role. Additional radio noise sources, such as atmospheric noise and the noise emitted by the temperature of the electronic components have to be suppressed. The received signal can be disturbed by transient signals, e.g. from radio or television broadcast, so-called radio frequency interference (RFI), in particular at S-band, which may ultimately lead to complete loss of one or more channels. Such cases have been increasingly reported depending on the environment, e.g. by Sorgente & Petrov (1999) at Matera, Italy. To keep the thermal noise as small as possible parts of the electronic are cooled down to a few degrees Kelvin, e.g. at the Radio Telescope Wettzell, Germany, using liquid Helium. For a low-loss focus of voltage the surface of the antenna has to be manufactured with a very high precision. The requirement for precision is about 0.05 of the wave length (Nottarp & Kilger, 1982).

Figure 2. The RTW (Radio Telescope Wettzell), a 20m diameter geodetic VLBI antenna at the Geodetic Observatory Wettzell, Germany (BKG/FESG), has observed the largest number of observations within IVS campaigns

Azimuth-elevation, X-Y, and polar or equatorial mounts can be found for geodetic VLBI systems. Most of the polar mounted antennae belonged to other programs and were used for other purposes before. The deep space network antenna at Hartebeesthoek, South Africa, for example, was constructed by NASA for tracking deep space vessels but later equipped with geodetic receivers and a precise timing unit and thus rearranged for geodetic purposes. The geodetic schedules require relatively large rotations of the telescopes switching among widely separated radio sources covering large azimuth and elevation angle distances. Therefore and to achieve a sufficient SNR, mid-size, about 20 m diameter, telescopes have proven to be optimal for geodetic schedules. Dishes with such dimensions of course considerably deform depending on thermal and wind-driven environmental conditions as well as due to gravitational sacking at various elevation angles. Variations of the telescope geometry may cause defocussing and thus the loss of the signal. In particular large steerable telescopes, such as Effelsberg, Germany, need to move the feed horn and with it the focal point of the reflector according to the elevation angle of the observation to remain focused. With smaller telescopes the deformations may not lead to the total loss of signal but the environmental effects may significantly distort the VLBI observable. There are empirical models, material constants, and antenna dependent data available for thermal deformations (Nothnagel, 2008), so that for most of the geodetic antennae this effect can be corrected. Only those antennae covered by a radome need to be treated individually, for the inside radome temperature is usually not available via IVS. For gravitational deformation there are models available, too, e.g. published by Sarti et al. (2010), but there are not all relevant antenna-dependent data collected for the consistent application. Gravitational deformations are larger for primary focus antennae, i.e. for those antennae, where the relatively heave receiver is located in the focus of the main reflector, and those are rather the minority of geodetic antennae. There are no models for wind induced deformations available. In case of strong winds, the telescope usually stops observing and moves to a safety position.

The VLBI reference point, where the VLBI measurements actually refer to, is usually an immaterial invariant point located at the intersection of the telescope axes. Since this point is usually not directly accessible, it needs to be eccentrically realized, e.g. through indirect measurements from external reference markers (Vittuari et al., 2001; Dawson et al., 2007). Those eccentric reference markers allow the access and maintenance of the VLBI reference point for other space-geodetic or engineer surveying techniques. The later are usually applied to determine the distance between reference points of various space-geodetic techniques, called local ties. The local ties, after transformation into the Cartesian geocentric system of the space-geodetic techniques, are one of the most important issues for the realization of multi-technique terrestrial reference frames, such as the current conventional International Terrestrial Reference Frame, ITRF2008 (Altamimi et al., 2011). The primary axes of the telescopes do practically not intersect. The size of the antenna axis offset can be only a few millimeters or can reach up to several meters (Nothnagel & Steinforth, 2005). For its determination, in the best case, there are measurements by precise engineer surveying methods available; otherwise it has to be estimated from VLBI observations.

The antennae are driven and controlled by the field system, a LINUX-based software for the movement of radio antennae (Himwich, 2000), which allows the automatic control during a VLBI-experiment scheduled in advance. The cable winding needs to be considered as well, since the antenna can only limitedly turn into one direction. New approaches of automized, semi-unattended antenna control are under investigation as well (Neidhardt et al., 2011).

With the aforementioned antennae it is in principle possible to receive frequencies between about 0.4 and 22 GHz. For geodetic and astrometric VLBI, the application of the so-called NASA-frequencies of about 8.4 GHz (X-band) and 2.3 GHz (S-band) became accepted. With the dual frequency reception, channels of a few to several hundreds of MHz around the mid-band frequencies are band pass filtered from the continuum noise of the radio source and then individually processed. The incoming radio signal, the so-called received frequency, is initially polarized and then amplified. Thereafter it is down converted to an intermediate frequency of about 300 MHz at the front end of the antenna (Whitney et al., 1976). To keep the data rate small, not the complete bandwidths but several channels of a few MHz are processed only. The individual channels are later on synthesized to a larger effective bandwidth applying bandwidth synthesis (Rogers, 1970; Hinteregger et al., 1972). Via a coaxial cable the signal is propagated to a nearby control building, the so-called back end, where further data processing steps take place until the signal is finally recorded. After the conversion to base band frequency, the signal goes through a formatter, where time stamps of a local oscillator are superimposed. This video signal is then sampled and quantized, so that the digital value is represented by e.g. 1-bit sampling. It is also possible to digitize the signal with higher bit assignment. Until the Mark III VLBI-system the gained digital signals were recorded onto magnetic tapes. Thereafter, applying Mark IV or newer VLBI-systems, the data are saved on hard disks. The requirements for data recording rates were and are still very challenging.

The data storage media, magnetic tapes or hard disks, are then shipped to a central processing unit, the so-called correlator, for further processing and determination of observables. Still under development is the step away from storage media towards e-VLBI, i.e. real time VLBI with data send via broad band cables (Whitney & Ruszczyk, 2006), such as via the internet. Unfortunately, this method of data exchange has to compete with commercial users and can thus become very expensive. In addition, many VLBI-antennae were intentionally built at rather remote places, so that cables in particular at the last few kilometers are often not available and would have to be laid only for this purpose. That's why e-VLBI has been successfully tested, but is often not applied for routine IVS VLBI-experiments at all the participating sites.

For measuring effects through the electronic components and instrumentation, an artificial signal is injected at each antenna system, with which variations of the signal phase can be detected, so-called phase calibration. The calibration signal is induced by a local oscillator at the front end in form of equally spaced pulses of 1 μs separation. The artificial signal undergoes the same signal way than the received signal. Since amplitude, phase, and frequency of the artificial signal are known, it is possible to reconstruct the effects on the received signal. To detect frequency depending characteristics of the instrumental effects, the phase calibration is done for each frequency channel separately (Whitney et al., 1976; Corey, 1999). Besides the phase calibration, the cable delay is calibrated as well, i.e. the delay of the signal between the epoch, the signal passes the VLBI reference point and the epoch it is actually recorded. Another source of error on the VLBI observable is due to polarization leakage. Polarization leakage arises from unavoidable imperfections in the construction of the polarizer. It corrupts the observed phase in a way that can depend on frequency.. So far polarization has shown to affect the geodetic observable in the order of 1.6 ps for 90% of the observations, an effect, which can still be neglected (Bertarini et al., 2011). Nevertheless, for the VLBI2010 observing system, polarization will become an issue.

2.4. Interferometer and interferometric principle

The wave length of the center frequencies, i.e. the geometric mean of the lower and upper cutoff-frequencies, in X- and S-band are about $\lambda_X = 3.6$ cm and $\lambda_S = 13$ cm. Since the bandwidths are rather small compared to the center frequencies, the frequency bands are sufficiently represented by their center frequencies. If observations were restricted to be carried out by a single antenna, angle resolutions of about 100 as (seconds of arc) could be achieved. By connecting two or more antennae of similar type on a baseline (Figure 3.), it is possible to synthesize a much larger antenna diameter. The angle resolution on an average baseline of about 6000 km, e.g. Westford, USA to Wettzell, Germany, already reaches 1.2 mas (milliarc-seconds) and is thus several orders of magnitude more precise than the resolution obtained by a single antenna. Through such a connection of antennae, called interferometer, however, not radio images, but patterns of interference are provided. Interference is excited, if the received signals fulfill the coherence condition. Temporal coherence of equally polarized radiation expressed in simplified terms means that the phase is temporarily invariant. Besides temporal coherence, which depends on the different lengths of the signal paths, spatial coherence plays a significant role as well. Adhering spatial coherence the diameter of the radio source needs to be rather small, optimally point like. Variations of the interference caused by spatial extension of the radio source are the basis for investigations of the source structure in radio astronomy. To separate spatial and temporal coherence, the frequency bandwidth has to be much smaller than the observed frequency.

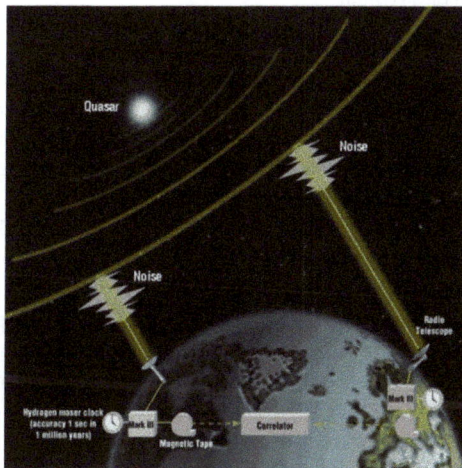

Figure 3. Scatch of a VLBI delay and the basic instrumental components of a Mark III VLBI system taken from the NASA/GSFC brochure "VLBI – measuring our changing Earth"

Considering the recorded bandwidths of a few hundreds of MHz compared to the observed frequencies of 2.3 and 8.4 GHz, the separation is in principle possible. For VLBI observations the coherence has to be realized through local frequency normals. Synchronization of the normals can be approximately achieved via time transfer e.g. by the GPS system. With local frequency normals, however, it is not possible to maintain the coherence during the whole observation session, yet during smaller time spans of a few minutes. This short coherent time span is usually sufficient to integrate a single observation, called a scan (Thompson et al., 2001). VLBI's requirements for frequency stability are very demanding, but only during these short time spans of up to about 1000 s (17 minutes). Hydrogen-maser normals have proven to deliver highest frequency stability over the required coherent time span and are thus installed at geodetic VLBI observatories. The imprecision through the synchronization and drift of the various masers is usually parameterized and estimated along with the other astrometric, geodetic and auxiliary parameters (see section 3). It is an inherent characteristic of the interferometric technique that only those quantities affecting the interfering signals in a different way are visible in the observation, i.e. the interferometer is independent of effects common to both signals.

2.5. Earth: Carrier of interferometer baselines

The temporal coherence condition is generally not fulfilled because of the motion of Earth during observation. The geometry of the baseline and the radio source is continuously changing, e.g. due to Earth rotation. The phase of the received signal, therefore, slowly varies with time. As a consequence, not a constant interference frequency, but a slowly varying fringe frequency mainly caused by the differential Doppler-effect of Earth rotation is observed during the finite duration of an observation.

The radio telescopes forming the baselines are quite stably attached to the underlying rock bed through their mount, a construction mainly made of steel and concrete. Earth's surface, however, is not stable. On the contrary, the lithosphere is subject to a variety of deformations. Some of the deformations are rather constant, secular, or periodical. Others are individual, episodic, and discontinuous, e.g. during and after a seismic event. Occasionally a significant antenna repair has to take place, where the antenna is lifted from its rail. In spite of the usually very carefully executed procedure, such a repair typically leads to a displacement of the VLBI reference point of a few millimeters. Local deformations have been observed at some sites, e.g. through increasing water abstraction or season-depending irrigation. Secular variations, such as the slow geodynamics of the lithosphere, in particular recent crustal motions, plate motions, and postglacial rebound, are the subject of plate motion models, such as the Actual Plate KInematic Model APKIM (Drewes, 2009). Those secular drifts of the plates are almost exact linear within time scales of thousands of years and are believed to be driven by the continuous process of sea floor spreading (Campbell et al., 1992). The station coordinate model of current terrestrial reference frames therefore contains at least a position and a linear velocity term for each site. Nevertheless, at the borders between plates, the plate boundaries, significant anomalies with respect to the linear velocities can be found. Stress and strain release at the plate boundaries is also a major origin of earthquakes. Co- and post-seismic lithosphere deformations can only be individually explained depending on the earthquake mechanism. For the very well observed M7.9 Denali earthquake with its epicenter close to Fairbanks,

Alaska, in November 2002, non-linear motions of the VLBI station at Gilmore Creek were successfully approximated by a combined logarithmic-exponential model. The exponential deformation held on for several years after the co-seismic positional jump (Heinkelmann et al., 2008).

Displacements of reference points through tidal and loading deformations occur on much shorter time scales, such as hours, days, months, or years. This group of effects is to such an extent sufficiently understood and described by geophysical models, that it is usually directly reduced from the observations and therefore no contributions appear in station coordinate residual time series. Deformations belonging to this group are:

i.   The solid Earth tides, caused mainly by the external torques of Moon and Sun, where deformations are related to the torques by a set of both, time and frequency dependent, constants, called Love and Shida numbers.

ii.   The loading deformations, which also comprise secondary effects on solid Earth due to interactions among the Earth system's spheres, mainly oceanic tidaly and atmosphere pressure induced, but also seasonal hydrological and snow induced loading deformations.

Besides the deformations of solid Earth and the displacements of attached reference points, the rotation axis of Earth is moving because of the inclination of the rotation axis with respect to the figure axis. The movement of the axis can be expressed with respect to the Earth's surface, i.e. the International Terrestrial Reference System (ITRS) or with respect to quasi-inertial space realized by the Geocentric Celestial Reference System (GCRS). The effects expressed with respect to Earth's surface are separated from those with respect to the celestial frame depending on frequency. By convention all terms around the retrograde diurnal band are addressed with respect to GCRS and the other terms, outside of the retrograde diurnal band, are attributed with respect to ITRS. Finally between ITRS and GCRS the diurnal Earth spin takes places. The terms with respect to ITRS are called polar motion. Besides the main periodically signals with 430 days (Chandler wobble) and annual periods, polar motion shows secular drifts, called polar wandering. Polar wandering is explained by large long-term mass variations inside the Earth system. Such a large mass variation occurred for example through the advance and following melting of glaciers during the last ice age at the end of the Pleistocene, where up to 3 km thick ice sheets covered large parts of Scandinavia, Greenland, and Canada. Since then Scandinavia for example has lifted about 300 m upwards and the global sea level raised about 120 to 130 m. Polar motion causes another group of reference point displacements:

iii.   The pole tides, a centrifugal effect due to the secular motion of the mean pole of Earth's rotation axis with respect to Earth's crust, and the ocean pole tides, a second order effect on solid Earth due to the equilibrium response of the oceans with respect to the main periodical signals within polar motion, the Chandler wobble and the annual term.

VLBI is an extraordinary technique for the determination of variations of the Earth rotation velocity. The velocity of Earth rotation shows for example tidally induced variations and a secular deceleration due to tidal friction in the two-body-system Earth-Moon. With respect to GCRS the orientation of Earth is varying, too, which is called precession-nutation. By continuous monitoring of these quantities models of the Earth interior could be improved. Free core nutation models were derived by Earth orientation parameters determined from VLBI observations. So far a period of about 430 days has been assumed for free core nutation, although some authors consider an interference of two signals with periods around 410 and 450 days (Malkin & Miller, 2007). For the determination of such models one is always bound to indirect methods, since the deepest drilling into Earth crust reached only about 15 km.

Even during the short delay between the receptions of the radio signals at the two VLBI antennae of about 20 ms as seen from a 6000 km baseline, the interferometer significantly moves due to Earth rotation and ecliptic motion. Consequently, the baseline is initially defined at those two epochs, called retarded baseline effect, and has to be referred to one epoch. The motion of the second antenna after the signal reception at the first antenna is accounted for in the theoretical VLBI delay model.

All those effects have a common consequence: each baseline formed by a pair of antennae is not constant but varies with time. Lengths and directions of the baseline vectors can change with respect to Earth's surface and with respect to the radio sources reference frame.

2.6. Correlation: Determination of observables

During correlation the individually recorded digital signals are superimposed for achieving interference whereby the observables are obtained. The equipment with which this process is done is called a correlator. In principle, one can distinguish between hardware and software correlators. While a hardware correlator is limited by the number of magnetic tapes or disks, which can be processed in parallel, the performance of a software correlator depends only on the available computing capacity. Applying modern concepts such as computer clusters or distributed systems and due to the steady improvements and developments of computer hardware, the technical borders of correlators are not yet reached (Kondo et al., 2004; Machida et al., 2006).

A variety of observables can be obtained depending on the purpose of the application. Astronomical VLBI primarily aims for high resolution image reconstruction. Therefore, the fringe amplitudes and phases are the desired observables. The contribution of each telescope is added in such a way as if the whole array would be one single antenna. For geodetic and astrometric VLBI precise group delay and delay rate observables are required. The group delay can be obtained by shifting the interfering signals in the time domain until the maximum of cross correlation is reached. In addition to the cross correlation of the bit streams the fringe rotation mainly caused by the Doppler effect due to Earth rotation, needs to be removed, so-called fringe stopping. This can be achieved by multiplication with sine and cosine terms, so-called quadrature mixing, where the fringe rotating cross correlation signal, which oscillates in the kHz range, is brought to a frequency close to zero. Fringe amplitude and phase can be obtained by summing up or dividing (tangent) the sine and cosine terms, respectively. The fringe frequency, the partial derivative of the phase delay with time, can be tracked for several minutes, as long as one radio source is continuously scanned. Switching among the radio sources, however, introduces ambiguities, which have to be solved for a phase measurement, called ambiguity solution. The ambiguity problematic can be avoided, if the group delay is used as observable, since the cross correlation function usually shows a unique maximum. The delay rate is the other observable used for geodetic VLBI to fix the ambiguities introduced by the broadband synthesis prior to the analysis of group delays. It can be obtained from the fringe frequency describing the phase drift due to Earth rotation. Since the precision of the delay rate in terms of geodetic target parameters is significantly worse compared to the group delay, it is usually not used itself for parameter determination.

The correlation procedure can be mathematically described through a cross correlation and a Fourier transformation. If the signals are at first cross correlated and then Fourier transformed, the correlator type is called XF-correlator, e.g. the Mark VLBI-systems. Otherwise, if the two mathematical procedures are applied vice versa, it is called FX-correlator, e.g. the VLBA-systems. Both correlator types have advantages and disadvantages (Moran, 1989; Alef, 1989; Whitney, 2000).

2.7. Precision of the group delay observable

The instrumental precision of the group delay from the correlation analysis

$$\sigma_\tau = \frac{1}{2\pi \cdot B_{eff} \cdot SNR} \tag{1}$$

primarily depends on the signal to noise ratio SNR and the effective bandwidth B_{eff} yielded by synthesizing the real observed and correlated channels

$$B_{eff} = \sqrt{\frac{\Sigma (f_i - f)^2}{N}} \tag{2}$$

where f_i are the channel frequencies ($i = 1,2,...,N$) and f is their mean frequency. The signal to noise ratio

$$SNR = \eta \sqrt{2 \cdot B_{eff}} \cdot t_{int} \frac{I}{2k} \sqrt{\frac{A_1 \cdot A_2}{T_{R,1} \cdot T_{R,2}}} b \tag{3}$$

is the criterion for a successful observation with $\eta \approx 0.73$ (1-bit sampling) an instrumental loss factor, t_{int} = 60 to 1000 s the coherent integration time, I the radio flux or intensity of the radio source, k the Boltzmann constant; $A_{1,2}$ denote the effective antenna areas and $T_{R,1}$ and $T_{R,2}$ are the noise temperatures of the electronic devices (Whitney et al., 1976). Above equations (1-3) describe the sensitivity of the measurement precision first of all to the effective bandwidth, but also to the intensity of the radio source. To observe radio sources with smaller intensities the coherent integration time needs to be quadratically increased to keep the measurement precision using the same equipment. The terms assume that a common system noise is implicitly present, independently at each channel and of the same size. There are, however, a number of instrumental negative effects, which affect individual channels only. Ray & Corey (1991) therefore discus an extension of the above model through two empirical variables: a scaling factor and an additive constant.

Besides the instrumental precision related to a single observation, the stochastic model of ensembles of VLBI observations can be described not only by the diagonal elements but also by off-diagonal elements in the weight or covariance matrices, respectively. A large variety of covariance models are possible in principle to describe the correlations of the observations among each other (Tesmer, 2004). Among them the modeling of station dependent noise has shown to significantly improve the stochastic of the VLBI equation systems (Gipson, 2007).

3. Analysis of group delays

After the ambiguities from broadband synthesis have been fixed incorporating the delay rate observable, the mathematical model of the analysis of group delays deals with

i.   the functional model, i.e. the creation of theoreticals. Simulated observations are calculated for each real observation applying the theoretical VLBI group delay model and a variety of correction models. The theoretical VLBI model used today is called the consensus model and includes all necessary terms to achieve 1 ps precision (about 0.3 mm). As presented at the IAU General Assembly 2012, the IAU Commission 52, Relativity in Fundamental Astronomy, is going to present a new model with 0.1 ps precision, soon. The current conventionally applied correction models are specified by IERS Conventions (2010).

ii.   The determination of parameters, which also includes the stochastic of the over-determined problem by modeling the committed errors and the neglected deterministic systematics. In geodetic VLBI several techniques have been applied for parameter determination: first of all least-squares estimation (Koch, 1997), then least-squares collocation (Titov & Schuh, 2000), Kalman filtering (Herring et al., 1990), and square-root-information filtering (Bierman, 1977).

The deterministic part of the mathematical model, the functional model of the group delay observation τ_{gd}, can be symbolically written as

$$\tau_{gd} = -\frac{1}{c} b' WRQk + \delta\tau_{rel} + \delta\tau_{iono} + \delta\tau_{neut} + \delta\tau_{cable} + \dots \qquad (4)$$

It contains the speed of light c, the baseline vector b', the polar motion matrix W, diurnal spin matrix R, the frame bias, precession-nutation matrix Q, and the radio source position vector k, which depends on the declination and right ascension at epoch J2000.0. The first expression at the right hand side before the relativistic delay correction $\delta\tau_{rel}$, is called geometric delay. Together with the relativistic delay correction the geometric delay is called theoretical delay model; specified e.g. by the consensus model (Eubanks, 1991) in its current conventional form (IERS Conventions, 2010). With the current precision it is sufficient to model the group delay in a Newtonian way and to add the relativistic implications, the retarded baseline and other special and general relativistic effects, in form of corrections to the Newtonian delay. The first order ionosphere correction

$$\delta\tau_{iono} = \frac{f_S^2}{f_X^2 - f_S^2}(\tau_X - \tau_S)$$ (5)

is derived by a linear combination of the single band delays at X- and S-band multiplied by a factor, the ratio of the square of the lower band frequency to the separation of the squared frequencies. Higher order terms can be neglected for the desired precision, (Hawarey et al., 2005). In analogy to this term the measurement error in S-band, which is due to the lower resolution usually larger than the one in X-band, propagates into the X-band group delay. The smaller the factor, the smaller is the error contribution from S-band on the X-band group delay. A wide frequency separation decreases the factor and is, thus, favorable for precision.

Additionally explicitly mentioned are the neutrosphere delay $\delta\tau_{neut}$ and the cable delay $\delta\tau_{cable}$ corrections. The three dots at the end of the above equation denote that there can be further corrections considered. Each of these delays contain two terms, one contribution from the first and one from the second station forming a baseline. If the radio signal first reaches the first antenna, the contribution to the group delay is positive for the second antenna and negative for the first antenna and vice versa. The neutrosphere correction of the i-th station is given by

$$\delta\tau_{neut,i} = \frac{1}{c} \bullet mf_h(e) \bullet zhd + mf_g(e)[cos(a)G_N + sin(a)G_E]$$ (6)

where mf are mapping functions of the hydrostatic delay (index h), or the gradients (index g), depending on the elevation angle e. The horizontal asymmetry is modeled by the two gradients in north-south G_N and east-west directions G_E depending on the azimuth angle a. The above term is multiplied with the factor -1, if i = 1. State-of-the-art mapping functions are derived by raytracing through numerical weather models, e.g. VMF1 (Böhm et al., 2006). The hydrostatic delay in zenith direction zhd is direct proportional to the surface air pressure at the VLBI reference point of the specific instrument. An appropriate apriori gradient model should be applied for geodetic VLBI analyses, because the estimated gradient parameters are usually constrained and can thus depend on the apriori values. Besides the apriori gradient model specified by IERS Conventions (2010), the usage of the apriori gradient model determined from the Data Assimilation Office weather model as provided by the IVS Analysis Center at NASA Goddard Space Flight Center (MacMillan & Ma, 1998) can be recommended. The cable delay is measured at each antenna and can be directly applied as taken from the IVS database. The theoreticals calculated in the above described way are subtracted from the observed group delays forming the o-c (observed minus computed) vector.

Before the parameter estimation process, the analyst has to decide on the parameterization, i.e. the definition of the parameters, which shall be determined. According to this decision, the design or Jacobi matrix relating the observations to the parameters is fixed in its overall structure. The entries of the design matrix are the partial derivatives of the observations with respect to the individual parameters. A large variety of partial derivatives of geodetic and astrometric parameters can be found in Nothnagel (1991) or in Sovers & Jacobs (1996). The row and column ranks of the equation system are extended by the geodetic datum and the pseudo-observations, respectively. The pseudo-observations are constraints on auxiliary parameters, which can be necessary to prevent singularities. The impact of the constraint is governed by its weight, which is chosen by the analyst. With the precision of the group delay as the basis for the weight matrix, the mathematical model is complete and the parameters can be determined.

The aim of the analysis is to produce normally distributed residuals. This implies that all present significant systematics are identified and sufficiently modeled and that the stochastic part contains pure white noise, an assumption of the estimation methods. By robust estimation the normal distribution can be forced to a certain extent. This method, however, practically discards observations violating certain robustness criteria and thus decreases the redundancy of the problem. It also requires some additional operations, what increases the runtime. Nevertheless, both disadvantages are usually accepted considering the advantages of robust estimation (Kutterer et al., 2003). The derived formal errors of the parameters are the outcome of observation and model errors projected into the parameter space by error propagation. Not all the possible error sources are necessarily included and thus the existence of further neglected errors, called omitted errors, can be assumed. Consequently, to obtain meaningful accuracies the type and size of omitted errors have to be assessed and added to the formal errors as well. If a significant error source is known but not considered, the formal errors, which are then actually smaller than the real errors, nust be inflated e.g. by a factor or a constant depending on the assumed characteristics of the omitted errors. Due to the inherent characteristics of VLBI, the baseline repeatability presents a reliable and often-used quality criterion for the performance of the parameter estimation, since it is independent from the geodetic datum. A more detailed overview of the geodetic VLBI analysis is presented by Schuh (2000).

3.1. Coordinates and Earth orientation parameters

VLBI for geodesy and astrometry group delay analyses primarily aim for the determination of terrestrial baselines (see Figure 4.), radio source positions, and the Earth orientation parameters (EOP). These parameter groups can be estimated at the same time. Nevertheless, the design of the observation sessions, specified through the scheduling, which is carried out in either geodetic or astrometric mode, usually optimizes one or two of those three parameter groups. In the standard estimation approach for a single about 24 h IVS session, the auxiliary parameters for clock synchronization and neutrosphere delay and gradient modeling are estimated at the same time. The clocks are already synchronized e.g. via the GPS system time. During the experiment the hydrogen maser normals show some drift and occasionally jumps with respect to each other. A reference clock needs to be defined and the drift of the other clocks can be usually approximated by a second order polynomial with respect to the reference clock. Additional auxiliaries are parameterized as piece-wise linear functions with a temporal resolution of about 30 minutes for stochastic fluctuations of clock and neutrosphere delays and several hours for neutrosphere gradients, respectively. To discriminate between common rotations of the radio sources and the Earth orientation parameters it is necessary to include no net rotation (NNR) condition equations for the celestial coordinates. Also the polyhedron of terrestrial baselines can be subject to overall rotations, which by convention are to be expressed by the EOP and thus, NNR conditions need to be applied at the terrestrial side as well. To derive network station coordinates instead of baselines both, an apriori set of coordinates and another geodetic datum have to be provided. For deriving geocentric coordinates the apriori values have to refer to the geocenter. This is the case for conventional terrestrial reference frames, such as a version of ITRF. The datum has to constrain the adjustments to the apriori values at least in such a way that the inherent translational ambiguities are appropriately fixed, and then no singularities appear in the equation system.

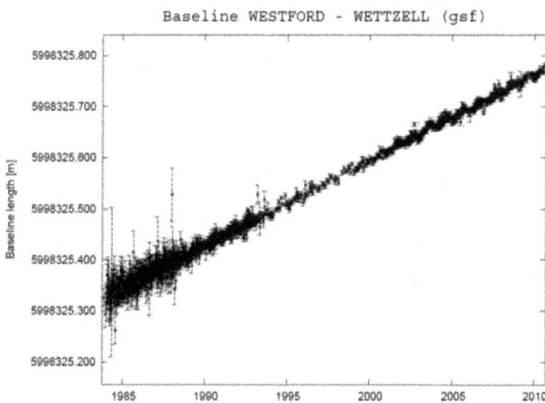

Figure 4. With its almost 6000 km the baseline WESTFORD (Westford, USA) – WETTZELL (Wettzell, Germany) shows a constant linear increase of about 1.7 cm yr⁻¹, which is due to the motion of the North American w.r.t. the Eurasian plate. The small annual signal is due to unmodelled geophysical effects. The figure is provided by IVS (http://vlbi.geod.uni-bonn.de/baseline-project/) and made of results from the IVS AC NASA/GSFC.

This set of condition equations is called no net translation (NNT). The NNR and NNT equations are the current best non-deforming datum conditions, because in contrast to fixing specific sets of coordinates, all coordinates are adjusted by this approach called free network adjustment. The origin and the orientation of the apriori frames are conserved but only in a kinematic sense. Consequently, due to the imperfect representation of the system through coordinates of its measured objects, there can still be some frame rotations present which cannot be suppressed by this approach. The imperfections of the kinematic non-rotating approach may lead to small differences between kinematically and dynamically non-rotating conditions, which may have to be considered by corrections.

3.2. Reference frames

In contrast to the analysis of a single VLBI session, the complete history of VLBI data are usually analyzed for the determination of reference frames and time series of EOP. If longer time spans are analyzed, velocities of terrestrial network stations need to be parameterized as well. Consequently, it is necessary to include additional datum conditions for the temporal evolution of both terrestrial NNR and NNT conditions. In principle, a datum gets more reliable, the more datum points are included (Baarda, 1968). For the frame determination, the choice of reference stations, or reference radio sources, respectively, is one of the major tasks. Stations are usually applied as a reference, if they sufficiently fulfill the linear station model. If a station shows a significant derivation from linearity, e.g. after a seismic event, it is usually excluded from the set of datum points. The current best realization of a terrestrial reference system is ITRF2008 (see Figure 5.). For the celestial reference points astrometric and astrophysical stability criteria are considered, if available (Heinkelmann et al., 2007a). Since the number of observed radio sources is much larger than the number of available network stations, this set of datum points can be selected applying more stringent criteria. For the current conventional realization, the ICRF2 (see Figure 6.), besides the stability, the geometrical distribution of radio reference sources was considered as well (IERS, 2009). The metric of the VLBI baselines depends only on the inserted time scale and the speed of light, which is one of the most precisely known natural constants. The precise network scale of the terrestrial coordinates is truly a strength of the VLBI technique. If the time scales of the VLBI model are correctly handled in particular during the Lorentz transformation from barycentric coordinate time (TCB) to geocentric coordinate time (TCG), the resulting geocentric VLBI metric is one of the utmost accurate among the space geodetic techniques. The scale of space-geodetic techniques incorporating satellites, such as GNSS and SLR, additionally depends on the imperfections of Earth's gravity field models and the uncertainty of the geocentric gravitational constant GM. Those techniques are, however, needed for the realization of the geocenter, in particular SLR. While the satellite orbit configurations directly refer to the center of mass of the Earth system, VLBI is independent from those dynamics and thus it is not possible to refer to the center of mass with VLBI alone.

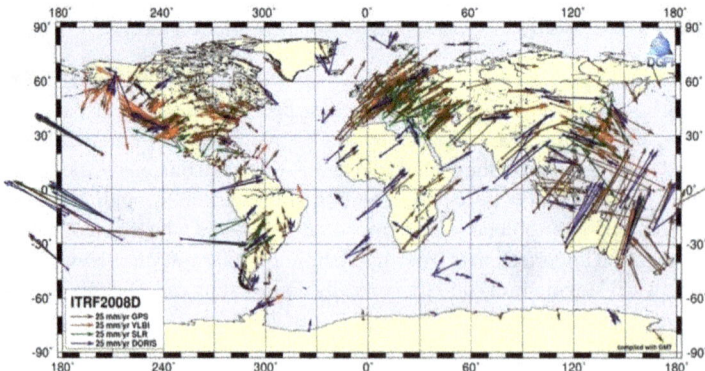

Figure 5. Kinematics of observatories of various space-geodetic techniques including VLBI (orange) as given by the ITRF2008 terrestrial reference frame solution determined at DGFI (http://www.dgfi.badw.de/), courtesy of M. Seitz

EOP are defined as the rotation parameters between the GCRS and the ITRS. VLBI provides a direct access to both systems. Thus, VLBI is the only technique for a direct determination of EOP. Satellite techniques require an additional transformation of their dynamic satellite orbit configurations to GCRS, liable to a number of error sources, and thus only derivatives of some of the EOP achieve a sufficient precision. Since five angles (EOP) are defined for a rotation, which could be achieved by only three independent angles, e.g. Euler angles, the EOP are by definition correlated with each other. Earth orientation parameters determined by VLBI have been used to determine a variety of quantities, such as the free core nutation (IERS Conventions, 2010) or ocean tidal terms (Englich et al., 2008).

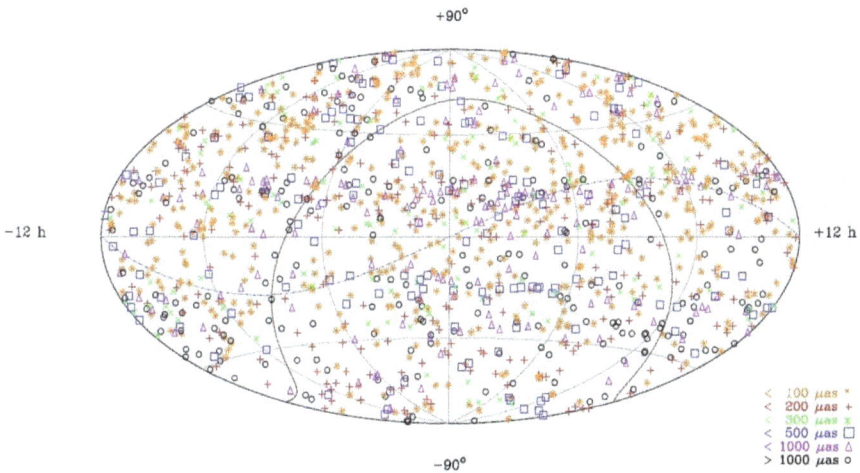

Figure 6. Mean positional errors of the multi-session radio sources of ICRF2 observed by IVS. Larger positional errors (> 1000 µas) can almost only be found close to the galactic plane (the omega-shaped line). Figure is taken from IERS Tech. Note 35, courtesy of C. Jacobs

3.3. Atmospheric quantities

The main goal of atmospheric analyses is the determination of atmospheric water vapor. Atmospheric water vapor, e.g. described by precipitable water, can be obtained by multiplying a factor on the estimated zenith wet delays (Heinkelmann et al., 2007b). In principle, the neutrosphere parameters are optimally decorrelated at low elevations. If the elevation angle of the observation is too small, however, the scatter gets too large and the observation is likely to get lost. As a tradeoff, an elevation cutoff angle is applied, which is in the case of VLBI usually 5°. Small elevation cutoff angles are possible for VLBI, because VLBI antennae are directional antennae and there are no multi-path effects comparable to those present with non-directional e.g. GNSS antennae. Another criterion for optimal neutrosphere parameter determination is the spatio-temporal sampling. Since about the beginning of the 1990s, the geodetic VLBI schedules require the antennae to point at very different directions within relative short time intervals. With such a spatio-temporal sampling the neutrosphere is observed with a good geometrical distribution within relative short time. Consequently, neutrosphere parameters can be successfully estimated with high temporal resolution.

Besides the standard analysis models and observation schedules, the precise determination of zenith wet delays depends on accurate surface air pressure values at the location of the network stations during the observation. The effects on the determination of zenith delays have been quantified and investigated. In particular the terrestrial reference frame, if not adjusted as well, can introduce large systematics into the zenith delay estimates. The weaknesses in the definition of the origin in z-direction of ITRF2000, for example, can cause apparent trends in atmospheric water vapor climatologies (Heinkelmann, 2008). In-situ measurements of air pressure have shown potential to be the best available data for this purpose. During the more than 30 years of observations, however, in-situ measurement time series of air pressure are often subject to inhomogeneities caused by a variety of possible effects. If those inhomogeneities are homogenized by appropriate procedures (Heinkelmann et al., 2005), the time series of zenith wet delay can be used to reliably estimate trends of atmospheric water vapor and other climate quantities. Besides, zenith wet delays can be used as a quality criterion for internal consistency (intra-technique comparison) and external validation (inter-technique comparison). Inter-technique comparisons with respect to other space-geodetic techniques at radio wavelengths have been carried out and published by a large number of authors. The neutrosphere delays determined by those techniques should equal each other in principle, if the reference points of the local instruments are situated at the same height. To adjust neutrosphere delays in terms of height a correction is necessary, which has to consider the height dependence of the symmetric neutrosphere model. Besides the aforementioned hydrostatic and non-hydrostatic (wet) zenith delays, also the mapping functions vary with height and contribute to the neutrospheric tie. The other type of neutrosphere parameter, the gradients should equal each other without a correction. In reality, however, neutrosphere parameters of various space-geodetic techniques differ (Teke et al., 2011), even if ties are applied. The reasons might be inherent simplifications of the neutrosphere model and neglections of cicumstances, e.g. the radomes above the instruments, which are not considered in the neutrosphere model. In addition, the wet zenith delay estimates, unfortunately, do not only describe the neutrosphere conditions, but also additional noise and systematics from unconsidered effects outside of the neutrosphere model, which show comparable elevation dependent characteristics.

The other examples of atmospheric quantities obtained by geodetic VLBI are less popular: Jin et al. (2008) showed that the hydrostatic component inside the total neutrosphere delay can be used to determine amplitudes of atmospheric tides and Tesmer et al. (2008) estimated coefficients of a model for atmospheric pressure loading deformations with an approach based on station height time series.

3.4. Further parameters obtained from group delays

Besides the aforementioned quantities, geodetic VLBI has shown its ability to precisely determine further parameters. These parameters are usually estimated by individually optimized analyses designed for the specific purpose, while other parameters are not considered or constrained. Further parameters obtained by geodetic VLBI are

i. Love and Shida numbers. Love and Shida numbers have been recently estimated by the IVS Analysis Center at the Insitute of Geodesy and Geophysics (IGG), TU Vienna, Austria (Spicakova et al., 2010). The values agree very well with the values reported in the IERS conventions (2004) and can be considered as an improvement over those values estimated by VLBI before (Haas & Schuh, 1996).

ii. Eccentricities of network stations. For the determination of eccentricities the geodesist usually prefers a measurement, e.g. obtained by engineer surveying methods. Nevertheless, whenever such measurements are unavailable, it is possible to insert eccentricities into the parameter space. The largest eccentricities for geodetic VLBI antennae were measured at the mobile observatories roughly during the 1990s. If different equipment is installed at the same platform, the eccentricity at a site is likely to change between two mobile occupations. At stationary antennae large eccentricities are usually avoided. When precisely measured eccentricities are available, it is in principle possible to compare the measured ones with the estimated ones. The transformation from local measurement system into the geocentric Cartesian system of the apriori catalogue may, however, significantly degrade the precision of the measured eccentricity and thus the comparison.

iii. The γ-parameter of the parameterized post-Newtonian theory. The γ-parameter describes how much unit mass deforms space-time and equals unity in Einstein's theory of gravity. For the γ-parameter determination, VLBI competes with ranging measurements to space crafts, which have nowadays shown to provide estimates with slightly better repeatability. Space-craft ranging measurements are, however, carried out at much smaller number of epochs and in almost one direction of the universe. VLBI, with its more than 30 years of continuous observations in quasi all directions of the universe, can in addition prove that this universal constant is indeed invariant to the measurement epoch and the directions in our universe. A review of the latest estimates and earlier VLBI results can be found in Heinkelmann & Schuh (2010).

iv. Velocities of radio sources and other advanced astrometric parameters. Extra-galactic radio sources should in principle be subject to galactic rotation of about 5.4 µas per year (Kovalevsky, 2003). An exact determination of the size of this effect is, however, very difficult. One difficulty is the correlation with the precession rate, which is about 50 as per year. The models for precession are also adjusted using VLBI (Capitaine et al., 2003). In spite of the fact that galactic rotation takes place in the galactic plane and precession is defined along the celestial equator, galactic aberration may propagate into the precession rate estimated by VLBI (Malkin, 2011). Furthermore, the investigation of radio source velocities led Titov et al. (2011) to an estimate of the Hubble constant describing an anisotropic expansion of the universe or low frequency gravitational waves, and thus, cosmology has become an issue for geodetic and astrometric VLBI as well.

4. Analysis of ionosphere delays

The analysis of VLBI ionosphere delays presents an independent branch,

$$\tau_{iono} = \frac{1.34 \bullet 10^{-7}}{f_X^2}\left[mf_i(e_2)VTEC_2 - mf_i(e_1)VTEC_1\right] + \tau_{offset,2} - \tau_{offset,1} \qquad (7)$$

enabling the determination of integrated electron density expressed by vertical total electron content $VTEC$. To derive $VTEC$ from slant total electron content (STEC) the application of an ionosphere mapping function mf_i is required. Ionosphere mapping functions are e.g. specified by IERS Conventions (2010) and can have significant effects on the ionosphere parameters (Dettmering et al., 2011b). The total electron content is usually specified in TECU (1 TECU := 10^{16} electrons per m^2). The ionosphere delays are derived from the real observed frequency channels around the two center frequencies in X- and S-band (equation 5) and are stored together with their formal errors in IVS databases for correction of the group delay observable (see section 3). The difference is that for the group delay analysis ionosphere delays are applied as a correction, whereas for this approach they will be used as primary observations. A small part of the ionosphere delay analysis is in common with the group delay analysis: the geometry of the network stations and the radio source needs to be known. Ionosphere parameters are usually referred to geocentric systems, such as the conventional terrestrial system or the geomagnetic equatorial system. Thus, the position of the network stations and the radio source need to be known in a geocentric system at the epoch of each observation. After the radio source coordinates have been transformed, the geometry is given, e.g. through pairs of azimuth and elevation angles, in a geocentric system. The ionosphere delay describes the difference of ionization along the observed ray paths. In addition there is an unknown instrumental offset included for each network station $\tau_{offset,i}$. Each delay contains at least three unknowns. Thus, without additional assumptions it would not be possible to estimate ionosphere parameters. The first assumption is that instrumental offsets are constant during an observation session, i.e. over 24 h. In this case, it would be possible to determine parameters, if some observations could be projected onto the same parameter. If $VTEC$ is estimated instead of STEC, it is possible to group the delays observed in various directions together within certain time intervals. Then, if the variation of the observations with time can be appropriately modeled, it is possible to reduce the number of unknowns. If short time spans are defined for the parameters, it is valid to assume that the ionosphere remains constant within these time spans. Since the maximum of ionization of Earth's ionosphere follows the movement of the Sun along the Earth's magnetic equator, this movement is accounted for in a simplified way

$$VTEC(t) = \left(1 + (\varphi' - \varphi)\begin{bmatrix} G_N \\ G_S \end{bmatrix}\right)VTEC[t + (\lambda' - \lambda)/15] \qquad (8)$$

Where φ' and λ' are the latitude and longitude of the ionospheric pierce point and φ and λ are the latitude and longitude of the specific VLBI antenna. With this approach and the respective assumptions it is possible to obtain an over-determined problem wherein $VTEC(t)$ with a certain temporal resolution and instrumental offsets at each network station can be determined. Spatial asymmetry in north-south direction can be considered by parameterizing additional gradient parameters (G_N, G_S). This method for ionosphere parameter estimation was introduced by Hobiger et al. (2006) and applied in a slightly refined way by us (Dettmering et al., 2011b).

Comparing VTEC obtained by various space-geodetic techniques reveals small differences depending on the technique (Dettmering et al., 2011a). Among GNSS and VLBI the mean differences observed during the continuous observation campaign CONT08 are about 1 TECU, with a slightly larger formal error. Thus, no significant dispersive offsets were found. The result proves that the rather simple model assumptions provide results of acceptable quality.

5. The current and future role of VLBI

Geodetic and astrometric VLBI under the umbrella of the International VLBI Service for Geodesy and Astrometry (IVS) provided observations of very good quality in a consistent way since more than 30 years (Schlüter & Behrend, 2007). Since 1999 the IVS comprises an adequate platform for the work packages of geodetic VLBI, which need to be shared. Besides the international observation campaigns, various components of the work flow are carried out by different institutions, world-wide, what requires a certain amount of organization. The IVS organizes conferences, proceedings, technical and analysis workshops and schools and the procedures, which can be done in an operational way. IVS is a service of the International Association of Geodesy (IAG) and as such it contributes in a unique way to IAG's flagship, the Global Geodetic Observing System (GGOS). Part of VLBI's contribution to GGOS is given through its infrastructure. The infrastructure is usually expensive and immobile and thus, the installation of VLBI equipment is rather a long-term investment. For the determination of geodetic and astrometric reference frames this is on the one hand an advantage, because, once installed, VLBI observations are usually carried out for many years by the same equipment, but on the other hand a disadvantage, because the absolute number of geodetic VLBI antenna is too small for many geoscientific applications, which require higher spatial resolutions. VLBI is able to measure the longest baselines on Earth limited only by the necessary common view of extra-galactic radio sources. The satellite-based space-geodetic techniques are restricted for example by the common view of a satellite or rely on networks of satellites or stations and can only indirectly reach comparable distances. By VLBI the differences at very remote locations can be determined directly and thus it is a truly global and consistent measurement technique.

In general and through projects, such as the VLBI2010, the development of technique and in the following observational strategies, analysis procedures and software, will go on and will continue to improve geodetic VLBI. Besides the VLBI receiver improvements, the antenna development is an ongoing process. Within project VLBI2010 the antenna specifications for diameter have decreased to 10 to 12 m (Niell et al., 2005; Shield & Godwin, 2006). The lower requirements for antenna diameters are possible because other receiver components, such as the effective bandwidth or the data recording rate, got significantly improved in recent years.

Several countries already have built or plan to build VLBI2010 compatible new VLBI telescopes, e.g. Australia, Spain, and Germany. The number of individual members in the IVS increases and we are looking faithfully into future.

In face of the upcoming GAIA optical astrometry mission (GAIA stands for global astrometric interferometer for astrophysics), VLBI will be needed to link the GAIA astrometry catalogue via the radio sources with Earth. The GAIA mission will operate only a limited number of years; five years are foreseen by now. The extrapolations of star positions of the GAIA catalogue outside of the temporal measurement range of its mission still foresee very high precision, but VLBI observations will go on in future and thus it will be possible to maintain VLBI based frames long after the GAIA mission will have ended.

Author details

Robert Heinkelmann

Deutsches Geodätisches Forschungsinstitut (DGFI), Bayerische Akademie der Wissenschaften, Munich, Germany

References

[1] Haas M. and K. Meisenheimer (2003) Sind Radiogalaxien und Quasare dasselbe? Die Antwort des Infrarotsatelliten ISO. Sterne und Weltraum, Nr. 11 (2003), 24-32

[2] Tateyama C.E., K.A. Kingham, P. Kaufmann, B.G. Piner, A.M.P. de Lucena, and L.C.L. Botti (1998) Observations of BL Lacertae from the geodetic VLBI archive of the Washington correlator. The Astrophysical Journal, Vol. 500, 810-815

[3] Preuss E. (1982) Zu Stand und Entwicklung der Radiointerferometrie in der Astronomie. Die Sterne, Vol. 58, No. 4 (1982), 232-251

[4] Charlot P. (1990) Radio-source structure in astrometric and geodetic very long baseline interferometry. The Astronomical Journal, Vol. 99, No. 4 (1990), 1309-1326

[5] Sovers O.J., P. Charlot, A.L. Fey, and D. Gordon (2002) Structure corrections in modeling VLBI delays for RDV data. In: Proceedings of the IVS 2002 General Meeting, N.R. Vandenberg & K.D. Baver (edts.), NASA/CP-2002-210002, 243-247

[6] Charlot P. (2002) Modeling radio source structure for improved VLBI data analysis. In: Proceedings of the IVS 2002 General Meeting, N.R. Vandenberg & K.D. Baver (edts.), NASA/CP-2002-210002, 233-242

[7] Moor A., S. Frey, S.B. Lambert, O.A. Titov, and J. Bakos (2011) On the connection of the apparent proper motion and the VLBI structure of compact radio sources. arxiv.org/pdf/1103.3963

[8] Dettmering D., R. Heinkelmann, and M. Schmidt (2011a) Systematic differences between VTEC obtained by different space-geodetic techniques during CONT08. Journal of Geodesy, DOI 10.1007/s00190-011-0473-z

[9] Dettmering D., M. Schmidt, R. Heinkelmann, and M. Seitz (2011b) Combination of different space-geodetic observations for regional ionosphere modeling. Journal of Geodesy, DOI 10.1007/s00190-010-0423-1

[10] Dettmering D., R. Heinkelmann, M. Schmidt, and M. Seitz (2010) Die Atmosphäre als Fehlerquelle und Zielgröße in der Geodäsie. Zeitschrift für Vermessung und Geoinformation, Nr. 2/2010, 100-105

[11] Saosao T. and M. Morimoto (1991) Antennacluster-antennacluster VLBI for geodesy and astrometry. In: Proceedings of the AGU Chapman Conference on Geodetic VLBI: Monitoring Global Change. NOAA Technical Report, No. 137, NGS 49, 48-62

[12] Clark T.A., D. Gordon, W.E. Himwich, C. Ma, A. Mallama, and J.W. Ryan (1987) Determination of relative site motions in the western united states using Mark III very long baseline interferometry. Journal of Geophysical Research, Vol. 92, No. B12, 12741-12750

[13] Ma C., J.M. Sauber, L.J. Bell, T.A. Clark, D. Gordon, W.E. Himwich, and J.W. Ryan (1990) Measurement of horizontal motions in Alaska using very long baseline interferometry. Journal of Geophysical Research, Vol. 95, No. B13, 21991-22011

[14] Hase H. (1999) Theorie und Praxis globaler Bezugssysteme. Mitteilungen des Bundesamtes für Kartographie und Geodäsie. Nr. 13, Verlag des Bundesamtes für Kartographie und Geodäsie, 177

[15] Rogers A.E.E. (1991) Instrumentation improvement to achieve millimeter accuracy. In: Proceedings of the AGU Chapman Conference on Geodetic VLBI: Monitoring Global Change. NOAA Technical Report, No. 137, NGS 49, 1-6

[16] Sorgente M. and L. Petrov (1999) Overview of performance of European VLBI geodetic network in Europe campaigns in 1998. In: proceedings of the 13th Working Meeting on European VLBI for Geodesy and Astrometry. W. Schlüter and H. Hase (edts.), Bundesamt für Kartographie und Geodäsie, 95-100

[17] Nottarp K. and R. Kilger (1982) Design criteria of a radio telescope for geodetic and astrometric purpose. Techniques d'Interférométrie à très grande Base, CNES, Toulouse

[18] Nothnagel A. (2008) Conventions on thermal expansion modeling of radio telescopes for geodetic and astrometric VLBI. Journal of Geodesy, DOI 10.1007/s00190-008-0284-z

[19] Sarti P., C. Abbondanza, L. Petrov, and M. Negusini (2010) Height bias and scale effect induced by antenna gravitational deformations in geodetic VLBI data analysis. Journal of Geodesy, DOI 10-1007/s00190-010-0410-6

[20] Vittuari L., P. Sarti, and P. Tomasi (2001) 2001 GPS and classical survey at Medicina observatory: local tie and VLBI antenna's reference point determination. In: Proceed-

ings of the 15th Working Meeting on European VLBI for Geodesy and Astrometry. D. Behrend and A. Rius (edts.), 161-167

[21] Dawson J., P. Sarti, G. Johnston, and L. Vittuari (2007) Indirect approach to invariant point determination for SLR and VLBI systems: an assessment. Journal of Geodesy, Vol. 81, Nos. 6-8, 433-441

[22] Altamimi Z., X. Collilieux, and L. Métivier (2011) ITRF2008: an improved solution of the international terrestrial reference frame. Journal of Geodesy, DOI 10.1007/s00190-011-0444-4

[23] Nothnagel A and C. Steinforth (2005) Analysis Coordinator Report. In: IVS 2004 Annual Report, D. Behrend and K.D. Baver (edts.), NASA/TP-2005-212772, 28-30

[24] Himwich W.E. (2000) Introduction to the field system for non-users. In: Proceedings of the IVS 2000 General Meeting. N.R. Vandenberg and K.D. Baver (edts.), NASA/CP-2000-209893, 86-90

[25] Neidhardt A., M. Ettl, H. Rottmann, C. Plötz, M. Mühlbauer, H. Hase, W. Alef, S. Sobarzo, C. Herrera, C. Beaudoin, W.E. Himwich (2011) New technical observation strategies with e-control (new name: e-RemoteCtrl). In: Proceedings of the 20th Working Meeting on European VLBI for Geodesy and Astrometry. W. Alef, S. Bernhart, and A. Nothnagel (edts.), 26-30

[26] Whitney A.R., A.E.E. Rogers, H.F. Hinteregger, C.A. Knight, J.L. Levine, S. Lippincott, T.A. Clark, I.I. Shapiro, and D.S. Robertson (1976) A very-long-baseline interferometer system for geodetic applications. Radio Science, Vol. 11, No. 5, 421-432

[27] Rogers A.E.E. (1970) Very long baseline interferometry with large effective bandwidth for phase-delay measurements. Radio Science, Vol. 5, No. 10, 1239-1247

[28] Hinteregger H.F., I.I. Shapiro, D.S. Robertson, C.A. Knight, R.A. Ergas, A.R. Whitney, A.E.E. Rogers, J.M. Moran, T.A. Clark, and B.F. Burke (1972) Precision geodesy via radio interferometry. Science, Vol. 178, No. 4059, 396-398

[29] Whitney A.R. and C.A. Ruszczyk (2006) e-VLBI development at Haystack observatory. In: Proceedings of the IVS 2006 General Meeting, D. Behrend and K.D. Baver (edts.) NASA/CP-2006-214140, 211-215

[30] Corey B. (1999) Sputious phase calibration signals: how to find them and how to cure them. In: Proceedings of VLBI chief meetings held in Haystack Observatory, 1-5

[31] Bertarini A., A.L. Roy, B. Corey, R.C. Walker, W. Alef, and A. Nothnagel (2011) Effects on geodetic VLBI measurements due to polarization leakage in S/X receivers. Journal of Geodesy, DOI 10.1007/s00190-011-0478-7

[32] Thompson A.R., J.M. Moran, and G.W. Swenson Jr. (2001) Interferometry and synthesis in radio astronomy. 2nd edition. John Wiley & Sons, 692

[33] Campbell J., A. Nothnagel, and H. Schuh (1992) VLBI-Messungen für geodynamische Fragestellungen. Zeitschrift für Vermessungswesen, No. 4 (1992), 214-227

[34] Heinkelmann R., J. Freymueller, and H. Schuh (2008) A postseismic relaxation model fort he 2002 Denali earthquake from GPS deformation analysis applied to VLBI data. In: Proceedings of the 2008 IVS General Meeting, Nauka, 335-340

[35] Malkin Z. and N. Miller (2007) An analysis of celestial pole offset observations in the free core nutation frequency band. Arxiv.org/abs/0704.3252v1

[36] Kondo T., M. Kimura, Y. Koyama, and H. Osaki (2004) Current status of software correlators developed at Kashima Space Research Center. In: Proceedings of the 2004 General Meeting, N.R. Vandenberg and K.D. Baver (edts.), NASA/CP-2004-212255, 186-190

[37] Machida M., M. Ishimoto, K. Takashima, T. Kondo, and Y. Koyama (2006) K5/VSSP data processing system of small cluster computing at Tsukuba VLBI Correlator. In: Proceedings of the IVS 2006 General Meeting, D. Behrend and K.D. Baver (edts.), NASA/CP-2006-214140, 117-126

[38] Moran J.M. (1989) Introduction to VLBI. In: Very Long Baseline Interferometry. Techniques and Applications. M. Felli and R.E. Spencer (edts.). Kluwer Academic Publishers, 27-45

[39] Alef W. (1989) Scheduling, correlating, and post-processing of VLBI observations. In: Very Long Baseline Interferometry. Techniques and Applications. M. Felli and R.E. Spencer (edts.). Kluwer Academic Publishers,97-139

[40] Whitney A.R. (2000) How do VLBI correlators work? In: Proceedings of the IVS 2000 General Meeting, N.R. Vandenberg and K.D. Baver (edts.), NASA/CP-2000-209893, 187-205

[41] Ray J.R. and B.E. Corey (1991) Current precision of VLBI multi-band delay observables. In: Proceedings of the AGU Chapman Conference on Geodetic VLBI: Monitoring Global Change. NOAA Technical Report, No. 137, NGS 49, 123-134

[42] Tesmer V. (2004) Das stochastische Modell bei der VLBI-Auswertung. DGK Reihe C Nr. 573. Verlag der Bayerischen Akademie der Wissenschaften in Kommission beim Verlag C.H. Beck. 97

[43] Gipson J.M. (2007) Incorporating correlated station dependent noise improves VLBI estimates. In: Proceedings of the 18th European VLBI for Geodesy and Astrometry Working Meeting, J. Böhm, A. Pany, and H. Schuh (edts.), 129-134

[44] IERS Conventions (2010) IERS Technical Note No. 36, G. Petit and B. Luzum (eds.), Verlag des Bundesamtes für Kartographie und Geodäsie, 179

[45] Koch K.R. (1997) Parameterschätzung und Hypothesentests, 3rd edition, Dümmler, 368

[46] Titov O.A. and H. Schuh (2000) Short periods in Earth rotation seen in VLBI data analysed by the least-squares collocation method. IERS Technical Note No. 28, Observatoire de Paris, 33-41

[47] Herring T.A., J.L. Davis, and I.I. Shapiro (1990) Geodesy by Radio Interferometry: The application of Kalman Filtering to the analysis of Very Long Baseline Interferometry data. Journal of Geophysical Research, Vol. 95, No. B8, 12561-12581

[48] Bierman G.J. (1977) Factorization methods for discrete sequential estimation. Mathematics in Science and Engineering, Vol. 128. Academic Press. Inc., 237

[49] Eubanks T.M.A. (1991) A consensus model for relativistic effects in geodetic VLBI. In: Proceedings of the USNO Workshop on Relativistic Models for use in space geodesy, 60-82

[50] Hawarey M., T. Hobiger, and H. Schuh (2005) Effects of the 2nd order ionospheric terms on VLBI measurements. Geophysical Research Letters, Vol. 32, No. 11, L11304

[51] Böhm J., B. Werl, and H. Schuh (2006) Troposphere mapping functions for GPS and very long baseline interferometry from European Centre for Medium-Range Weather Forecasts operational analysis data. Journal of Geophysical Research, Vol. 111, B02406. DOI 10.1029/2005JB003629

[52] MacMillan D.S. and C. Ma (1998) Using meteorological data assimilation models in computing tropospheric delays at microwave frequencies. Physics and Chemistry of the Earth, Vol. 23, No. 1, 97-102

[53] Nothnagel A. (1991) Radiointerferometrische Beobachtungen zur Bestimmung der Polbewegung unter Benutzung langer Nord-Süd-Basislinien. DGK Reihe C, Nr. 368, Verlag des Instituts für Angewandte Geodäsie, 93

[54] Sovers O.J. and C.S. Jacobs (1996) Observation model and parameter partials for the JPL VLBI parameter estimation software "MODEST" – 1996. JPL Publications 83-39, Rev. 6, 150

[55] Kutterer H., R. Heinkelmann, and V. Tesmer (2003) Robust outliers detection in VLBI data analysis. In: Proceedings of the 16th Working Meeting on European VLBI for Geodesy and Astrometry, W. Schwegmann and V. Thorandt (eds.) Bundesamt für Kartographie und Geodäsie, 247-256

[56] Schuh H. (2000) Geodetic analysis overview. In: Proceedings of the IVS 2000 General Meeting, N.R.Vandenberg and K.D. Baver (eds.), NASA/CP-2000-209893

[57] Baarda W. (1968) A testing procedure for use in geodetic networks. Publications on Geodesy, New Series, Vol. 2, No. 5, Netherlands Geodetic Commission, 97

[58] Heinkelmann R., J. Böhm, and H. Schuh (2007a) Effects of geodetic datum definition on the celestial and terrestrial reference frmaes determined by VLBI. In: Proceedings of the 18th European VLBI for Geodesy and Astrometry Working Meeting, J. Böhm, A. Pany, and H. Schuh (edts.), 200-205

[59] IERS (2009) The second realization of the International Celestial Refernce Frame by Very Long Baseline Interferometry. IERS Technical Note No. 35, Verlag des Bundesamts für Kartographie und Geodäsie, 204

[60] Englich S., R. Heinkelmann, and H. Schuh (2008) Re-assessment of ocean tidal terms in high-frequency Earth rotation variations observed by VLBI. In: Proceedings of the IVS 2008 General Meeting, A. Finkelstein and D. Behrend (eds.), Nauka, 314-318

[61] Heinkelmann R., M. Schmidt, J. Böhm, and H. Schuh (2007b) Determination of water vapor trends from VLBI observations. Vermessung & Geoinformation, 2/2007, 73-79

[62] Heinkelmann R. (2008) Bestimmung des atmosphärischen Wasserdampfes mittles VLBI als Beitrag zur Klimaforschung. Geowissenschaftliche Mitteilungen, Heft Nr. 82, 212

[63] Heinkelmann R., J. Böhm, and H. Schuh (2005) Homogenization of surface pressure recordings and ist impact on long-term series of VLBI tropospheric parameters. In: Proceedings of the 17th Working Meeting on European VLBI for Geodesy and Astrometry, M. Vennebusch and A. Nothnagel (eds.), INAF – Istituto di Radioastronomia – Sezione di NOTO – Italy, 74-78

[64] Teke K., J. Böhm, T. Nilsson, H. Schuh, P. Steigenberger, R. Dach, R. Heinkelmann, P. Willis, R. Haas, S. García-Espada, T. Hobiger, R. Ichikawa, and S. Shimizu (2011) Multi-technique comparison of troposphere zenith delays and gradients during CONT08. Journal of Geodesy, DOI 10.1007/s00190-010-0434-y

[65] Jin S., Y. Wu, R. Heinkelmann, and J. Park (2008) Diurnal and semidiurnal atmospheric tides observed by co-located GPS and VLBI measurements. Journal of Atmospheric and Solar-Terrestrial Physics, Vol. 70, 1366-1372

[66] Tesmer V., J. Böhm, B. Meisel, M. Rothacher, and P. Steigenberger (2008) Atmospheric loading coefficients determined from homogeneously reprocessed GPS and VLBI height time series. In: Proceedings of the IVS 2008 General Meeting, A. Finkelstein and D. Behrend (eds.), Nauka, 307-313

[67] Spicakova H., J. Böhm, S. Böhm, T. Nilsson, A. Pany, L. Plank, K. Teke, and H. Schuh (2010) Estimation of geodetic and geodynamical parameters with VieVS. In: Proceedings of the 2010 IVS General Meeting, D. Behrend and K.D. Baver (eds.) NASA/CP-2010-215864, 202-206

[68] IERS conventions (2004) IERS Technical Note No. 32, D.D. McCarthy and G. Petit (eds.) Verlag des Bundesamts für Kartographie und Geodäsie, 127

[69] Haas R. and H. Schuh (1996) Determination of frequency dependent Love and Shida numbers from VLBI data. Geophysical Research Letters, Vol. 23, No. 12, 1509-1512

[70] Heinkelmann R. and H. Schuh (2010) Very Long Baseline Interferometry (VLBI): Accuracy limits and relativistic tests. In: Proceedings of the IAU Symposium, No. 261, S. Klioner, P.K. Seidelmann, M. Soffel (eds.), 286-290

[71] Kovalevsky J. (2003) Aberration in proper motions. Astronomy & Astrophysics, Vol. 404, 743-747

[72] Capitaine N., P.T. Wallace, and J. Chapront (2003) Expressions for IAU 2000 precession quantities. Astronomy & Astrophysics, Vol. 412, 567-586

[73] Malkin Z. (2011) The influence of galactic aberration on precession parameters determined from VLBI observations. arxiv.org/pdf/1109.0514.pdf

[74] Titov O.A., S.B. Lambert, and A.-M. Gontier (2011) VLBI measurement of the secular aberration drift. arxiv.org/pdf/1009.3698v4.pdf

[75] Hobiger T., T. Kondo, and H. Schuh (2006) Very long baseline interferometry as a tool to probe the ionosphere. Radio Science, Vol. 41, DOI 10.1029/2005RS003297

[76] Schlüter W. and D. Behrend (2007) The International VLBI Service for Geodesy and Astrometry (IVS): current capabilities and future prospects", Journal of Geodesy, DOI 10.1007/s00190-006-0131-z

[77] Niell A.E., A.R. Whitney, B. Petrachenko, W. Schlüter, N.R. Vandenberg, H. Hase, Y. Koyama, C. Ma, H. Schuh, and G. Tuccari (2005) VLBI2010: current and future requirements for geodetic VLBI systems. Final Report of Working Group 3 to the IVS Directing Board, 21

[78] Shield P. and M. Godwin (2006) A new lower cost 12m full motion antenna. In: Proceedings of the IVS 2006 General Meeting, D. Behrend and K.D. Baver (eds.) NASA/CP-2006-214140, 77-82

Chapter 2

Signal Acquisition and Tracking Loop Design for GNSS Receivers

Kewen Sun

Additional information is available at the end of the chapter

1. Introduction

1.1. GNSS Signal Model

The signal received at the input of a Global Navigation Satellite System (GNSS) receiver, in a one-path additive Gaussian noise environment, can be represented by:

$$y_{RF}(t) = \sum_{i=1}^{N_s} r_{RF,i}(t) + \eta_{RF}(t) \tag{1}$$

that is the sum of N_s useful signals, emitted by N_s different satellites, and of a noise term $\eta_{RF}(t)$. The expression of the Signal in Space (SIS) transmitted by the ith satellite and received at the receiver antenna usually assumes the following structure:

$$r_{RF,i}(t) = A_i c_i(t - \tau_i) d_i(t - \tau_i) \cos[2\pi(f_{RF} + f_{d,i})t + \varphi_{RF,i}] \tag{2}$$

where

- A_i is the amplitude of the ith useful signal;
- τ_i is the code phase delay introduced by the transmission channel;
- $c_i(t - \tau_i)$ is the spreading sequence which is given by the product of several terms and it is assumed to take value in the set $\{-1, 1\}$;
- $d_i(t - \tau_i)$ is the navigation message, Binary Phase Shift Keying (BPSK) modulated, containing satellite data, each binary unit is called *bit*;

- $f_{d,i}$ is the Doppler frequency shift affecting the ith useful signal and $\varphi_{RF,i}$ is the initial carrier phase offset;
- f_{RF} is the carrier frequency and it depends on the GNSS signal band under analysis. For the Galileo E1 Open Service (OS) signal, $f_{RF} = 1575.420$ MHz.

The spreading sequence $c_i(t)$ can be expressed as

$$c_i(t) = c_{1,i}(t)c_{2,i}(t)s_{b,i}(t) \tag{3}$$

where $c_{1,i}$ is the periodic repetition of the primary spreading code, $c_{2,i}(t)$ is the secondary code and $s_{b,i}(t)$ is the subcarrier signal. The subcarrier $s_{b,i}(t)$ is the periodic repetition of a basic wave that determines the spectral characteristics of $r_{RF,i}(t)$. Two examples of subcarrier signals are BPSK and Binary Offset Carrier (BOC). The BPSK is adopted by the Global Positioning System (GPS) Coarse Acquisition (C/A); with the advent of new GNSSs, such as the European Galileo [1] and the modernized GPS [2], more complex modulations have been adopted.

The primary spreading code $c_{1,i}(t)$ consists of a unique sequence of chips which exhibits the orthogonal property necessary to avoid interference among signals. Denoting with T_c the chip interval, $c_{1,i}(t)$ can be expressed as $c_{1,i}(t) = \sum_k c_{1,i,k}P_{T_c}(t - kT_c)$, where $c_{1,i,k}$ is the kth chip of the Pseudo Random Noise (PRN) sequence for the ith satellite with a chip rate $R_c = 1/T_c$, and $P_{T_c}(t)$ is a unitary rectangular window with the duration T_c. $d(t)$ is a sequence of data bits, whose duration is T_b. T_b is much higher than T_c, with $T_b = 4$ ms in the Galileo E1 OS signal case and $T_b = 20$ ms in the GPS case. In case of Galileo E1 OS the primary spreading code $c_{1,i}(t)$ is a PRN with the chip rate R_c of 1.032 MHz and the repetition period $T_p = T_b = 4$ ms, which means that there is a potential bit sign transition in each primary code period, while there is always a sequence of at least 10 primary code periods without bit sign transition in case of GPS.

It is also known that Galileo E1 OS pilot channel (E1-C) signal is based on primary and secondary codes, by using the so called tired codes construction. Tired codes are generated modulating a short duration primary code by a long duration secondary code sequence. The secondary code has a code length of 100 ms including 25 chips. The secondary code acts exactly as the navigation message for the Galileo E1 OS data channel (E1-B) signal and it can be the cause of a sign reversal in the correlation operation over the integration time interval. From this point of view the impact of bit sign transitions on the primary spreading code has no difference between the data channel and the pilot channel signals.

The noise $\eta_{RF}(t)$ is assumed to be a zero-mean stationary additive white Gaussian noise (AWGN) process with power spectral density (PSD) $N_0/2$. In reality the noise will be neither Gaussian nor white, however the Gaussian approximation is justified by the central limit theorem, and is found to be accurate in practice. In addition, the sampled noise process is not white, as successive noise samples are correlated, so the white noise assumption is only an approximation. Each useful signal is characterized by power

$$C_i = \frac{A_i^2}{2} \tag{4}$$

and the overall signal quality is quantified by the carrier-to-noise-power-density ratio (C_i/N_0).

The input signal $y_{RF}(t)$, defined in (1), is received by the receiver antenna, down-converted and filtered by the receiver front-end. In this way, the received signal before the Analog to Digital (A/D) conversion is given by

$$
\begin{aligned}
y(t) &= \sum_{i=1}^{N_s} r_i(t) + \eta(t) \\
&= \sum_{i=1}^{N_s} A_i \tilde{c}_i(t - \tau_i) d_i(t - \tau_i) \cos[2\pi(f_{IF} + f_{d,i})t + \varphi_i] + \eta(t)
\end{aligned}
\tag{5}
$$

where f_{IF} is the receiver intermediate frequency (IF). The term $\tilde{c}_i(t - \tau_i)$ represents the spreading sequence after filtering of the front-end and here the simplifying condition

$$
\tilde{c}_i(t) \approx c_i(t)
\tag{6}
$$

is assumed and the impact of the front-end filter is neglected. $\eta(t)$ is the down-converted and filtered noise component.

In a digital receiver the IF signal is sampled through an Analog-to-Digital Converter (ADC). The ADC generates a sampled sequence $y(nT_s)$, obtained by sampling $y(t)$ at the sampling frequency $f_s = 1/T_s$. From now on the notation $x[n] = x[nT_s]$ will be adopted to indicate a generic sequence $x[n]$ to be processed in any digital platform. After the IF signal of Eq. (5) is sampled and digitized, neglecting the quantization impact, the following signal model is obtained:

$$
y[n] = \sum_{i=1}^{N_s} A_i \tilde{c}_i[n - \tau_i/T_s] d_i[n - \tau_i/T_s] \cos(2\pi F_{D,i} n + \varphi_i) + \eta[n]
\tag{7}
$$

where $F_{D,i} = (f_{IF} + f_{d,i})T_s$.

The spectral characteristics of the discrete-time random process $\eta[n]$ depends on the type of filtering, and the sampling and decimation strategy adopted in the front-end. If the choice on the sampling frequency $f_s = 2B_{IF} = 4f_{IF}$ is adopted, the IF signal and noise are sampled at Nyquist rate, where B_{IF} is the bandwidth of the front-end. In this case, it is easy to know that the noise variance becomes

$$
\sigma_{IF}^2 = E\{\eta^2[t]\} = E\{\eta^2[nT_s]\} = \frac{N_0}{2} f_s = N_0 B_{IF}
\tag{8}
$$

Another important parameter for the noise characterization is its auto-correlation function

$$
R_{IF}[m] = E\{\eta[n]\eta[n + m]\} = \sigma_{IF}^2 \delta[m]
\tag{9}
$$

which implies that the discrete-time random process $\eta[n]$ is a classical independent and identically distributed (i.i.d.) wide sense stationary (WSS) random process, or a white sequence. $\delta[m]$ is the Kronecker delta function.

Due to the orthogonality property of the spreading code sequence, the different GNSS signals are analyzed separately by the receiver, and only a single satellite is considered and the the index i of a satellite is dropped. The resulting signal is written as

$$y[n] = r[n] + \eta[n] = Ac[n - \tau/T_s]d[n - \tau/T_s]\cos(2\pi F_D n + \varphi) + \eta[n] \qquad (10)$$

1.2. GNSS Signal Acquisition Basic Concepts

In a code division multiple access (CDMA) based GNSS system, each satellite continuously transmits a periodic code signal, which is modulated by information symbols. The code signal is a spreading sequence made up of L_c chips and the sequence length (or repetition period) is denoted by T_p. Each satellite is characterized by an unique PRN code sequence. The cross correlation properties of such codes allow the GNSS receiver to efficiently separate received satellite signals which are superposed in the time domain.

It is well known that the first task performed by any GNSS receiver is to detect the presence of a generic satellite and to perform a global search for approximate values of the code phase delay τ and Doppler shift f_d of the SIS of each detected satellite. This stage, known as signal acquisition, provides an estimation $\hat{\tau}$ and $\hat{f}_d$ of the SIS parameters τ and f_d to the following signal tracking stage. The first parameter, code delay τ, is the time alignment of the PRN code in the current block of data, which contains the basic range and time information required to compute user position and clock offset. It is necessary to know the code phase delay in order to generate a local PRN code replica that is perfectly aligned with the incoming code. Only when this is the case, the incoming code can be removed from the received signal. PRN codes have high auto correlation only for zero lag. That is, the two signals must be perfectly aligned to remove the incoming code. The carrier frequency, which in case of down conversion corresponds to IF. The IF should be known for example from the Galileo E1 carrier frequency of 1575.42MHz and from the mixers in the down converter. However, the frequency can deviate from the expected value. The line-of-sight (LOS) velocity of the satellite (with respect to the receiver) causes a Doppler effect f_d resulting in a higher or lower frequency. In the worst case, the frequency can deviate up to ± 10 kHz. This unknown Doppler shift f_d is the second parameter needed to be estimated in the signal acquisition stage. It is important to know the frequency of the signal to be able to generate a local carrier signal, which is used to remove the incoming carrier from the signal.

All the acquisition systems for GNSS applications are based on the evaluation and processing of the cross ambiguity function (CAF) that, in the discrete time domain, can be defined as

$$R_{y,r}(\bar{\tau}, \bar{f}_d) = \sum_{n=0}^{L-1} y[n]c_{Loc}[nT_s - \bar{\tau}]e^{-j2\pi(f_{IF} + \bar{f}_d)nT_s} \qquad (11)$$

where $y[n]$ is the received signal; $c_{Loc}[nT_s - \bar{\tau}]$ is the the local code replica reproducing the PRN code and the subcarrier; $\bar{\tau}$ and $\bar{f}_d$ are the code delay and the Doppler frequency tested

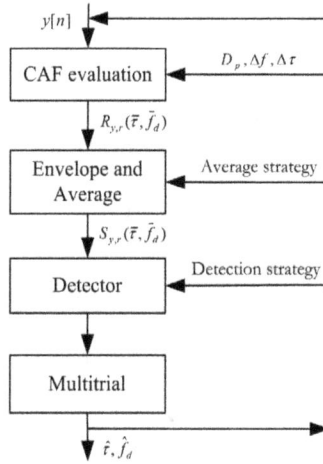

Figure 1. Functional blocks of an acquisition system.

by the receiver. Ideally the CAF should present a sharp peak that corresponds to the values of $\bar{\tau}$ and $\bar{f}_d$ matching the code delay and the Doppler shift of the SIS. However the phase of the incoming signal, the noise and other impairments can degrade the readability of the CAF and further processing is needed. For instance, in a non-coherent acquisition block only the envelope of the CAF is considered, avoiding the phase dependence. Moreover coherent and non-coherent integrations can be employed in order to reduce the noise impact. When the envelope of the averaged CAF is evaluated, the system can make a decision on the presence of the satellite. Different detection strategies can be employed. Some strategies are only based on the partial knowledge of the CAF and interactions among the different acquisition steps may be required. The detection can be further enhanced by using multi-trial techniques that require the use of CAFs evaluated on subsequent portions of the incoming signal.

In Fig. 1 the general scheme of an acquisition system is reported, which usually consists of four functional blocks:

- CAF evaluation;
- Envelope and Average;
- Detection and Decision;
- Multitrial and Verification.

The first two stages are devoted to the evaluation of the CAF in Eq. 11. The last two determine the signal presence and verify if the decision that has been taken is correct.

1.3. CAF Evaluation

The first stage of the acquisition block consists in the evaluation of the CAF of Eq. (11). The received signal $y[n]$ is multiplied by two orthogonal sinusoids at the frequency $\bar{F}_D =$

$(f_{IF} + \bar{f}_d)T_s$, which are called the in-phase and quadrature reference signals, respectively. In this way two new signals from the in-phase and quadrature channels are generated:

$$Y_c(n, \bar{F}_D) = y[n] \cos(2\pi \bar{F}_D n)$$
$$Y_s(n, \bar{F}_D) = -y[n] \sin(2\pi \bar{F}_D n) \tag{12}$$

The multiplication by these two orthogonal sinusoids is aimed at translating the received signal into baseband, removing the effect of the Doppler shift. This process is sometimes called carrier wipeoff because the signal is no longer modulated by the carrier frequency or any of of the intermediate frequencies. The normalized frequency

$$\bar{F}_D = (f_{IF} + \bar{f}_d)T_s = (f_{IF} + \bar{f}_d)/f_s$$

is given by two terms:

- the intermediate frequency, f_{IF},
- the local Doppler frequency, $\bar{f}_d$.

The intermediate frequency f_{IF} can be known from the Galileo E1 carrier frequency of 1575.42 MHz and from the mixers in the downconverter, whereas $\bar{f}_d$ is chosen from a finite set of the type:

$$\bar{f}_d = f_{d,\min} + l\Delta f \quad \text{for } l = 0, 1, \cdots, N_{f_d} - 1 \tag{13}$$

where Δf is called Doppler frequency bin size. Different Doppler frequencies are tested in order to determine the Doppler shift of the incoming signal. For low dynamic applications, $-5\text{KHz} \leq \bar{f}_d \leq 5\text{KHz}$; and for a high speed craft, $-10\text{KHz} \leq \bar{f}_d \leq 10\text{KHz}$.

The signals $Y_c(n, \bar{F}_D)$ and $Y_s(n, \bar{F}_D)$ are then multiplied by a local signal replica $c_{Loc}(nT_s)$ that includes the primary PRN code and the subcarrier. The local signal replica is delayed by $\bar{\tau}$ and the signals

$$Y_c'(n, \bar{\tau}, \bar{F}_D) = y[n] \cos(2\pi \bar{F}_D n) c_{Loc}(nT_s - \bar{\tau})$$
$$Y_s'(n, \bar{\tau}, \bar{F}_D) = -y[n] \sin(2\pi \bar{F}_D n) c_{Loc}(nT_s - \bar{\tau}) \tag{14}$$

are obtained. The delay $\bar{\tau}$ is taken from a set

$$\bar{\tau} = \tau_{\min} + h\Delta\tau \quad \text{for } l = 0, 1, \cdots, N_\tau - 1 \tag{15}$$

where $\Delta\tau$ is called delay bin size. By testing the different code delays, the acquisition block is able to estimate the delay of the received signal $y[n]$.

The signals $Y_c'(n, \bar{\tau}, \bar{F}_D)$ and $Y_s'(n, \bar{\tau}, \bar{F}_D)$ are then integrated, leading to the in-phase and quadrature components $Y_I(\bar{\tau}, \bar{F}_D)$ and $Y_Q(\bar{\tau}, \bar{F}_D)$:

$$Y_I(\bar{\tau}, \bar{F}_D) = \frac{1}{N} \sum_{n=0}^{N-1} Y_c'(n, \bar{\tau}, \bar{F}_D)$$

$$Y_Q(\bar{\tau}, \bar{F}_D) = \frac{1}{N} \sum_{n=0}^{N-1} Y_s'(n, \bar{\tau}, \bar{F}_D)$$

(16)

In Eq. (16), N represents the number of samples used for evaluating the in-phase and quadrature components and is used to define the coherent integration time (in discrete time sense):

$$T_{\mathrm{coh}} = N T_s$$

(17)

which is usually chosen as a multiple of the primary PRN code period. In general, N_τ can be different from N since only a subset of all possible delays need to be tested. The two components of Eq. (16) represent the real and the imaginary parts of the CAF which is finally given by

$$R_{y,r}(\bar{\tau}, \bar{F}_D) = Y(\bar{\tau}, \bar{F}_D) = Y_I(\bar{\tau}, \bar{F}_D) + jY_Q(\bar{\tau}, \bar{F}_D)$$

(18)

In Fig. 2, the operations previously described are highlighted. The CAF is a two-dimensional function that depends on the Doppler frequency $\bar{F}_D$ and on the code delay $\bar{\tau}$. Since both $\bar{F}_D$ and $\bar{\tau}$ are evaluated on the discrete sets represented by Eqs. (13) and (15), the CAF results are represented over a two-dimensional grid that is usually referred to as search space. It is clear that the search space is a plane, containing $N_\tau \times N_{f_d}$ cells, N_τ delay bins and N_{f_d} Doppler bins (or frequency bins), which covers the whole delay-frequency uncertainty zone. Each value of $\bar{F}_D$ and $\bar{\tau}$ corresponds to a cell of the search space, where a random variable for deciding the presence of the useful signal is evaluated.

A macro classification of the classical acquisition methods used to evaluate the CAF is described as follows.

1.3.1. Serial scheme

In the GNSS community the term serial acquisition scheme is used to denote CAF evaluation without any block processing accelerator algorithm, which is illustrated in Fig. 2. In this scheme CAF is evaluated at each instant n. The input vector y can be updated instant by instant by adding a new input value and by discarding the former one. With this approach the delay $\bar{\tau}$ moves throughout the vector y at each new instant. Therefore the local code replica $c_{Loc}[n]$ is always the same and the CAF is given by the expression

$$Y(\bar{\tau}, \bar{F}_D) = \frac{1}{N} \sum_{m=0}^{N-1} y[\bar{\tau} - N + m + 1] c_{Loc}(mT_s) e^{-j2\pi \bar{F}_D m}$$

(19)

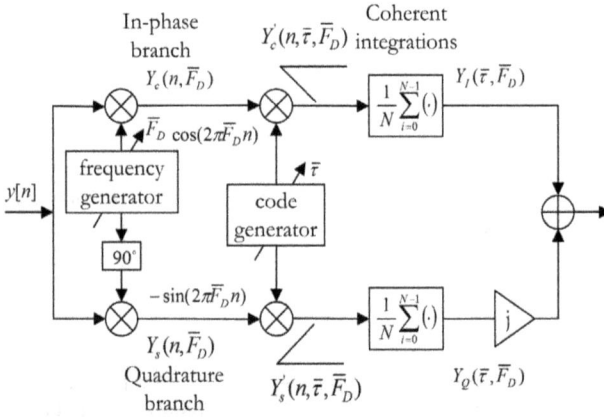

Figure 2. Conceptual scheme for the evaluation of the CAF. The received signal is multiplied by two orthogonal sinusoids and a local signal replica. The resulting signals are then integrated, generating the real and imaginary parts of the CAF.

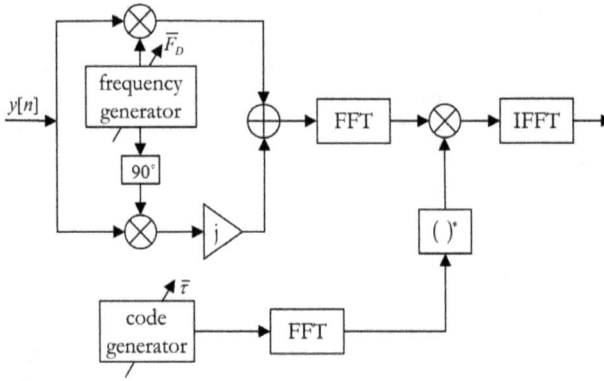

Figure 3. Time parallel acquisition scheme: the CAF is determined by using a circular convolution employing efficient FFT's.

1.3.2. FFT in the time domain

In this scheme the vector y of N samples is extracted by the incoming SIS and multiplied by $e^{-j2\pi \bar{F}_D n}$, so obtaining a sequence

$$q_l[n] = y[n]e^{-j2\pi \bar{F}_D n} \tag{20}$$

for each frequency bin $\bar{f}_d$ value. At this point the term

$$Y(\bar{\tau}, \bar{F}_D) = \frac{1}{N} \sum_{n=0}^{N-1} q_l[n] c_{Loc}(nT_s - \bar{\tau}) \tag{21}$$

assumes the form of a cross-correlation function (CCF), which can be evaluated by means of a circular cross-correlation defined as

$$Y(\bar{\tau}, \bar{F}_D) = \frac{1}{N} \text{IDFT}\{\text{DFT}[q_l[n]] \cdot \text{DFT}[c_{Loc}(nT_s)]^*\} \tag{22}$$

where DFT and IDFT stand for the well-known Discrete Fourier Transform and Inverse Discrete Fourier Transform, and * indicates the complex conjugate operator. The scheme based on DFT is shown in Fig. 3, where the Fast Fourier Transform (FFT) is used to evaluate the DFT. A CCF evaluated with a moving window and the circular CCF coincide only in presence of periodic sequences. This is the case when $\bar{f}_d = f_d$, except for the noise contribution and a negligible residual term due to a double frequency ($2f_d$) component contained in the term $q_l[n]$. In the other frequency bins or in the right bin but with $\bar{f}_d$ not exactly equal to f_d ($\bar{f}_d \approx f_d$), the presence of a sinusoidal component could alter the periodicity of the sequence. These aspects are generally negligible, while it is evident that the presence of a bit sign transition in the vector y completely destroys the code periodicity, so leading to serious CAF peak impairments in the search space.

1.3.3. FFT in the Doppler domain

This parallel fast acquisition employing FFT in the Doppler domain algorithm aims to compute a subset or the whole set of the frequency bins in a single step. This technique was first introduced in [3] for a high-dynamic GPS receiver, to solve the problem of the carrier frequency synchronization in presence of severe Doppler shift. Later on the method was also adopted in presence of no critical Doppler shifts to speed-up the acquisition process [4]. In this scheme a moving vector y can be extracted by the incoming SIS instant by instant, as in the method 1, and multiplied by the local code $c_{Loc}[n]$, so obtaining a sequence

$$q[m] = y(\bar{\tau} + mT_s)c_{Loc}[m] \tag{23}$$

for each delay bin $\bar{\tau}$. A similar result can be obtained by extracting an input vector y every N samples, and multiplying it by a delayed version of the local code $c_{Loc}[n]$. As mentioned

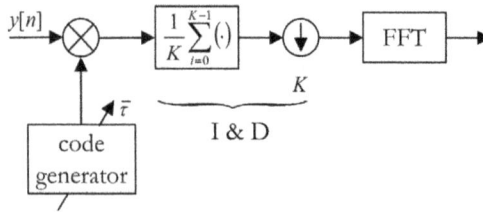

Figure 4. Frequency parallel acquisition scheme: the CAF is evaluated by using efficient FFT.

before, this delay has to be obtained by applying a circular shift to the samples of $c_{Loc}[n]$. At this point the term

$$Y(\bar{\tau}, \bar{F}_D) = \frac{1}{N} \sum_{m=0}^{N-1} q[m] \exp\{-j2\pi \bar{F}_D m\} \tag{24}$$

assumes the form of a Discrete-Time Fourier Transform (DTFT). It is well known that a DTFT can be evaluated by using a FFT if the normalized frequency $\bar{F}_D$ is discretized with a frequency interval

$$\Delta F = \frac{1}{N} \tag{25}$$

in the frequency range $(0, 1)$, which corresponds to the analog frequency range $(0, f_s)$. The evaluated frequency points become

$$\bar{f}_d T_s = \frac{l}{N} - f_{IF} T_s, \quad \text{for } l = 0, \cdots, N-1 \tag{26}$$

and the CAF can be written as

$$R_{y,r}(\bar{\tau}, \bar{F}_D) = \frac{1}{N} \sum_{m=0}^{N-1} q[m] \exp\left(-j\frac{2\pi l m}{N}\right) \tag{27}$$

With this method the search space along the frequency axis and the frequency bin size depend on the sampling frequency f_s and on the integration time N. If the same support and bin size used in methods 1 and 2 are adopted, the integration time has to be changed, and some decimation (with pre-filtering) has to be adopted before applying the FFT. This modifies the input signal, degrading its quality and introducing some losses [5]. In Fig. 4, the schematic block of the fast acquisition approach in frequency domain is reported. An "Integrate and Dump" (I & D) block followed by a decimation unit is inserted in order to reduce the number of samples on which the FFT is evaluated. This operation reduces the computational load but introduces a loss in the CAF quality.

1.4. Acquisition Is an Estimation Problem

Before GNSS signals can be used for obtaining the receiver's position the signals must be acquired. The signals arrived at the GNSS receiver's antenna have a corresponding code phase delay due to the transmission distance between the satellite transmitter and the receiver; in addition, the line-of-sight velocity of the satellite with respect to the receiver can cause Doppler shift effect resulting in a higher or lower frequency. For low mobility applications, such as a terrestrial receiver, the Doppler range is ±5 KHz; in high dynamic applications, such as a high-speed aircraft, the range becomes ±10 KHz.

The first task of a GNSS receiver is to detect the presence or not of a generic satellite and to perform a global search for approximate values of the parameters $(\tau_i, f_{d,i})$ of the corresponding SIS. This stage, known as signal acquisition, provides estimates $\hat{\tau}_i^{(A)}$ and $\hat{f}_{d,i}^{(A)}$ of the SIS parameters τ_i and $f_{d,i}$. It is necessary to estimate the code phase delay in order to generate a local code replica which is perfectly aligned with the incoming code. Only in this case the incoming code can be wiped off from the received SIS. It is important to obtain the Doppler shift of the received signal to generate a local carrier replica for the removal of the incoming carrier from the received SIS.

Once the signal is acquired, the estimated parameters $\hat{\tau}_i^{(A)}$ and $\hat{f}_{d,i}^{(A)}$ can be passed to signal tracking representing the second stage of the GNSS receiver, which is a fine tuning process performing a local search for accurate estimates of τ_i and $f_{d,i}$. In the coherent tracking case, the estimation of the carrier phase may be evaluated. Once the GNSS signals of the detected satellites are tracked, the navigation data message can be demodulated, the pseudo-ranges can be measured, and then the position, velocity and time(PVT) can be calculated.

The purpose of the signal acquisition stage is not confined to accessing the presence or absence of a given satellite. The important task of the acquisition system is to provide the subsequent tracking system with rough estimates of the received signal parameters. The mathematical discipline performed by the acquisition system is the parameter estimation theory. Therefore, signal acquisition process in a GNSS receiver can be formulated as a parameter estimation problem.

Considering that a signal is transmitted by a source (that is, a satellite transmitter), denoting by a vector $y = [y[0], y[1], \cdots, y[N-1]]$ of N measured samples of a realization $y[n]$ of a random process $Y[n]$ of the type $Y[n] = r[n] + W[n]$, where $W[n]$ is a zero-mean *white Gaussian noise* (WGN) random process, and $r[n]$ is an effective signal which contains K unknown parameters forming a vector $p = [p[0], p[1], \cdots, p[K-1]]$. The parameter point $\hat{p}$ which minimizes some suitably chosen cost function can be regarded as the best estimate of p given y.

Almost all the important parameters estimation theories and techniques rely on knowledge of the probability density function (p.d.f.) of the received signal conditioned on the true parameters, which can be denoted $f_{y|p}(y|p)$. This p.d.f. denotes the a priori probability of observing the vector y given that the signal parameters are provided by p. Thus, $f_{y|p}(y|p)$ can be viewed as a function of y parameterized by p. When receiving the signal, the problem is inverted: y is known, but p becomes unknown. Now look at this function at a different perspective: let the observed vector y be fixed "parameter" of this function, whereas the value of p is allowed to vary freely. In this case, $f_{y|p}(y|p)$ is considered to be a function of

p parameterized by y. $f_{y|p}(y|p)$ is a measure of the likelihood given that the observation vector is y. From this point of view, $f_{y|p}(y|p)$ is known as the likelihood function.

Among all the signal estimation theories, one form of optimal estimator is the minimum mean square error (MMSE) estimator. A MMSE estimator describes the approach which minimizes the mean square error (MSE). Let p denote an unknown random vector, and let Y denote a known random vector (the measurement or observation). An estimator $\hat{p}(y)$ of p is any function of the measurement Y. Let $e = \hat{p} - p$ denote the estimation error vector between the estimated and true signal parameters, then the MMSE estimator can be defined as the estimator achieving minimal MSE, given by $E[e^T e]$, where $E[\cdot]$ denotes the expected value of the random variable (r.v.) and $[\cdot]^T$ denotes the transpose of the vector. This estimator depends on knowledge of the a posteriori distribution of the true parameter vector p given the observation vector y, which is denoted $f_{p|y}(p|y)$. Under some weak regularity assumptions, the MMSE estimator is uniquely defined, and is given by [6]

$$\hat{p}_{MMSE}(y) = E[p|Y = y] \tag{28}$$

In other words, the MMSE estimator is the conditional expectation of p given the observed value of the measurements.

Another related estimator is the maximum a posteriori (MAP) estimator. In Bayesian statistics, a MAP estimate is a mode of the posterior distribution. The MAP can be adopted to achieve a point estimate of an unobserved quantity on the basis of empirical data. The MAP estimate $\hat{p}_{MAP}(y)$ maximizes the posterior probability that the estimate is correct on the basis of observations y. Assume that a prior distribution g over p exists. This allows us to treat p as a random vector as in Bayesian statistics. Then the posterior distribution of p can be written as follows [7]

$$f_{p|y}(p|y) = \frac{f_{y|p}(y|p)g_p(p)}{\int_{p' \in P} f_{y|p'}(y|p')g_p(p')dp'} = \frac{f_{y|p}(y|p)g_p(p)}{f_y(y)} \tag{29}$$

where g is the density function of p, P is the domain of g. In Eq. (29), the denominator of the posterior distribution (so-called partition function) does not depend on p and therefore plays no role in the optimization. The MAP estimate of p coincides with the Maximum likelihood(ML) estimate when the prior g is uniform (that is, a constant function).

The method of MAP estimation estimates p as the mode of the posterior distribution of this random vector, therefore, given by the solution to the equation:

$$\hat{p}_{MAP}(y) = \arg\max_{p} f_{p|y}(p|y) \tag{30}$$

The MAP estimate is a limit of Bayes estimators under under the 0-1 loss function, but it is generally not very representative of Bayesian methods.

The difficulty with MAP estimation is the reliance on the availability of $f_{p|y}(p|y)$. In practice, we normally have an expression for $f_{y|p}(y|p)$, and these two distributions are related in Eq. (29) by Bayes' theorem. However, usually we do not have expressions for $f_y(y)$ or $g_p(p)$, and this is the real case in GNSS signal acquisition problem. In such cases an alternative estimation, the maximum likelihood estimation (MLE), is commonly adopted [7]. MLE was recommended, analyzed and popularized by R. A. Fisher between 1912 and 1922. In statistics, MLE corresponds to many well-known estimation methods, which is popularly used for providing estimates for the statistical model's parameters. In general, considering a fixed set of observed data and underlying statistical model, the MLE method chooses the set of estimated values of the statistical model parameters which maximizes the likelihood function.

Suppose that there is a vector y of independent and identically distributed observations, with an unknown probability density $f_0(\cdot)$. It is however known that the function f_0 belongs to a certain family of distributions $\{f_{\cdot|p}(\cdot|p), p \in P\}$, called the parametric model, so that f_0 corresponds to $p = p_t$. The value p_t is unknown which is referred to as the true value of the parameter. It is desirable to find the value $\hat{p}_{ML}$ (the ML estimator) which would be as close to the true value p_t as possible.

To make use of the method of maximum likelihood, the joint density function for all observations should be specified firstly. For an independent and identically distributed observed sample, this joint density function is $f_{y|p}(y_1, y_2, \cdots, y_N|p) = f(y_1|p) \times f(y_2|p) \times \cdots \times f(y_N|p)$. Considering the observed values $y_1, y_2, \cdots, y_N$ to be fixed "parameters" of this function, whereas p becomes the function's variable which is allowed to vary freely; therefore, this function is called the likelihood:

$$L(p|y_1, y_2, \cdots, y_N) = f_{y|p}(y_1, y_2, \cdots, y_N|p) = \prod_{i=1}^{i=N} f(y_i|p) \tag{31}$$

The likelihood function of the parameter point p is simply the a priori probability of the received data signal y given p. The method of maximum likelihood estimates p_t by finding the value of p that maximizes likelihood function. This is the maximum likelihood estimator of p_t. The ML estimate is therefore given as follows

$$\hat{p}_{ML} = \arg \max_p f_{y|p}(y|p) \tag{32}$$

if any maximum exists.

MLE gives a unified approach to parameters estimation problem, which is well-defined in the case of the normal distribution and many other problems. For many statistical models, a ML estimator can be found as an explicit function of the observed data $y_1, y_2, \cdots, y_N$. For many other models, however, closed-form solution to the maximization problem is not available, and an MLE has to be obtained numerically adopting optimization methods. There may

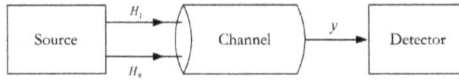

Figure 5. Binary hypothesis detector model.

exist multiple estimates that maximize the likelihood for some problems, whereas in some complicated problems, ML estimators are unsuitable or do not exist.

In addition, a ML estimator coincides with the most probable Bayesian estimator given a uniform prior distribution on the parameters. Assigning a uniform distribution to p is essentially equivalent to having zero a priori information about p.

1.5. Acquisition Is a Detection Problem

The operation performed by an acquisition system can be identified as a *detection* problem. In our application attention is limited to a situation where a data vector $y = [y[0], y[1], \cdots, y[N-1]]$ of N measured samples of the signal $y[n]$ is known, where $r[n]$ can be either present or absent. The purpose of the acquisition block is to detect whether or not a signal from a given satellite is present at the receiver antenna. In reality, GNSS signal acquisition is a combined detection/estimation problem. In this section an overview of the basic concepts of detection theory has been provided. The simplest problem in detection theory is called the binary hypothesis test [8], and is a useful example to illustrate the principle of detection. For example, in a digital communication system, a string of zeros and ones may be transmitted over some channel. At the receiver, the received signals representing the zeros and ones are corrupted in the channel by some additive noise and by the receiver noise. The receiver does not know which signal represents a zero and which signal represents a one, but it must make a decision as to whether the received signals represent zeros or ones. The process that the receiver undertakes in selecting a decision rule is implemented by a decision making device, or detector.

The situation above may be described by a source emitting two possible outputs at various instants of time. The outputs are referred to as hypotheses. The detector must choose between two hypotheses: the first hypothesis is called the null hypothesis denoted H_0 which represents a zero (target not present) while the second hypothesis is termed the alternate hypothesis denoted H_1 which represents a one (target present), as shown in Fig. 5. The detector, therefore, make one of the two decisions:

- D_0: The decision that H_0 is true,
- D_1: The decision that H_1 is true.

Thus, there are four possible outcomes to the simple binary hypothesis test, as tabulated in Table 1.

	H_0	H_1
D_0	Correct Rejection	False Dismissal
D_1	False Alarm	Correct Detection

Table 1. The Four Possible Outcomes of a Binary Hypothesis Test

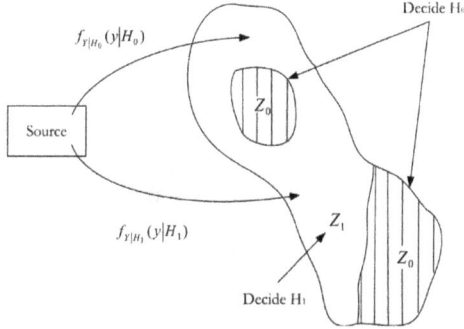

Figure 6. Decision regions.

Each hypothesis corresponds to one or more observations that are represented by random variables. Based on the observation values of these random variables, the receiver decides which hypothesis (H_0 or H_1) is true. Assume that the receiver is to make a decision based on N observation samples of the received signal which together form the observation vector y. The set of all possible observation values that the random vector Y takes constitutes an observation space Z. The observation space is partitioned into two distinct regions Z_0 and Z_1, such that if y lies in Z_0 ($y \in Z_0$), the receiver decides in favor of H_0; while if y lies in Z_1 ($y \in Z_1$), the receiver decides in favor of H_1, as shown in Fig. 6. The observation space Z is the union of Z_0 and Z_1; that is, $Z = Z_0 \cup Z_1$. The probability density functions of Y corresponding to each hypothesis are $f_{Y|H_0}(y|H_0)$ and $f_{Y|H_1}(y|H_1)$, where y is a particular value of the random vector Y.

The detector performance is, therefore, entirely dependent on how the decision regions Z_0 and Z_1 are chosen. Each time a decision is made, based on some criterion, for this binary hypothesis testing problem, four possible cases can occur, which can be measured by four parameters (conditional p.d.f.'s corresponding to the four outcomes of Table 1):

- The probability of correct detection: $P_d = P(D_1|H_1) = \int_{Z_1} f_{Y|H_1}(y|H_1)dy$;
- The probability of false alarm: $P_{fa} = P(D_1|H_0) = \int_{Z_1} f_{Y|H_0}(y|H_0)dy$;
- The probability of a miss: $P_m = P(D_0|H_1) = \int_{Z_0} f_{Y|H_1}(y|H_1)dy$;
- The probability of correct rejection: $P_r = P(D_0|H_0) = \int_{Z_0} f_{Y|H_0}(y|H_0)dy$.

Optimization criteria for the detection problem are determined by the degree of prior knowledge available. In Bayes' criterion [9], two assumptions are made. First, assume the probability of occurrence of the two source outputs is known. They are the a priori probabilities $P(H_0)$ and $P(H_1)$. $P(H_0)$ is the probability of occurrence of hypothesis H_0, while $P(H_1)$ is the probability of occurrence of hypothesis H_1. Denoting the a priori probabilities $P(H_0)$ and $P(H_1)$ by P_0 and P_1 respectively, and since either hypothesis H_0 or H_1 will always occur, we have

$$P_0 + P_1 = 1 \tag{33}$$

The second assumption is that a cost is assigned to each possible decision. In situations where we have knowledge of P_0 and P_1 it is common to apply Bayes' criterion to the optimization of the detection problem. A cost is assigned to each possible decision. The cost is due to the fact that some action will be taken based on a decision made. The consequences of one decision are different from the consequences of another. For example, in a radar detection problem, the consequences of miss are not the same as the consequences of false alarm. If we let D_i, $(i = 0,1)$, where D_0 denotes "decide H_0" and D_1 denotes "decide H_1", we can define C_{ij}, $(i,j = 0,1)$, as the cost of making decision D_i given that the true hypothesis is H_j. That is,

$$P(\text{incurring cost } C_{ij}) = P(\text{decide } D_i, H_j \text{ true}), \quad i,j = 0,1 \tag{34}$$

Bayes' criterion is to determine the decision rule so that the average cost $E[C]$, also known as risk $\Re$, is minimized. The operation $E[C]$ denotes expected value. It is also assumed that the cost of making a wrong decision is greater than the cost of making a correct decision. That is,

$$\begin{aligned} C_{01} &> C_{11} \\ C_{10} &> C_{00} \end{aligned} \tag{35}$$

Given $P(D_i, H_j)$, the joint probability that we decide D_i, and that the hypothesis H_j is true, the average cost is

$$\Re = E[C] = C_{00}P(D_0, H_0) + C_{01}P(D_0, H_1) + C_{10}P(D_1, H_0) + C_{11}P(D_1, H_1) \tag{36}$$

In terms of the defined four conditional p.d.f.'s corresponding to the four outcomes in Table 1, the average cost in Eq. (36) is rewritten as

$$\begin{aligned} \Re = P_0 C_{00} \int_{Z_0} f_{Y|H_0}(y|H_0)dy + P_1 C_{01} \int_{Z_0} f_{Y|H_1}(y|H_1)dy + \\ P_0 C_{10} \int_{Z_1} f_{Y|H_0}(y|H_0)dy + P_1 C_{11} \int_{Z_1} f_{Y|H_1}(y|H_1)dy \end{aligned} \tag{37}$$

Using the fact that

$$\int_{Z_0} f_{Y|H_j}(y|H_j)dy + \int_{Z_1} f_{Y|H_j}(y|H_j)dy = 1 \tag{38}$$

where $f_{Y|H_j}(y|H_j)$, $(j = 0,1)$, is the p.d.f. of Y corresponding to each hypothesis, substituting for Eq. (38) in Eq. (37), we obtain

$$\Re = P_0 C_{10} + P_1 C_{11} + \int_{Z_0} \left\{ \left[P_1(C_{01} - C_{11}) f_{Y|H_1}(y|H_1) \right] - \left[P_0(C_{10} - C_{00}) f_{Y|H_0}(y|H_0) \right] \right\} dy \tag{39}$$

The quantity $P_0 C_{10} + P_1 C_{11}$ is constant, independent of how we assign points in the observation space Z, and that the only variable quantity is the region of integration Z_0.

From Eq. (35), the terms inside the brackets of Eq. (39) $P_1(C_{01} - C_{11})f_{\boldsymbol{Y}|H_1}(\boldsymbol{y}|H_1)$ and $P_0(C_{10} - C_{00})f_{\boldsymbol{Y}|H_0}(\boldsymbol{y}|H_0)$, are both positive. Consequently, the risk $\Re$ is minimized by selecting the decision region $\boldsymbol{Z_0}$ to include only those points of $\boldsymbol{Y}$ for which the second term is larger, and hence the integrand is negative. Specifically, we assign to the region $\boldsymbol{Z_0}$ those points for which

$$P_1(C_{01} - C_{11})f_{\boldsymbol{Y}|H_1}(\boldsymbol{y}|H_1) < P_0(C_{10} - C_{00})f_{\boldsymbol{Y}|H_0}(\boldsymbol{y}|H_0) \tag{40}$$

The values for which the two terms are equal do not affect the risk, and can be assigned to either $\boldsymbol{Z_0}$ or $\boldsymbol{Z_1}$. Consequently, we decide H_1 if

$$P_1(C_{01} - C_{11})f_{\boldsymbol{Y}|H_1}(\boldsymbol{y}|H_1) > P_0(C_{10} - C_{00})f_{\boldsymbol{Y}|H_0}(\boldsymbol{y}|H_0) \tag{41}$$

It can be shown that fulfilling Bayes' criterion, the decision rule is equivalent to performing the test [10]:

$$\frac{f_{\boldsymbol{Y}|H_1}(\boldsymbol{y}|H_1)}{f_{\boldsymbol{Y}|H_0}(\boldsymbol{y}|H_0)} \underset{H_0}{\overset{H_1}{\underset{<}{>}}} \frac{P_0(C_{10} - C_{00})}{P_1(C_{01} - C_{11})} \tag{42}$$

where this notation means: if the left hand side is greater than the right hand side, then decide H_1; otherwise, decide H_0. The ratio of $f_{\boldsymbol{Y}|H_1}(\boldsymbol{y}|H_1)/f_{\boldsymbol{Y}|H_0}(\boldsymbol{y}|H_0)$ is called the likelihood statistic, denoted $\Lambda(\boldsymbol{y})$, which is a random variable since it is a function of the random vector $\boldsymbol{y}$. The threshold is

$$\beta = \frac{P_0(C_{10} - C_{00})}{P_1(C_{01} - C_{11})} \tag{43}$$

Therefore, Bayes' criterion, which minimizes the average cost, results in the likelihood ratio test (LRT):

$$\Lambda(\boldsymbol{y}) \underset{H_0}{\overset{H_1}{\underset{<}{>}}} \beta \tag{44}$$

We have seen that for the Bayes' criterion we require knowledge of the a priori probabilities and cost assignments for each possible decision. In many other physical situations, such as radar detection, it is very difficult to assign realistic costs and meaningful a priori probabilities. To overcome this difficulty, we use the conditional probabilities of false alarm, P_{fa}, and detection P_d. The Neyman-Pearson (N-P) test requires that P_{fa} be fixed to some value α while P_d is maximized. Since $P_m = 1 - P_d$, maximizing P_d is equivalent to minimizing P_m.

In order to minimize P_m (maximize P_d) subject to the constraint that $P_{fa} = \alpha$, we use the calculus of extrema, and form the objective function J to be

$$J = P_m + \lambda(P_{fa} - \alpha) \tag{45}$$

where λ ($\lambda \geq 0$) is the Lagrange multiplier. Given the observation space Z, there are many decision regions Z_1 for which $P_{fa} = \alpha$. The question is to determine those decision regions for which P_m is minimum. Consequently, we rewrite the objective function J in terms of the decision region to obtain

$$
\begin{aligned}
J &= \int_{Z_0} f_{Y|H_1}(y|H_1)dy + \lambda \left[\int_{Z_1} f_{Y|H_0}(y|H_0)dy - \alpha \right] \\
&= \lambda(1 - \alpha) + \int_{Z_0} \left[f_{Y|H_1}(y|H_1) - \lambda f_{Y|H_0}(y|H_0) \right] dy
\end{aligned}
\tag{46}
$$

Hence, J is minimized when values for which $f_{Y|H_1}(y|H_1) \geq \lambda f_{Y|H_0}(y|H_0)$ are assigned to the decision region Z_1. The decision rule is, therefore,

$$
\Lambda(y) = \frac{f_{Y|H_1}(y|H_1)}{f_{Y|H_0}(y|H_0)} \underset{H_0}{\overset{H_1}{\gtrless}} \lambda
\tag{47}
$$

The threshold β derived from the Bayes' criterion in Eq. (43) is equivalent to λ, the Lagrange multiplier in the Neyman-Pearson test for which the probability of false alarm is fixed to the value α. If we define the conditional density of Λ given that H_0 is true as $f_{\Lambda|H_0}(\lambda|H_0)$, then $P_{fa} = \alpha$ may be rewritten as

$$
P_{fa} = \int_{Z_1} f_{Y|H_0}(y|H_0)dy = \int_{\lambda}^{+\infty} f_{\Lambda(y)|H_0}(\lambda(y)|H_0)d\lambda
\tag{48}
$$

1.6. The Detector / Estimator

The optimal form of the detector/estimator (in the absence of data modulation) was derived by Hurd et al. [3], and is referred to here as the ML detector. The structure of this detector is illustrated in Fig. 7. This single cell detector calculates the metric of the decision statistic $S(\bar{\tau}, \bar{F}_D)$ for a given satellite and parameter estimate $\hat{\theta} = [\hat{\tau}, \hat{f}_d]$. The parameters of this detector are:

- The number of samples per primary code period, denoted as N_c;
- The number of code periods coherently integrated, denoted as M;
- The decision threshold, denoted as β.

We have assumed that an integer number of code periods are used in the observation vector y, this guarantees that the auto and cross correlations properties of the spreading codes are maintained. We have also assumed an integer number of samples per code period. If the detector metric $S(\bar{\tau}, \bar{F}_D)$ exceeds the decision threshold β, then a "hit" is declared and $\hat{\theta}$ is chosen as the optimal parameter estimate.

The detector performance is measured by the two parameters: false alarm probability and detection probability, which are defined in the following

$$
P_{fa}(\beta) = P(S(\bar{\tau}, \bar{F}_D) > \beta|H_0) = P(S(\bar{\tau}, \bar{F}_D) > \beta|\bar{\tau} \neq \tau \cup \bar{F}_D \neq F_D)
\tag{49}
$$

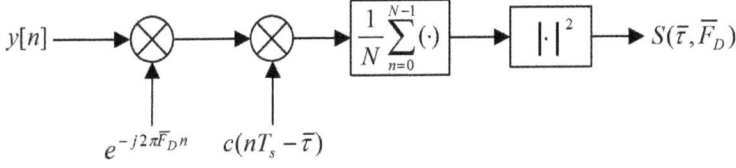

Figure 7. The ML detector.

$$P_d(\beta) = P(S(\bar{\tau}, \bar{F}_D) > \beta|H_1) = P(S(\bar{\tau}, \bar{F}_D) > \beta|\bar{\tau} = \tau \cap \bar{F}_D = F_D) \tag{50}$$

When only coherent integration is used, as in the scheme depicted in Fig. 2, the decision statistic for each cell in the search space is formed:

$$\begin{aligned} S(\bar{\tau}, \bar{F}_D) &= \left| \frac{1}{N} \sum_{n=0}^{N-1} r[n]c(nT_s - \bar{\tau})\exp\{-j2\pi\bar{F}_d n\} \right|^2 \\ &= |Y_I(\bar{\tau}, \bar{F}_d) + jY_Q(\bar{\tau}, \bar{F}_d)|^2 \\ &= Y_I^2(\bar{\tau}, \bar{F}_d) + Y_Q^2(\bar{\tau}, \bar{F}_d) \end{aligned} \tag{51}$$

The decision statistic $S(\bar{\tau}, \bar{F}_d)$ is obtained as the square absolute value of a complex Gaussian r.v. with independent real and imaginary parts. Moreover

$$\begin{aligned} \mathrm{Var}\left[Y_I(\bar{\tau}, \bar{F}_D)\right] &= \mathrm{Var}\left\{ \mathcal{Re}\left[\frac{1}{N} \sum_{n=0}^{N-1} r[n]c(nT_s - \bar{\tau})\exp\left(-j2\pi\bar{F}_D n\right) \right] \right\} \\ &= \mathrm{Var}\left\{ \frac{1}{N} \sum_{n=0}^{N-1} r[n]c(nT_s - \bar{\tau})\cos(2\pi\bar{F}_D n) \right\} \\ &= \frac{1}{N^2} \sum_{n=0}^{N-1} \mathrm{Var}\left\{ r[n]c(nT_s - \bar{\tau})\cos(2\pi\bar{F}_D n) \right\} \\ &= \frac{1}{N^2} \sum_{n=0}^{N-1} \sigma_{IF}^2 \cos^2(2\pi\bar{F}_D n) \\ &\approx \frac{\sigma_{IF}^2}{2N} \end{aligned} \tag{52}$$

Similarly,

$$
\begin{aligned}
\mathrm{Var}\left[Y_Q(\bar{\tau}, \bar{F}_d)\right] &= \mathrm{Var}\left\{ \mathcal{I}m\left[\frac{1}{N} \sum_{n=0}^{N-1} r[n]c(nT_s - \bar{\tau})\exp\left(-j2\pi\bar{F}_D n\right)\right]\right\} \\
&= \mathrm{Var}\left\{ -\frac{1}{N} \sum_{n=0}^{N-1} r[n]c(nT_s - \bar{\tau})\sin(2\pi\bar{F}_D n)\right\} \\
&= \frac{1}{N^2} \sum_{n=0}^{N-1} \mathrm{Var}\left\{ r[n]c(nT_s - \bar{\tau})\sin(2\pi\bar{F}_D n)\right\} \\
&= \frac{1}{N^2} \sum_{n=0}^{N-1} \sigma_{IF}^2 \sin^2(2\pi\bar{F}_D n) \\
&\approx \frac{\sigma_{IF}^2}{2N}
\end{aligned}
\tag{53}
$$

Thus, we can denote

$$
\mathrm{Var}\left[Y_I(\bar{\tau}, \bar{F}_D)\right] = \mathrm{Var}\left[Y_Q(\bar{\tau}, \bar{F}_D)\right] = \frac{\sigma_{IF}^2}{2N} = \sigma_n^2
\tag{54}
$$

Under null hypothesis H_0, $E[Y_I(\bar{\tau}, \bar{F}_D)] = 0$, $E[Y_Q(\bar{\tau}, \bar{F}_D)] = 0$. $S(\bar{\tau}, \bar{F}_D)$ is the sum the squares of two independent, zero-mean Gaussian r.v.'s. Thus, $S(\bar{\tau}, \bar{F}_D)$ has a central χ^2 distribution with two degrees of freedom (d.o.f.'s) [11], and its probability density function (p.d.f.) is given by:

$$
f_{S(\bar{\tau}, \bar{F}_D)|H_0}(x|H_0) = \begin{cases} \dfrac{1}{2\sigma_n^2}\exp(-\dfrac{x}{2\sigma_n^2}) & x \geq 0 \\[2mm] 0 & x \leq 0 \end{cases}
\tag{55}
$$

The probability of false alarm is then given by:

$$
P_{fa} = \int_{\beta}^{\infty} f_{S(\bar{\tau}, \bar{F}_D)|H_0}(x|H_0)\, dx = \exp\left(-\frac{\beta}{2\sigma_n^2}\right)
\tag{56}
$$

Under alternative hypothesis H_1, $y[n]$ contains both signal and noise components: $y[n] = r[n] + \eta[n]$, thus $Y_I(\bar{\tau}, \bar{F}_D)$ and $Y_Q(\bar{\tau}, \bar{F}_D)$ are no longer zero mean, and in particular:

$$
\begin{aligned}
E\left\{ Y_I(\bar{\tau}, \bar{F}_D)\right\} &= E\left\{ \frac{1}{N} \sum_{n=0}^{N-1} y[n]c(nT_s - \bar{\tau})\cos(2\pi\bar{F}_D n)\right\} \\
&= \frac{1}{N} \sum_{n=0}^{N-1} E\left\{ r[n] + \eta[n]\right\} c(nT_s - \bar{\tau})\cos(2\pi\bar{F}_D n) \\
&= \frac{1}{N} \sum_{n=0}^{N-1} r[n]c(nT_s - \bar{\tau})\cos(2\pi\bar{F}_D n)
\end{aligned}
\tag{57}
$$

By using the signal model in Eq. (10) and by assuming that $\bar{F}_D = F_D$ and $\bar{\tau} = \tau$, Eq. (57) becomes

$$
\begin{aligned}
\mathrm{E}\left\{Y_I(\bar{\tau}, \bar{F}_D)\right\} &= \frac{A}{N} \sum_{n=0}^{N-1} c^2(nT_s - \tau) \cos(2\pi F_D n + \varphi) \cos(2\pi F_D n) \\
&= \frac{A}{2N} \sum_{n=0}^{N-1} \left\{ \cos(4\pi F_D n + \varphi) + \cos\varphi \right\} \\
&\approx \frac{A}{2} \cos\varphi
\end{aligned}
\tag{58}
$$

Eq. (58) has been evaluated by neglecting the quantization effect, the impact of the front-end filter and code delay and frequency residual errors. Similarly, $\mathrm{E}\left\{Y_Q(\bar{\tau}, \bar{F}_D)\right\}$ is given by

$$
\mathrm{E}\left\{Y_Q(\bar{\tau}, \bar{F}_D)\right\} \approx \frac{A}{2} \sin\varphi
\tag{59}
$$

The variance of $Y_I(\bar{\tau}, \bar{F}_D)$ and $Y_I(\bar{\tau}, \bar{F}_D)$ is not influenced by the presence of the useful signal, which is considered as a deterministic component. Thus

$$
\begin{aligned}
Y_I(\bar{\tau}, \bar{F}_D)|H_1 &\sim \mathcal{N}(\frac{A}{2}\cos\varphi, \sigma_n^2) \\
Y_Q(\bar{\tau}, \bar{F}_D)|H_1 &\sim \mathcal{N}(\frac{A}{2}\sin\varphi, \sigma_n^2)
\end{aligned}
\tag{60}
$$

The sum of the square of two non-zero mean independent Gaussian r.v.'s leads to a non-central χ^2 r.v. with two d.o.f.'s [12]

$$
S(\bar{\tau}, \bar{F}_D)|H_1 = Y_I^2(\bar{\tau}, \bar{F}_D) + Y_Q^2(\bar{\tau}, \bar{F}_D)|H_1 \sim \chi_{nc,2}^2(\lambda, \sigma_n^2)
\tag{61}
$$

where

$$
\lambda = \mathrm{E}^2\left[Y_I(\bar{\tau}, \bar{F}_D)\right] + \mathrm{E}^2\left[Y_Q(\bar{\tau}, \bar{F}_D)\right] = \frac{A^2}{4}
\tag{62}
$$

is the non-centrality parameter. The p.d.f. of $S(\bar{\tau}, \bar{F}_D)$ under alternative hypothesis H_1 is given by

$$
f_{S(\bar{\tau}, \bar{F}_D)|H_1}(x|H_1) =
\begin{cases}
\dfrac{1}{2\sigma_n^2} \exp\left\{-\dfrac{x+\lambda}{2\sigma_n^2}\right\} I_0\left(\dfrac{\sqrt{x\lambda}}{\sigma_n^2}\right) & x \geq 0 \\
0 & x \leq 0
\end{cases}
\tag{63}
$$

where $I_\nu(\cdot)$ is the νth order modified Bessel function of the first kind, defined by:

$$I_\nu(x) = \sum_{k=0}^{\infty} \frac{(-1)^k}{k!\Gamma(k+\nu+1)} \left(\frac{x}{2}\right)^{2k+\nu} \tag{64}$$

where $\Gamma(K)$ is the gamma function, defined by:

$$\Gamma(x) = \int_0^{+\infty} \exp(-t)t^{x-1}\, dt \tag{65}$$

which, for $x \in Z^+$, is related to the factorial function by: $\Gamma(x) = (x-1)!$. The probability of detection P_d is given by:

$$P_d = \int_\beta^\infty f_{S(\tilde{\tau}, \tilde{F}_D)|H_1}(x|H_1)\, dx = Q_1\left(\frac{\sqrt{\lambda}}{\sigma_n}, \frac{\sqrt{\beta}}{\sigma_n}\right) \tag{66}$$

where $Q_K(a, b)$ is the Kth order generalized Marcum Q-function [13, 14], defined by

$$Q_K(a,b) = \frac{1}{a^{K-1}} \int_b^{+\infty} x^K \exp\left\{-\frac{a^2+x^2}{2}\right\} I_{K-1}(ax)\, dx \tag{67}$$

Non-coherent combining is a technique for increasing detection probability for a given false alarm probability in the presence of a randomly varying phase offset (for example, due to a carrier Doppler offset, or data modulation effects) at the receiver antenna. This approach dates back to the early days of radar detection theory [13] and the detection of unknown signals in noise [15]. It has found wide applications in the acquisition of Direct-Sequence Spread-Spectrum (DS/SS) signals in the presence of Doppler offset and data modulation [16–21] and it is also widely used in the acquisition of GNSS signals [22, 23]. In Fig. 8, the acquisition scheme with non-coherent integrations is is illustrated. This technique operates by simply accumulating a number, denoted K, of sequential instances of the output of the basic ML detector. The squaring blocks remove the phase dependence and the CAFs are non-coherently summed. The final decision variable is obtained as

$$S_K(\tilde{\tau}, \tilde{F}_D) = \sum_{k=0}^{K-1} S_k(\tilde{\tau}, \tilde{F}_D) \tag{68}$$

where the subscript K indicates that K non-coherent integrations have been used. The index k has been used in the right side of Eq. (68) in order to distinguish different realizations of the basic squared CAF envelope $S_k(\tilde{\tau}, \tilde{F}_D)$. Those realizations have been evaluated by using non-overlapping portions of the input signal $y[n]$.

Under null hypothesis H_0, it has already been shown that, for $K = 1$, i.e. in absence of non-coherent integration, the output of the ML detector is a central χ^2 distributed r.v. with

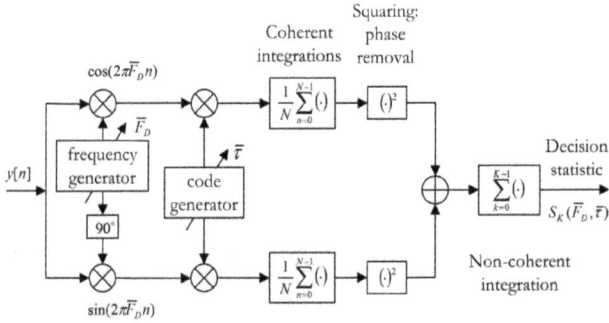

Figure 8. Non-coherent acquisition scheme.

two d.o.f.'s. The non-coherently integrated decision statistic $S_K(\bar{\tau}, \bar{F}_D)$ is given by the sum of K independent r.v.'s with 2 d.o.f.'s. From [11], we know that the sum of K independent central χ^2 distributed r.v.'s with n d.o.f.'s is itself a central χ^2 r.v., with nK d.o.f.'s. By using the properties of χ^2 r.v.'s, the distribution of $S_K(\bar{\tau}, \bar{F}_D)$ under null hypothesis H_0 is, therefore, the distribution of a central χ^2 r.v. with $2K$ d.o.f.'s and variance of the underlying Gaussian distribution σ_n^2, whose conditional p.d.f. under H_0 is given by:

$$f_{S_K(\bar{\tau},\bar{F}_D)|H_0}(x|H_0) = \begin{cases} \dfrac{1}{2\sigma_n^2} \dfrac{1}{\Gamma(K)} \left(\dfrac{x}{2\sigma_n^2}\right)^{K-1} \exp\left(-\dfrac{x}{2\sigma_n^2}\right) & x \geq 0 \\ 0 & x < 0 \end{cases} \tag{69}$$

The probability of false alarm P_{fa} is then given by:

$$P_{fa,K}(\beta) = \int_{\beta}^{+\infty} f_{S_K(\bar{\tau},\bar{F}_D)|H_0}(x|H_0)\,dx = \frac{\Gamma_K\left(\frac{\beta}{2\sigma_n^2}\right)}{\Gamma(K)} \tag{70}$$

where $\Gamma_K(x)$ is the complementary incomplete Gamma function of order K:

$$\Gamma_K(x) = \int_x^{+\infty} \exp(-t)\,t^{K-1}\,dt \tag{71}$$

Hence, using Eq. (71), it is easy to obtain

$$P_{fa,K}(\beta) = \frac{1}{2^K(K-1)!\,\sigma_n^{2K}} \int_{\beta}^{+\infty} x^{K-1}\exp\left(-\frac{x}{2\sigma_n^2}\right) dx \tag{72}$$

The final expression can be then obtained integrating by part $K-1$ times. After some algebraic manipulations it is possible to obtain the relationship between the false alarm probability

$P_{fa,K}(\beta)$ and the decision threshold β in the following form:

$$
\begin{aligned}
P_{fa,K}(\beta) &= \frac{1}{2^K(K-1)!\,\sigma_n^{2K}}\left[\sum_{i=1}^{K}(2\sigma_n^2)^i\frac{(K-1)!}{(K-i)!}\beta^{K-i}\exp\left(-\frac{\beta}{2\sigma_n^2}\right)\right] \\
&= \frac{\exp\left(-\frac{\beta}{2\sigma_n^2}\right)\beta^K}{2^K\sigma_n^{2K}}\sum_{i=1}^{K}\frac{1}{(K-i)!}\left(\frac{2\sigma_n^2}{\beta}\right)^i \\
&= \exp\left(-\frac{\beta}{2\sigma_n^2}\right)\sum_{i=0}^{K-1}\frac{1}{i!}\left(\frac{\beta}{2\sigma_n^2}\right)^i
\end{aligned}
\tag{73}
$$

which has to be be computed numerically.

Under alternative hypothesis H_1, when the code and the Doppler frequency shift of the local signal replica match the ones of the incoming signal, the output of the ML detector $S_k(\tilde{\tau}, \tilde{F}_D)$ is a non-central χ^2 distributed r.v. with two d.o.f.'s and non-centrality parameter λ. Denoting by λ_i the non-centrality parameter of the ith output of the ML detector, then the metric $S_K(\tilde{\tau}, \tilde{F}_D)$ is a non-central χ^2 distributed r.v. with $2K$ d.o.f.'s and non-centrality parameter λ_K:

$$
\lambda_K = \sum_{i=1}^{K}\lambda_i = K\lambda = K\frac{A^2}{4}
\tag{74}
$$

The distribution of $S_K(\tilde{\tau}, \tilde{F}_D)$ is then given by:

$$
f_{S_K(\tilde{\tau},f_D)|H_1}(x|H_1) = \begin{cases} \dfrac{1}{2\sigma_n^2}\left(\dfrac{x}{\lambda}\right)^{\frac{K-1}{2}}\exp\left(-\dfrac{x+\lambda}{2\sigma_n^2}\right)I_{K-1}\left(\dfrac{\sqrt{x\lambda}}{\sigma_n^2}\right) & x \geq 0 \\[3mm] 0 & x < 0 \end{cases}
\tag{75}
$$

where $I_K(\cdot)$ is the modified Bessel function of the first kind of order K, as defined by Eq. (64).

Therefore, the probability of detection is given by:

$$
\begin{aligned}
P_{d,K}(\beta) &= \int_{\beta}^{+\infty}f_{S_K(\tilde{\tau},f_D)|H_1}(x|H_1)\,dx \\
&= Q_K\left(\frac{\sqrt{\lambda_K}}{\sigma_n},\frac{\sqrt{\beta}}{\sigma_n}\right) \\
&= Q_K\left(\frac{\sqrt{K\lambda}}{\sigma_n},\frac{\sqrt{\beta}}{\sigma_n}\right)
\end{aligned}
\tag{76}
$$

where $Q_K(a,b)$ is the Kth order Marcum Q-function, σ_n^2 is the variance of the in-phase and quadrature outputs defined in Eq. (54).

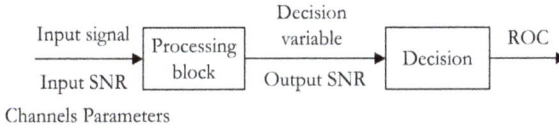

Figure 9. General detection scheme. The input signal is processed in order to produce a decision variable for establishing the presence of a desired signal.

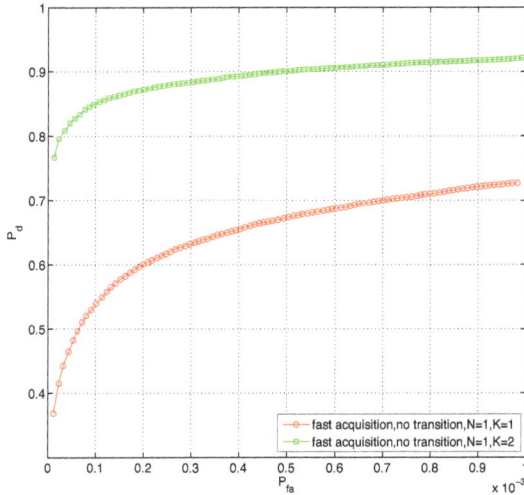

Figure 10. ROC comparison with Galileo E1 OS BOC(1,1) signal for a case of $C/N_0 = 35$ dB-Hz, coherent integration time 4 ms.

1.7. Receiver Operating Characteristics

A general detection process consists in determining the presence or absence of a desired signal from a set of noisy data[9]. A general detection process is depicted in Fig. 9: the noisy input signal is processed and a decision variable derived. The decision variable is then used for deciding the presence or absence of the desired signal. The input signal is characterized by an input signal-to-noise ratio (*input SNR*), which is the ratio between the desired signal and noise powers. The desired signal can be further degraded by the presence of additional impairments, such as clutter, fading and interference. All these impairments are accounted for by specific models [24–26] and characterized by a set of parameters describing the channel characteristics responsible for the degradation of the useful signal.

The processing block is aimed at enhancing the desired signal by combining its samples and by exploiting a priori information available at the detector. The processing block is at first a detection process aimed to determine the presence or absence of the signal transmitted by a specific satellite. The processing block can be characterized by the same parameters exploited to characterize a general detector [27]. Different processing techniques can be adopted, such as coherent, non-coherent [28] and differentially coherent integrations [29].

The processing block is characterized by a set of parameters, such as the coherent integration time and the number of non-coherent integrations. The output of the processing block is a r.v., namely the decision statistic, characterized by two p.d.f.'s referring to the presence or absence of the desired signal. These p.d.f.'s and, in particular, the corresponding complementary cumulative distributions, completely determine the detector performance. The probability that the decision statistic passes a threshold β is called the detection probability P_d if the desired signal is present, and false alarm probability P_{fa} if it is absent. In signal detection theory, the performance of a generic detector may be completely expressed in a graphical plot of detection probability P_d versus false alarm probability P_{fa}. The curves which depict P_d versus P_{fa} for various values of SNR are known as Receiver Operating Characteristic (ROC), or simply ROC curves [30, 31]. ROC analysis provides tools to select possibly optimal detectors.

As an example, Fig. 10 depicts an ideal optimal ROC comparison between the Galileo E1 OS BOC(1,1) signals when employing different number of non-coherent integrations. The results have highlighted how better performance can be achieved by means of non-coherent integrations.

In general the ROC curves are used to compare the performances of different detectors operating on the same input data. The curve closer to the upper left corner of the diagram generally identifies the best detector among the compared ones.

2. Bit Sign Transition Cancellation Method

2.1. Detection and Estimation Main Strategy

All the acquisition systems for GNSS applications described in the literature [32–34] are based on a well-known result of the ML estimation theory [35], which can be briefly summarized as follows.

Suppose that there is a vector $y = [y[0], y[1], \cdots, y[N-1]]$ which is given by measured samples of a realization of a random process of the type $Y[n] = r[n] + W[n]$, where $W[n]$ is a zero-mean *white Gaussian noise* (WGN) random process, and $r[n]$ is a signal which contains K unknown parameters which form a vector $p = [p[0], p[1], \cdots, p[K-1]]$. The ML estimation $\hat{p}_{ML}$ of p is

$$\hat{p}_{ML} = \arg\max_{\bar{p}} \sum_{n=0}^{N-1} y[n]\bar{r}[n] \tag{77}$$

where $\bar{r}[n]$ is a test signal with the similar structure to $r[n]$ and with parameter vector $\bar{p}$ which consists of variables $\bar{p}_i$, $i = 1, \cdots, K$, defined in a proper support D_p which contains all the possible values that can be assumed by the unknown parameters p_i. Eq. (77) is a special case of ML estimation when the parameters to be estimated are contained in a signal affected by additive white Gaussian noise, and the energy of the test signal $\bar{r}[n]$ does not depend on $\bar{p}$. In this case, the ML estimation only depends on the scalar product between the test signal and the received signal, defined in the form:

$$R_{y,r}(\bar{p}) = \sum_{n=0}^{N-1} y[n]\bar{r}[n] \tag{78}$$

In principle in GNSS applications, the parameter vector should contain several unknowns: the Doppler frequency shift, the code phase delay, the carrier phase and the value of the data bits. However, the task of the acquisition is to estimate only the code phase delay and Doppler frequency shift; therefore, the problem can be simplified by ignoring the carrier phase estimation and trying to bypass the data bit value issue. This makes the computational burden of the acquisition stage feasible in any practical GNSS receiver; otherwise a complete choice would lead to theoretical results which could not be even applied with the most sophisticated current technology. The classical acquisition approach used in GNSS receiver is to adopt a non-coherent acquisition scheme and to ignore the presence of the navigation data bits [32–34].

Therefore, in the GNSS applications the test signal $\tilde{r}[n]$ has not exactly the same structure with the received signal $r[n]$, the correlation function $R_{y,r}(\tilde{\tau}, \tilde{f}_d)$ becomes

$$R_{y,r}(\tilde{\tau}, \tilde{f}_d) = \sum_{n=0}^{N-1} y[n] c_{Loc}(nT_s - \tilde{\tau}) \exp\left\{ -j2\pi(f_{IF} + \tilde{f}_d)nT_s \right\} \tag{79}$$

where N denotes the coherent integration time (in the discrete time sense), the parameter vector $\tilde{p} = [\tilde{\tau}, \tilde{f}_d]$ contains only two elements, $\tilde{r}[n]$ is a constant-energy signal of the type

$$\tilde{r}[n] = c_{Loc}(nT_s - \tilde{\tau}) \exp\left\{ -j2\pi(f_{IF} + \tilde{f}_d)nT_s \right\} \tag{80}$$

In the non-coherent scheme, the search for the maximum is performed on the squared CAF envelope $S_{y,r}^2(\tilde{\tau}, \tilde{f}_d) = |R_{y,r}(\tilde{\tau}, \tilde{f}_d)|^2$, from which we obtain

$$\hat{p}_{ML} = \arg\max_{\tilde{p}} S_{y,r}^2(\tilde{\tau}, \tilde{f}_d) \tag{81}$$

The CAF evaluation is only the first step performed by a GNSS acquisition scheme. In the other steps, the envelope is evaluated, and some average operations are performed before envelope computation (coherent integration) or after envelope computation (noncoherent integration). The detection phase is activated at this point. All these operations can be repeated by adopting some multitrial techniques.

The CAF evaluation method by using the FFT's to perform circular correlation is adopted. It is well-known that FFT's can be used to perform fast circular correlations, so FFT-based methods are often adopted to evaluate the CAF. This method is extremely efficient because it works on vectors in a parallel way, however it is sensitive to CAF peak impairments due to bit sign transitions [36–39].

By applying the results of the ML estimation theory, it is possible to show that the best estimates of the code phase delay $\tilde{\tau}$ and the Doppler shift $\tilde{f}_d$ in the presence of AWGN, are based on the maximization of the CAF. In the FFT-based scheme, a signal vector $y = [y[0], y[1], \cdots, y[N-1]]$ of N samples is extracted from the incoming IF signal and

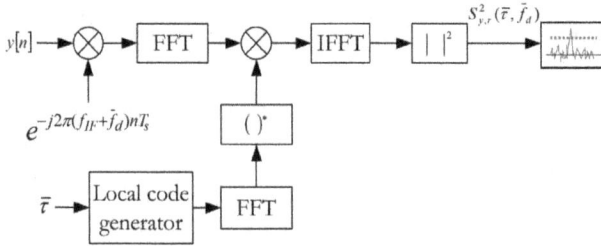

Figure 11. The parallel acquisition scheme: the CAF is determined by using a circular convolution employing efficient FFT's. The code generator also includes the subcarrier.

multiplied by a complex test signal $\exp\{-j2\pi(f_{IF} + \bar{f}_d)nT_s\}$, so as to obtain a sequence $q_l[n] = y[n]\exp\{-j2\pi(f_{IF} + \bar{f}_d)nT_s\}$ for each $\bar{f}_d$ value, that is for each Doppler bin in the search space.

In Fig. 11, the sequence $q_l[n]$ is then FFT-transformed and multiplied by the complex conjugate of the FFT of the local code replica $c_{Loc}[n]$ including primary PRN code and sub-carrier. Finally the inverse FFT is made so as to obtain the circular CCF $R_{y,r}(\bar{\tau}, \bar{f}_d)$, which can be evaluated in the form

$$R_{y,r}(\bar{\tau}, \bar{f}_d) = \text{IDFT}\left\{\text{DFT}\left[q_l[n]\right] \cdot \left[\text{DFT}\left[c_{Loc}(nT_s)\right]\right]^*\right\} \tag{82}$$

The FFT is used to evaluate the DFT. The local code generator includes the subcarrier.

A CCF evaluated by applying a classical serial scheme [40] and circular CCF coincide only in the presence of periodic sequences. The presence of a sign transition in the data vector completely destroys the code periodicity, so leading to serious peak impairments in the search space [36, 37, 39]. Since the FFT-based scheme takes advantage of parallel processing of the data vector, if we want to modify the scheme to solve the problem of the bit sign transition, we have to adopt a method which does not alter the benefits deriving from the block processing approach.

2.2. Bit Sign Transition Problem

Bit sign transition affects the CAF evaluation especially in the block processing methods working on an input vector y. Due to the existence of bit sign transitions in the input vector y, the fast acquisition method based on FFT suffers from the CAF peak splitting impairments since its intrinsic nature of processing blocks of data.

Bit sign transition modifies the shape of the CAF envelope. The approach used in GPS C/A code acquisition is to say that an acquisition system based on a 10 ms integration time (equivalent to ten primary code periods) works properly, since the bit sign transition could not occur in two consecutive signal segments of 10 ms, the bit duration being 20 ms. This means that there is at least one bit sign transition free search space in two consecutive signal segments of 10 ms. In the case of Galileo E1 OS signal, the sign transition could possibly

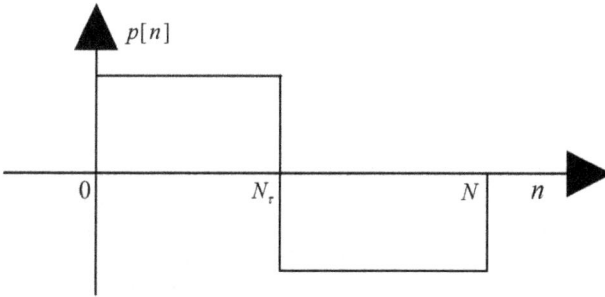

Figure 12. Function $p[n]$.

occur in any time interval of 4 ms (equivalent of a single primary code period), so the same approach does not apply.

The fast acquisition method based on FFT is extremely efficient, but since its intrinsic nature of processing blocks of data, it may suffer from the CAF peak splitting impairments due to the presence of bit sign transitions. In case of the Galileo E1 data channel (E1-B) signal, the bit sign transition could possibly occur in any time interval of 4 ms (equivalent to a single primary code period).

It is possible to show that the presence of bit sign transitions does not destroy the possibility of detecting the satellites in view, but it introduces an estimation error in the selection of the estimated pair $\hat{p} = [\hat{\tau}, \hat{f}_d]$, where $\hat{\tau}$ is the estimated code delay and $\hat{f}_d$ is the estimated Doppler shift in the acquisition stage. In fact when the local code replica matches the received signal perfectly, a code stripping process can be applied to $y[n]$, obtaining the signal:

$$
\begin{aligned}
x[n] &= Ad[n - \tau/T_s]\tilde{c}[n - \tau/T_s]c_{Loc}[n - \tau/T_s]\cos[2\pi(f_{IF} + f_d)nT_s + \varphi] \cdot \\
&\quad \exp\{-j2\pi(f_{IF} + \tilde{f}_d)nT_s\} \\
&= Ad[n - \tau/T_s]\cos[2\pi(f_{IF} + f_d)nT_s + \varphi]\exp\{-j2\pi(f_{IF} + \tilde{f}_d)nT_s\}
\end{aligned}
\tag{83}
$$

The CAF envelope becomes

$$
\begin{aligned}
S_{y,r}(\tau, \tilde{f}_d) &= \left| \sum_{n=0}^{N-1} x[n] \right| \\
&= \left| \sum_{n=0}^{N-1} Ad[n - \tau/T_s]\cos[2\pi(f_{IF} + f_d)nT_s + \varphi]e^{-j2\pi(f_{IF} + \tilde{f}_d)nT_s} \right|
\end{aligned}
\tag{84}
$$

where the term $d[n - \tau/T_s]\cos[2\pi(f_{IF} + f_d)nT_s + \varphi]$ can be written as

$$b_\tau[n] = p_N[n]d[n - \tau/T_s]\cos[2\pi(f_{IF} + f_d)nT_s + \varphi]$$
$$= p[n]\cos[2\pi(f_{IF} + f_d)nT_s + \varphi] \tag{85}$$

with the presence of a rectangular window function $p_N[n]$ in the interval $n \in [0, N-1]$ which has a unitary amplitude. In case of bit sign transition the function $p[n] = p_N[n]d[n - \tau/T_s]$ becomes a two-pulse signal which reverses the sign, as shown in Fig. 12, where $N_\tau = \lfloor \tau/T_s \rfloor$ is the delay expressed in the discrete time notation. Eq. (84) can be regarded as the Discrete Time Fourier Transform (DTFT) of a sinusoidal function modulated by $p[n]$, which behaves as a sort of subcarrier. The effect on the CAF peak is a split of its power into two different smaller side lobes along the Doppler shift axis.

By using the Euler formula $\cos\alpha = 1/2(e^{j\alpha} + e^{-j\alpha})$, and introducing the discrete-time function $p[n]$, the analytical expression of the spectrum can be obtained as

$$R_{y,r}(\tau, \bar{f}_d) = \sum_{n=0}^{N-1} Ap[n]\cos[2\pi(f_{IF} + f_d)nT_s + \varphi]\exp\{-j2\pi(f_{IF} + \bar{f}_d)nT_s\}$$
$$= \frac{1}{2}A\sum_{n=0}^{N-1} p[n]\left\{\exp\left\{j[2\pi(f_d - \bar{f}_d)nT_s + \varphi]\right\} + \right. \tag{86}$$
$$\left. \exp\left\{-j[2\pi(2f_{IF} + f_d + \bar{f}_d)nT_s + \varphi]\right\}\right\}$$

The second high frequency term can be neglected, so we can obtain

$$R_{y,r}(\tau, \bar{f}_d) \approx \frac{1}{2}Ae^{j\varphi}\sum_{n=0}^{N-1} p[n]\exp\left\{j2\pi(f_d - \bar{f}_d)nT_s\right\}$$
$$= \frac{1}{2}Ae^{j\varphi}\left\{\sum_{n=0}^{N_\tau-1}\exp\{j2\pi(f_d - \bar{f}_d)nT_s\} - \sum_{n=N_\tau}^{N-1}\exp\{j2\pi(f_d - \bar{f}_d)nT_s\}\right\} \tag{87}$$

The two terms in Eq. (87) are two truncated geometrical series, which can be easily summed giving the result [41]

$$R_{y,r}(\tau, \bar{f}_d) \approx \frac{1}{2}Ae^{j\varphi}\left\{e^{j\alpha_1}\frac{\sin(\pi(f_d - \bar{f}_d)N_\tau T_s)}{\sin(\pi(f_d - \bar{f}_d)T_s)} - e^{j\alpha_2}\frac{\sin(\pi(f_d - \bar{f}_d)(N - N_\tau)T_s)}{\sin(\pi(f_d - \bar{f}_d)T_s)}\right\} \tag{88}$$

where $\alpha_1 = \pi(f_d - \bar{f}_d)(N_\tau - 1)T_s$, and $\alpha_2 = \pi(f_d - \bar{f}_d)(N + N_\tau - 1)T_s$. In the correct Doppler cell ($f_d = \bar{f}_d$), Eq. (88) becomes

$$R_{y,r}(\tau, \bar{f}_d) = \frac{1}{2}Ae^{j\varphi}[N_\tau - (N - N_\tau)] \tag{89}$$

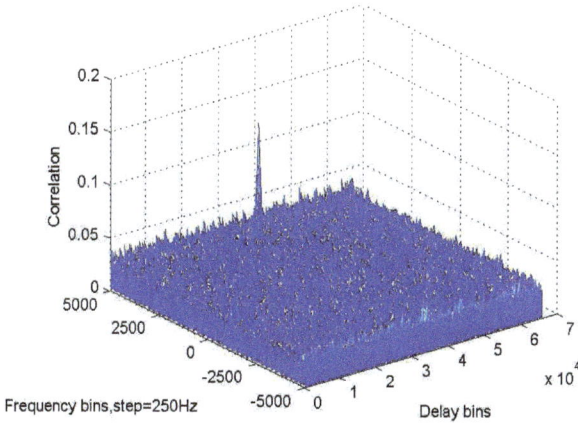

Figure 13. CAF envelope of the Galileo E1-B BOC(1,1) signal. The signal contains only the primary PRN code (no data sign transition presents). The SIS parameters are $f_d = 3500$ Hz, $\tau = 2$ ms, and $C/N_0 = 45$ dB-Hz. The coherent integration time is 4 ms.

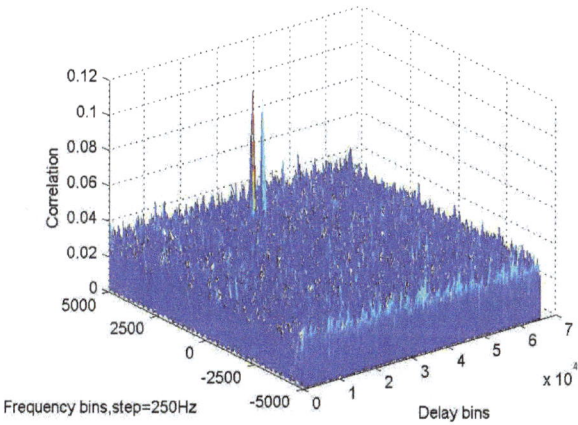

Figure 14. CAF envelope of the Galileo E1-B BOC(1,1) signal, with $C/N_0 = 45$ dB-Hz. The signal contains the primary code and the data (a bit sign transition is present). The coherent integration time is 4 ms.

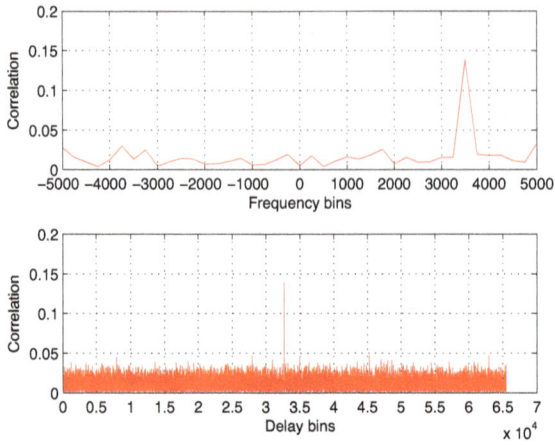

Figure 15. Curves (Energy spectrum and CCF) extracted from the CAF envelope in case of no bit sign transition.

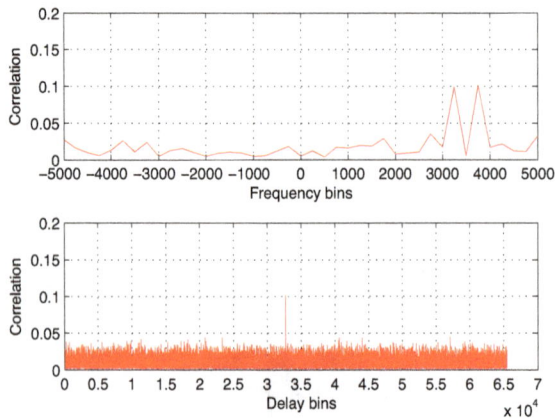

Figure 16. Curves (Energy spectrum and CCF) extracted from the CAF envelope when a bit sign transition is present in the primary code period.

which becomes zero for $N_\tau = \frac{N}{2}$. This means that the CAF peak completely disappears in the correct Doppler shift position when the bit transition occurs at the middle of the code period. However, the information is not lost as the correlation function in Eq. (87) exhibits side peaks, which can be properly exploited to recover the information on the code delay and Doppler frequency. In section 2.3, a two-step GNSS signal acquisition method will be described to recover the CAF peak in presence of bit sign transitions.

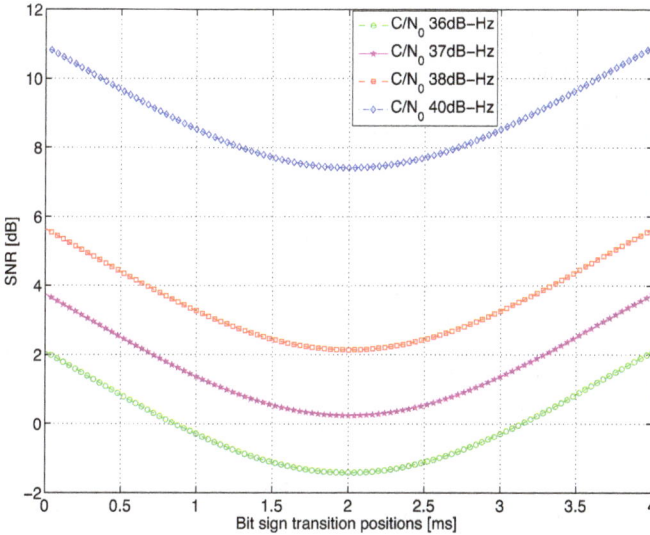

Figure 17. SNR evolutions dependent on the bit sign transition position. Three code periods are coherently integrated.

To show the presence of side lobes in the search space the Galileo Open Service E1 data channel (E1-B) signal containing navigation data bits has been simulated with the symbol rate of 250 symbols/s, which means that a potential bit sign transition exists in each primary code period. The acquisition experiments were performed with a Doppler shift f_d of 3500 Hz, a code phase delay τ of 2 ms and a *Carrier-to-Noise Ratio* (C/N_0) of 45 dB-Hz. This C/N_0 value represents a relatively optimistic situation, which was considered only to better show the spectrum splitting effect of the CAF and then to motivate the modification to the state of the art acquisition scheme. In Fig. 13, the CAF envelope is evaluated based on the fast acquisition scheme in case of no presence of bit sign transitions. When a bit sign transition is introduced to the signal, the splitting effect of the CAF main lobe can be clearly seen in Fig. 14. In this case the FFT-based fast acquisition scheme suffers much from the CAF peak loss caused by the presence of bit sign transition.

Two plots extracted from the CAF envelopes of Fig. 13 and Fig. 14 are provided, which are shown in Fig. 15 and Fig. 16, respectively. The upper curves represent the sections of the CAF envelopes in the Doppler shift domain (which are energy spectrum functions) at the correct code delay bin; the lower curves indicate the CCF in the code delay domain at the right Doppler shift bin. In Fig. 15, it is known that the CAF peak locates its position correctly along the code delay and Doppler shift axes respectively when bit sign transition is not present in the signal. When dealing with the bit sign transition case, in the upper plot of Fig. 16, it is clearly observed that the CAF peak is divided into two different smaller side lobes along the Doppler shift axis, leading to a wrong Doppler shift estimation; while from the lower plot of Fig. 16 it is evident that the presence of bit sign transition does not impair the CAF peak position which is located in the correct code delay bin. It is possible to have a conclusion that the CAF main peak position remains unchanged in the code delay domain even with the presence of bit sign transition.

In order to further evaluate the CAF peak splitting effect dependent on the bit sign transition position in the function $p[n]$ in Eq. (85), an appropriate metric known as Signal-to-Noise Ratio (SNR) is adopted, which is defined as

$$\mathrm{SNR} \overset{def}{=} \frac{|R_s(\hat{\tau}, \hat{f}_d)|^2}{E\{|R_n(\bar{\tau}, \bar{f}_d)|^2\}} \tag{90}$$

where $R_s(\hat{\tau}, \hat{f}_d)$ is the circular correlation function value specific for the CAF peak position when only useful signal is present, $E\{|R_n(\bar{\tau}, \bar{f}_d)|^2\}$ is the expected value of the squared CAF envelope due to only noise contribution. SNR is thus a measure of the signal power to the average noise power. To determine the SNR values dependent on different positions of bit sign transitions in the received signal segment $y[n]$, simulation campaigns have been performed with several cases of C/N_0 values. In the simulation tests three code periods are coherently integrated and bit sign transitions occur in the code periods alternatively. The simulation results are shown in Fig. 17, where the SNR value tends to decrease when the bit sign transitions move towards the middle position in a code period, resulting in about 3.5 dB loss.

2.3. Two-step Acquisition Scheme

In this section, a two-step acquisition algorithm is proposed to overcome the problem of CAF peak splitting caused by the presence of bit sign transitions. The idea is to exploit the fact that the CAF peak splitting only occurs in the Doppler shift domain, while in the code delay domain the CAF peak position remains almost unchanged. In the first acquisition step the code delay $\hat{\tau}$ is estimated so as to tentatively align the local code sequence with the bit sign transition position in the received signal segment, while in the second acquisition step the Doppler shift $\hat{f}_d$ is estimated. In other words, the estimated pair $\hat{p} = [\hat{\tau}, \hat{f}_d]$ is obtained in two consecutive steps. The first acquisition step aims to get estimated code delay value $\hat{\tau}_1$ by using the FFT-based fast acquisition method. The Doppler shift $\hat{f}_{d,1}$ is not estimated in this step because it could be erroneous due to the CAF peak splitting effect. Noise reduction techniques, such as coherent integration and non-coherent integration, can be adopted in order to increase the acquisition sensitivity. The coherently integrated CAF envelope $S_1(\bar{\tau}, \bar{f}_d)$ evaluated in the first acquisition step can be written as

$$S_1(\bar{\tau}, \bar{f}_d) = \left| \frac{1}{N_1} \sum_{n=1}^{N_1} R_n(\bar{\tau}, \bar{f}_d) \right| \tag{91}$$

where $R_n(\bar{\tau}, \bar{f}_d)$ is the nth contribution in the coherent integration process; N_1 is the number of code periods applied to the coherent integration process in the first step. Non-coherent integration can be used after the coherent integration operation is made. The non-coherently integrated CAF envelope $G_1(\bar{\tau}, \bar{f}_d)$ can be written as in the following

$$G_1(\bar{\tau}, \bar{f}_d) = \sqrt{\frac{1}{K_1} \sum_{k=1}^{K_1} S_{1,k}^2(\bar{\tau}, \bar{f}_d)} \qquad (92)$$

where $S_{1,k}(\bar{\tau}, \bar{f}_d)$ is the kth coherently integrated CAF envelope in the non-coherent integration process; K_1 is the number of periods adopted in the evaluation of the non-coherently integrated CAF envelope $G_1(\bar{\tau}, \bar{f}_d)$. In the first acquisition step the estimated pair $\hat{\boldsymbol{p}}_{ML,1} = [\hat{\tau}_1, \hat{f}_{d,1}]$

$$\hat{\boldsymbol{p}}_{ML,1} = [\hat{\tau}_1, \hat{f}_{d,1}] = \underset{\boldsymbol{p}_1}{\arg\max}\ G_1(\bar{\tau}, \bar{f}_d) \qquad (93)$$

is obtained, but only the estimated code delay value $\hat{\tau}_1$ is retained as valid. The estimated Doppler shift value $\hat{f}_{d,1}$ is discarded, as it could be possibly affected by the CAF peak splitting errors (as shown in Fig. 16).

In the second acquisition step the obtained code delay estimate $\hat{\tau}_1$ is used to extract a new signal vector aligned with the local code replica. In this way the effect of the bit transition practically disappears, even if the alignment is not perfect. Coherent and non-coherent integrations can be again employed in the second acquisition step. The coherently integrated CAF envelope $S_2(\bar{\tau}, \bar{f}_d)$ evaluated in the second acquisition step can be written as

$$S_2(\bar{\tau}, \bar{f}_d) = \left| \frac{1}{N_2} \sum_{n=1}^{N_2} R_i(\bar{\tau}, \bar{f}_d) \right| \qquad (94)$$

Similarly, the non-coherently integrated CAF envelope $G_2(\bar{\tau}, \bar{f}_d)$ in the second acquisition step can be written in the form

$$G_2(\bar{\tau}, \bar{f}_d) = \sqrt{\frac{1}{K_2} \sum_{k=1}^{K_2} S_{2,k}^2(\bar{\tau}, \bar{f}_d)} \qquad (95)$$

A new pair $\hat{\boldsymbol{p}}_{ML,2} = (\hat{\tau}_2, \hat{f}_{d,2})$ is now estimated as

$$\hat{\boldsymbol{p}}_{ML,2} = (\hat{\tau}_2, \hat{f}_{d,2}) = \underset{\boldsymbol{p}_2}{\arg\max}\ G_2(\bar{\tau}, \bar{f}_d) \qquad (96)$$

and only the Doppler frequency shift value $\hat{f}_{d,2}$ is retained. The code delay estimate $\hat{\tau}_2$ should give a null value now, due to the new signal alignment performed in the second acquisition step. Therefore the code delay estimate $\hat{\tau}_2$ can be now discarded or it could be further used to refine the estimated code delay value $\hat{\tau}_1$ obtained in the first acquisition step.

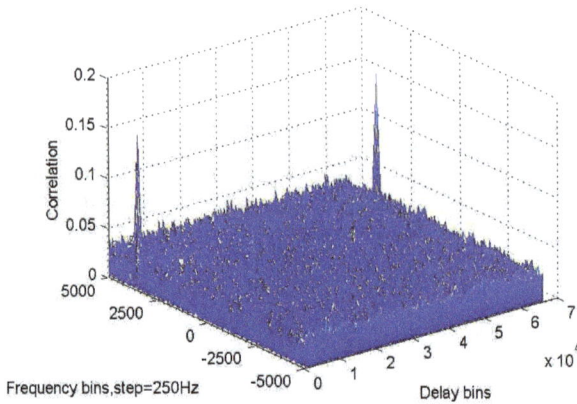

Figure 18. CAF envelope of the Galileo E1-B signal evaluated by the two-step acquisition method for a C/N_0 of 45 dB-Hz with bit sign transition.

The CAF envelope in the search space evaluated in the second step ($N_2 = 1$ and $K_2 = 1$) is shown in Fig. 18. Two CAF peaks appear at the correct Doppler frequency estimate value ($f_d = 3500$ Hz). This is due to the fact that the code delay is zero in the second step, the bit sign transition practically disappears, and then the typical correlation triangles are located at the beginning and end positions along the code delay axis where each correlation peak energy is correctly concentrated. This result is better highlighted in Fig. 19: the upper curve shows that the CAF peak is located at the correct Doppler shift position ($f_d = 3500$ Hz); the lower curve proves that the local code replica aligns perfectly to the bit sign transition position in the second acquisition step because of the available right recovery of the code phase delay $\hat{\tau}_1$ achieved in the first acquisition step.

In order to validate this proposed two-step acquisition technique, simulation campaigns have been performed on the simulated Galileo E1 OS BOC(1,1) signal, where the spreading code is modulated by fake data with correct rate. The behavior of the proposed technique is given in terms of histograms of the estimated Doppler shift and code phase delay; and in order to assess the acquisition performances ROC and SNR curves have been also addressed to compare the proposed two-step acquisition method with the state-of-the-art acquisition approach.

Firstly a preliminary performance analysis of the two-step acquisition method has been carried out by means of histogram plots of the estimated Doppler shift and the estimated code phase delay. The simulation scenario considered a Galileo E1 BOC(1,1) signal with a code delay τ of 2.5 ms, a Doppler shift f_d of 3500 Hz and a carrier to noise power density ratio C/N_0 of 30 dB-Hz. Six code periods have been coherently integrated ($N = 6$, $K = 6$) for both the classical fast acquisition approach and the proposed two-step acquisition technique. The Monte Carlo simulation experiments have been repeated for 1000 times and the histograms of the estimates of f_d and τ are given respectively.

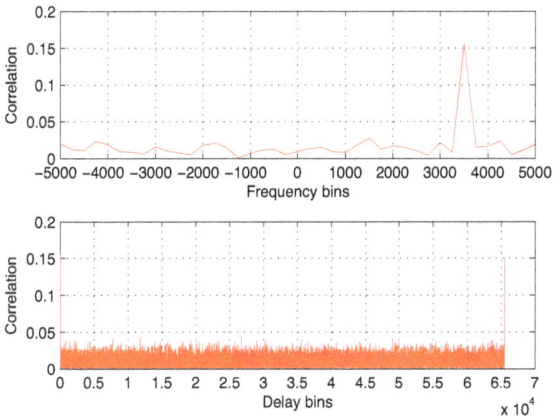

Figure 19. Curves (Energy spectrum and CCF) extracted from the CAF envelope evaluated with the two-step acquisition method with bit transition.

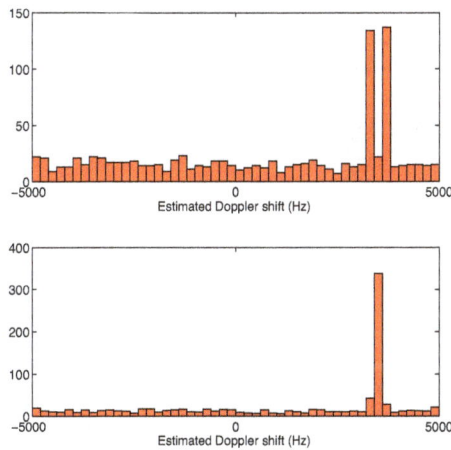

Figure 20. Histograms of the estimated Doppler shifts in two cases: the fast acquisition approach and the two-step acquisition method for a C/N_0 of 30 dB-Hz with bit sign transitions, and $N = 6$, $K = 6$.

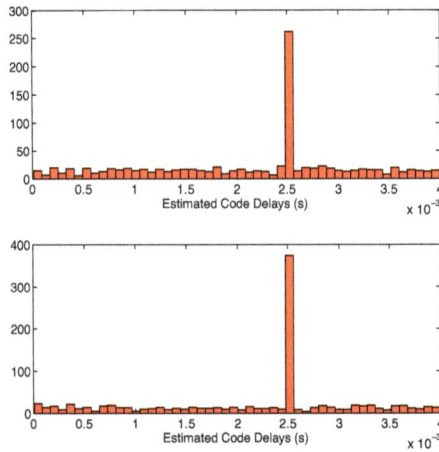

Figure 21. Histograms of the estimated code delays in two cases: the fast acquisition approach and the two-step acquisition method for a C/N_0 of 30 dB-Hz with bit sign transitions, and $N = 6$, $K = 6$.

Fig. 20 shows that the histograms of the Doppler shift estimates for the classical fast acquisition approach and the proposed two-step acquisition technique. The upper plot of Fig. 20 is the histogram of the Doppler shift estimates evaluated by the classical fast acquisition approach, which shows that the Doppler shift estimates deviate much from its true value ($f_d = 3500$ Hz) due to the CAF peak splitting effect. The classical fast acquisition approach shows inadequate performance when dealing with bit sign transition problem. The lower plot of Fig. 20 is the histogram of the Doppler shift estimates obtained by the proposed two-step acquisition technique, as it can be observed that the achieved Doppler shift estimates much more concentrate around the correct Doppler shift value. This proposed methodology is able to partially mitigate this CAF peak splitting effect and it outperforms the classical fast acquisition approach.

Fig. 21 shows the comparison of the histograms of the code delay estimates for both aforementioned acquisition techniques. The upper histogram of the code delay estimates is evaluated by the classical acquisition approach and the lower histogram is achieved by the proposed acquisition technique. It is easily known from Fig. 21 that the code phase delay estimates achieved by the proposed acquisition technique are usually not so sensitive to the CAF peak splitting effect as the classical fast acquisition scheme, and the proposed technique provides much improved detection rate of the code phase delay.

A more detailed analysis has been performed by evaluating the ROC curves [42]. ROC curve completely characterizes the acquisition system performance [43]. ROC provides a statistical characterization of the acquisition performance allowing comparative analysis for the different algorithms. The presence of bit sign transitions in the Galileo signals reduces the benefits deriving by coherently extending the integration time, for such a reason in the first acquisition step a combination strategy between the coherent and non-coherent integrations techniques over multiple code periods is suggested. In the simulation experiments different

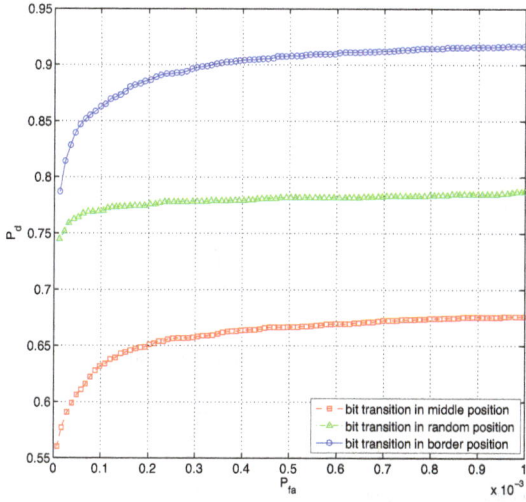

Figure 22. ROC comparison varying with the bit transition position by the two-step acquisition method for a case of C/N_0 value of 38 dB-Hz and $N = 1, K = 1$.

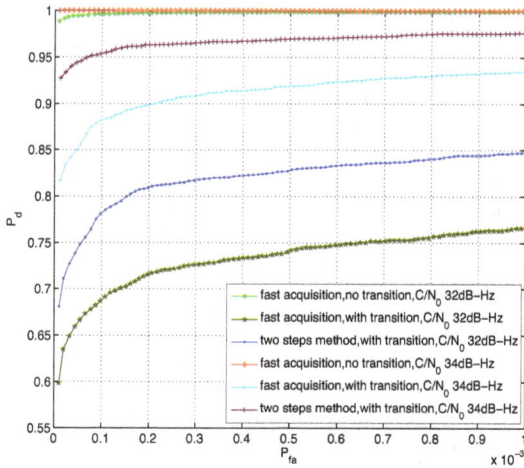

Figure 23. Comparison between the fast acquisition approach and the two-step method for two C/N_0 values of 32 and 34 dB-Hz, N=2, K=6.

code periods of the coherent and non-coherent integrations operations have been chosen to compare the performances between different acquisition schemes. Each simulated ROC curve reports the performance comparison of different acquisition schemes using the same number of code periods, coherently or non-coherently integrated. Monte Carlo simulation campaigns have been performed on the simulated Galileo E1 OS BOC(1,1) signals.

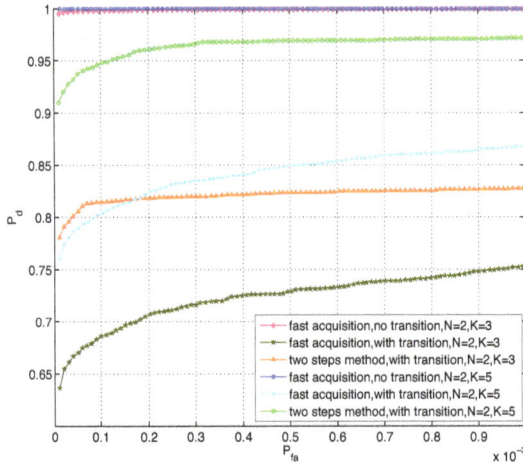

Figure 24. Comparison between the fast acquisition approach and the two-step acquisition method for a case of $C/N_0 = 35$ dB-Hz.

The CAF main peak splitting effect dependent on the bit sign transition position in the signal segment is presented here in terms of ROC curve. Simulation test is made for three typical bit sign transitions distributions cases: bit transition present in the middle or border positions, or randomly distributed in the signal segment, which is implemented by the proposed two steps acquisition algorithm for a C/N_0 value of 38 dB-Hz. The simulation result is shown in Fig. 22, which indicates that the acquisition performance degrades greatly when the bit transition occurs in the middle of the signal segment, while the acquisition system provides better performance when the bit transition is close to the border position of the signal segment. When dealing with the signal which presents bit transition randomly distributed in a code period, the acquisition performance lies in between the two above described cases. In the following performance analysis signals presenting bit sign transitions randomly distributed are simulated and adopted in the ROC evaluations.

Fig. 23 depicts the performance comparison among three acquisition cases: the fast acquisition approach with or without bit transitions and the proposed two-step acquisition technique with sign reversals during the correlation. The simulation has been made considering coherent integration of two Galileo BOC(1,1) code periods (N=2) and six non-coherent integration operations (K=6) for two C/N_0 values of 32 dB-Hz and 34 dB-Hz respectively. The results of Fig. 23 show that the two-step acquisition method provides improved acquisition performance in terms of detection probability over the classical fast acquisition approach when the received signal presents the well known problem of bit sign transitions. It is also obviously known that much improved detection probability can be achieved when the C/N_0 value increases while keeping the coherent and non-coherent operations unchanged. This can be verified from the simulation results in Fig. 23 when the C/N_0 value increases from 32 dB-Hz to 34 dB-Hz.

To achieve a reasonable estimation rate of the code phase delay in the first acquisition step to consequently recover the correct Doppler frequency shift in the second acquisition

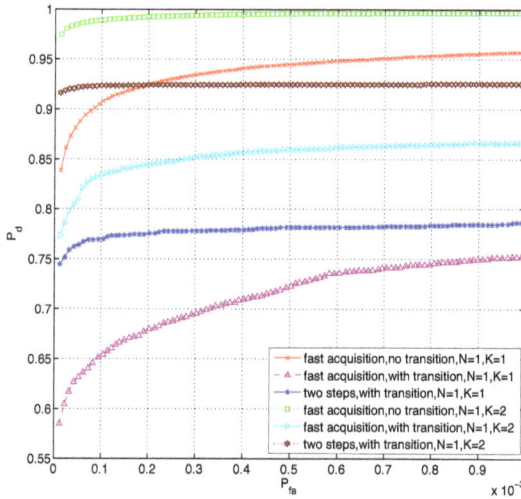

Figure 25. Comparison between the fast acquisition approach and the two-step acquisition method for a case of $C/N_0 = 38$ dB-Hz.

step, a combination strategy of coherent and non-coherent integrations operations is usually adopted in the first acquisition step. It is possible to know that even though the non-coherent integration technique adopted in the first acquisition step causes the side effect of a squaring loss, it is able to achieve a good estimate of the code phase delay to facilitate the second acquisition step, and the price to be paid is a longer acquisition time. In Fig. 24 the acquisition performance comparison is outlined varying the non-coherent integration periods and keeping the coherent integration code periods unchanged. The results shown in Fig. 24 highlight how better performance can be achieved by the two-step acquisition technique when the non-coherent integration code periods increase.

This trend is even more evident for high C/N_0 values. As the C/N_0 value increases, less non-coherent integration operations are required to be able to achieve a good estimate of the code phase delay at which the bit sign transition might occur in the first acquisition step in order to initialize the new signal alignment properly in the second acquisition step. Fig. 25 clearly shows that the two-step acquisition method provides much improved performance in comparison with the classical fast acquisition scheme, which could aid the acquisition phase of a GNSS receiver in real situations.

Finally the acquisition performances comparison is also presented in terms of SNR curve, which is the detection probability plotted versus the input value of C/N_0 for a fixed false alarm probability. Simulations for the SNR curve are made for a selected P_{fa} of 10^{-3}. In Fig. 26, it shows that the two-step acquisition method outperforms the classical fast acquisition scheme in presence of bit transitions when the combination strategy of coherent and non-coherent integrations operations is adopted in the first acquisition step. The analysis results have proved the validity and effectiveness of the proposed two-step acquisition technique, which is able to mitigate the CAF peak splitting effect caused by the bit sign transitions.

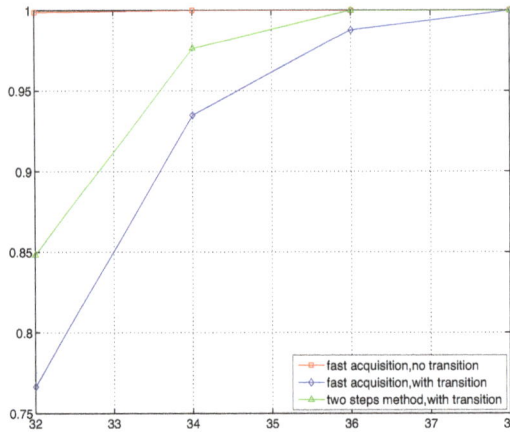

Figure 26. SNR curve comparison between the fast acquisition approach and the two-step acquisition method for the Galileo BOC(1,1) signal, $N = 2$, $K = 6$ and $P_{fa} = 10^{-3}$.

The two-step acquisition method can generally provide better performance when the signal modulation presents a potential bit sign transition in each code period, but the price to be paid for this enhanced performance is its increased computational complexity.

3. Differential Detection Technique

3.1. Differential Detection Basic Concepts

The simplest acquisition strategy is a coherent integration, where different CAF's are averaged before evaluating the envelope. From the noise point of view the coherent integration corresponds to increase the integration time of the correlator, performing de facto a longer coherent integration (in other term, a correlation with a local code replica containing more code periods). When coherent integration is adopted, the effect of the noise variance of the correlator output will be reduced. The price to be paid for this improvement of the acquisition performance is a modification of the CAF shape in the right bins. The coherent integration generally affects the main lobe of the CAF in the correct bins. The peak becomes narrower if different CAF's in the accumulation are evaluated without altering the phase relationships of all the signals involved in the CAF evaluation. The width of the main lobe of the CAF deceases as the integration time increases. In order to limit the frequency loss, the frequency bin size of the search space has to be reduced proportionally to the inverse of the coherent integration time [33]. As a consequence the number of the Doppler shift bins to be evaluated and analyzed in the search space increases. Moreover, the achievable maximum performance of extending the coherent integration time for improving the acquisition performance has received significant attenuation in the presence of bits (or chips of the secondary codes), which could seriously degrade the CAF modifying the main lobe characteristics and result in a CAF main peak splitting effect as discussed in section 2.2.

When non-coherent integration strategy is employed, the decision variable $S_K(\bar{\tau}, \bar{F}_D)$ is obtained by squaring $Y_{k,I}(\bar{\tau}, \bar{F}_D)$ and $Y_{k,Q}(\bar{\tau}, \bar{F}_D)$ and summing K independent realizations

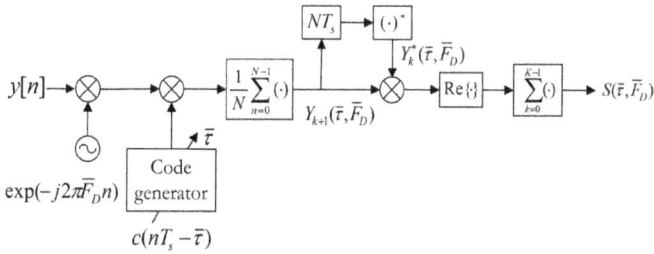

Figure 27. Differentially coherent combining detector. * denotes complex conjugation.

of those r.v.'s

$$S_K(\bar{\tau}, \bar{F}_D) = \sum_{k=0}^{K-1} \left[Y_{k,I}^2(\bar{\tau}, \bar{F}_D) + Y_{k,Q}^2(\bar{\tau}, \bar{F}_D) \right] \qquad (97)$$

The non-coherent integration scheme works by ignoring the residual phase effects that depend on the unknown phase of the input signal. This phase dependence is removed by squaring the coherent correlator outputs in Eq. (97), thus, signal degradation due to phase errors, such as Doppler shift offset and bit sign transitions, is reduced. However, in this way, noise component is also squared, leading to a definite positive process whose mean is no longer zero, and the post-correlation averaging is less effective since the noise components do not cancel out any longer, thus a residual noise term still remains, this effect is called squaring loss.

The coherent combining scheme is optimal only for static channels. When the channel varies as results of fading and frequency offset, the correlator outputs have different phases. The coherent combining scheme is not optimal in such conditions, due to the absence of phase compensation for correlator outputs. The non-coherent combining eliminates the need for phase compensation. However, its performance may be poor due to non-coherent combining squaring loss.

In order to circumvent the limitations with the coherent integration and non-coherent integration methods, as a phase compensation method, a differential detection technique may be considered to achieve better acquisition sensitivity for a fixed SNR. A phase reference of the current correlator output is provided by the previous correlator output in the differential detection scheme. Noise reduction can be achieved by using differential integration scheme since cross-correlated noise samples can be assumed to be independent and their accumulation leads to an equivalent noise term with asymptotically null power. The main idea behind the differential detection is that there will be a high degree of correlation between the phases of successive correlator outputs when the useful signal is present, but they will be essentially independent under the influence of noise alone. Denoting by $Y_k(\bar{\tau}, \bar{F}_D)$ the kth output of the coherent correlator, the differentially coherent product is formed as:

$$R_k(\bar{\tau}, \bar{F}_D) = Y_{k+1}(\bar{\tau}, \bar{F}_D) Y_k^*(\bar{\tau}, \bar{F}_D) \qquad (98)$$

where $Y_k^*(\bar{\tau}, \bar{F}_D)$ denotes the complex conjugate of $Y_k(\bar{\tau}, \bar{F}_D)$. The differential detector forms a decision variable by accumulating a number (say K) of these differentially coherent products. The differentially coherent combining detector is illustrated in Fig.27.

There are different kinds of differential integrations [44, 45], depending on how the correlator outputs are combined together and how the final decision variable $S(\bar{\tau}, \bar{F}_D)$ is computed. For example, differential post detection integration-real (DPDI-Real) and differential post detection integration-absolute (DPDI-Abs) are depicted in Fig. 28. The DPDI-Real decision statistic is written as follows

$$S_K(\bar{\tau}, \bar{F}_D) = \mathcal{R}e\left\{ \sum_{k=0}^{K-1} \left[Y_{k+1,I}(\bar{\tau}, \bar{F}_D) + jY_{k+1,Q}(\bar{\tau}, \bar{F}_D) \right] \cdot \left[Y_{k,I}(\bar{\tau}, \bar{F}_D) + jY_{k,Q}(\bar{\tau}, \bar{F}_D) \right]^* \right\} \quad (99)$$

The DPDI-Abs decision statistic is

$$S_K(\bar{\tau}, \bar{F}_D) = \left| \sum_{k=0}^{K-1} \left[Y_{k+1,I}(\bar{\tau}, \bar{F}_D) + jY_{k+1,Q}(\bar{\tau}, \bar{F}_D) \right] \cdot \left[Y_{k,I}(\bar{\tau}, \bar{F}_D) + jY_{k,Q}(\bar{\tau}, \bar{F}_D) \right]^* \right|^2 \quad (100)$$

In order to simplify the following acquisition strategy analysis, the pairwise form of differential detector is mainly considered here, whose decision variable is given by:

$$S_K(\bar{\tau}, \bar{F}_D) = \mathcal{R}e\left\{ \sum_{k=0}^{K-1} Y_{2k+1}(\bar{\tau}, \bar{F}_D)Y_{2k}^*(\bar{\tau}, \bar{F}_D) \right\} \quad (101)$$

where the accumulation of every second differential product is considered, and only the real part is exploited as the decision statistic. This differential correlation scheme will be further analyzed in section 3.2 and it will be shown that the resulting r.v. can be expressed as the difference of two χ^2 r.v.'s. In this paper this kind of differential integration scheme is adopted in the following proposed two-step based differentially coherent acquisition technique.

The differential acquisition technique operates by maintaining differential phase information between successive correlator outputs, which has been proposed in CDMA literature [44–47] as a method to reduce the effects of phase fluctuations due to frequency offset and fading. The fading is slow enough so as to avoid significant variation over two consecutive correlation intervals. In GPS acquisition case, the data bit can be regarded as a slow varying fading (one possible bit sign transition within 20 PRN code periods), which can be compensated with a differential coherent integration scheme. For the case of Galileo, there is a potential sign transition of data bit / secondary code at each primary code period as the bit duration is equal to the code length. Therefore, this differential detection technique needs to be modified. In order to remove the dependence on the product of the navigation message and secondary codes, a possible solution is to take into account the absolute value of the differential block's output, thus the decision variable is formed as [48]

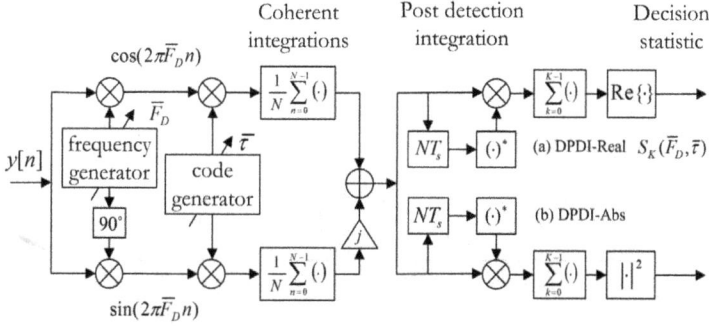

Figure 28. Differentially combining schemes: DPDI-Real and DPDI-Abs.

$$S_K(\bar{\tau}, \bar{F}_D) = \sum_{k=0}^{K-1} \left| \mathcal{Re}\{Y_{2k+1}(\bar{\tau}, \bar{F}_D)Y_{2k}^*(\bar{\tau}, \bar{F}_D)\}\right| \tag{102}$$

where $Y_{2k}^*(\bar{\tau}, \bar{F}_D)$ denotes the complex conjugate of $Y_{2k}(\bar{\tau}, \bar{F}_D)$.

3.2. Differentially Coherent Acquisition Strategy Analysis

Adopting the threshold-crossing criterion for the binary detection problem, the detection performance is characterized in terms of false alarm probability P_{fa} and detection probability P_d, defined as:

$$P_{fa}(\beta) = P(S(\bar{\tau}, \bar{F}_D) > \beta | H_0) = \int_{\beta}^{+\infty} f_{S(\bar{\tau}, \bar{F}_D)|H_0}(x|H_0)\, dx \tag{103}$$

$$P_d(\beta) = P(S(\bar{\tau}, \bar{F}_D) > \beta | H_1) = \int_{\beta}^{+\infty} f_{S(\bar{\tau}, \bar{F}_D)|H_1}(x|H_1)\, dx \tag{104}$$

where β is the decision threshold; $f_{S(\bar{\tau}, \bar{F}_D)|H_0}(x|H_0)$ and $f_{S(\bar{\tau}, \bar{F}_D)|H_1}(x|H_1)$ are the conditional probability density functions (c.p.d.f.'s) of the decision variable $S(\bar{\tau}, \bar{F}_D)$ under hypotheses H_0 and H_1 respectively. In the following, P_{fa} and P_d are analytically derived for the differentially coherent acquisition scheme.

It is able to prove that the real part of the product of the two independent Gaussian r.v.'s can be rewritten as the difference of two independent χ^2 r.v.'s:

$$
\begin{aligned}
&\mathcal{R}e\{Y_{2k+1}Y_{2k}^*\} \\
&= \left|\frac{Y_{2k+1}+Y_{2k}}{2}\right|^2 - \left|\frac{Y_{2k+1}-Y_{2k}}{2}\right|^2 \\
&= \underbrace{\left[\left(\frac{Y_{2k+1,I}+Y_{2k,I}}{2}\right)^2 + \left(\frac{Y_{2k+1,Q}+Y_{2k,Q}}{2}\right)^2\right]}_{\chi^2(2)} - \underbrace{\left[\left(\frac{Y_{2k+1,I}-Y_{2k,I}}{2}\right)^2 + \left(\frac{Y_{2k+1,Q}-Y_{2k,Q}}{2}\right)^2\right]}_{\chi^2(2)}
\end{aligned}
\tag{105}
$$

In particular, when the useful signal is absent or not correctly aligned, i.e. under null hypothesis H_0, each element $\frac{Y_{2k+1,I\{Q\}} \pm Y_{2k,I\{Q\}}}{2}$ has a Gaussian distribution:

$$
\frac{Y_{2k+1,I\{Q\}} \pm Y_{2k,I\{Q\}}}{2} \sim \mathcal{N}\left(0, \frac{\sigma_n^2}{2}\right)
\tag{106}
$$

Thus, $\left|\frac{Y_{2k+1}+Y_{2k}}{2}\right|^2$ and $\left|\frac{Y_{2k+1}-Y_{2k}}{2}\right|^2$ are two independent central χ^2 r.v.'s with two degrees of freedom (d.o.f.'s). Therefore, $\mathcal{R}e\{Y_{2k+1}Y_{2k}^*\}$ is a r.v. which is equal to the difference of two independent χ^2 r.v.'s.

Let $p_{\mathcal{R}_k|H_0}(x)$ denote the p.d.f. of $\mathcal{R}e\{Y_{2k+1}Y_{2k}^*\}$, in [49], $p_{\mathcal{R}_k|H_0}(x)$ is expressed as

$$
p_{\mathcal{R}_k|H_0}(x) =
\begin{cases}
\dfrac{1}{2\sigma_n^2}\exp\left(\dfrac{x}{\sigma_n^2}\right) & x < 0 \\[2ex]
\dfrac{1}{2\sigma_n^2}\exp\left(-\dfrac{x}{\sigma_n^2}\right) & x \geq 0
\end{cases}
\tag{107}
$$

Then the c.p.d.f. of the decision variable $S_k(\bar{\tau},\bar{F}_D)(= |\mathcal{R}e\{Y_{2k+1}Y_{2k}^*\}|)$ under H_0 can be derived in the following

$$
f_{S_k(\bar{\tau},\bar{F}_D)|H_0}(x|H_0) = p_{\mathcal{R}_k|H_0}(x) + p_{\mathcal{R}_k|H_0}(-x) = \frac{1}{\sigma_n^2}\exp\left(-\frac{x}{\sigma_n^2}\right) \quad x \geq 0
\tag{108}
$$

In Eq. (108), it has clearly shown that the decision variable $S_k(\bar{\tau},\bar{F}_D)$ is exponentially distributed under H_0, $S_k(\bar{\tau},\bar{F}_D) \sim \text{Exp}(\frac{1}{\sigma_n^2})$, which is a special case of a Gamma distribution, i.e. $S_k(\bar{\tau},\bar{F}_D) \sim \Gamma(1,\sigma_n^2)$.

From the false alarm probability defined in Eq. (103), the probability of false alarm $P_{fa}(\beta,1)$ is obtained as

$$
P_{fa}(\beta,1) = \exp\left(-\frac{\beta}{\sigma_n^2}\right)
\tag{109}
$$

The acquisition over several periods can be performed by directly accumulating K independent realizations of $S_k(\bar{\tau}, \bar{F}_D)$:

$$S_K(\bar{\tau}, \bar{F}_D) = \sum_{k=0}^{K-1} \left| \mathcal{R}e\{Y_{2k+1}(\bar{\tau}, \bar{F}_D) Y_{2k}^*(\bar{\tau}, \bar{F}_D)\} \right| \tag{110}$$

$S_K(\bar{\tau}, \bar{F}_D)$ is the sum of K independent Gamma distributed r.v.'s, thus

$$S_K(\bar{\tau}, \bar{F}_D) = \sum_{i=0}^{K-1} S_k(\bar{\tau}, \bar{F}_D) \sim \Gamma(K, \sigma_n^2) \tag{111}$$

The false alarm probability for the decision variable $S_K(\bar{\tau}, \bar{F}_D)$ assumes the following expression:

$$P_{fa}(\beta, K) = \exp\left(-\frac{\beta}{\sigma_n^2}\right) \sum_{i=0}^{K-1} \frac{1}{i!} \left(\frac{\beta}{\sigma_n^2}\right)^i \tag{112}$$

Under alternative hypothesis H_1, we have

$$\begin{aligned}
\frac{Y_{2k+1,I\{Q\}} - Y_{2k,I\{Q\}}}{2} &\sim \mathcal{N}\left(0, \frac{\sigma_n^2}{2}\right) \\
\frac{Y_{2k+1,I} + Y_{2k,I}}{2} &\sim \mathcal{N}\left(\sqrt{\lambda}\cos\varphi, \frac{\sigma_n^2}{2}\right) \\
\frac{Y_{2k+1,Q} + Y_{2k,Q}}{2} &\sim \mathcal{N}\left(\sqrt{\lambda}\sin\varphi, \frac{\sigma_n^2}{2}\right)
\end{aligned} \tag{113}$$

Therefore, from Eq. (105), $\mathcal{R}e\{Y_{2k+1}Y_{2k}^*\}$ can be seen as the difference of a non-central and a central χ^2 r.v.'s with two d.o.f.'s, and λ is the non-centrality parameter. In [49], the p.d.f. of $R_k(\bar{\tau}, \bar{F}_D)(= \mathcal{R}e\{Y_{2k+1}Y_{2k}^*\})$ under H_1 is obtained

$$p_{\mathcal{R}_k|H_1}(x) = \begin{cases} \dfrac{1}{2\sigma_n^2}\exp\left(\dfrac{x}{\sigma_n^2}\right)\exp\left(-\dfrac{\lambda}{2\sigma_n^2}\right) & x < 0 \\[3mm] \dfrac{1}{2\sigma_n^2}\exp\left(\dfrac{x}{\sigma_n^2}\right)\exp\left(-\dfrac{\lambda}{2\sigma_n^2}\right)Q_1\left(\sqrt{\dfrac{\lambda}{\sigma_n^2}}, \sqrt{\dfrac{4x}{\sigma_n^2}}\right) & x \geq 0 \end{cases} \tag{114}$$

Then the conditional p.d.f. of the decision variable $S_k(\bar{\tau}, \bar{F}_D)(= |\mathcal{R}e\{Y_{2k+1}Y_{2k}^*\}|)$ can be derived in the following

$$\begin{aligned}
f_{S_k(\bar{\tau},\bar{F}_D)|H_1}(x|H_1) &= p_{\mathcal{R}_k|H_1}(x) + p_{\mathcal{R}_k|H_1}(-x) \\
&= \frac{1}{2\sigma_n^2}\exp\left(-\frac{\lambda}{2\sigma_n^2}\right)\left[\exp\left(-\frac{x}{\sigma_n^2}\right) + \exp\left(\frac{x}{\sigma_n^2}\right)Q_1\left(\sqrt{\frac{\lambda}{\sigma_n^2}}, \sqrt{\frac{4x}{\sigma_n^2}}\right)\right]
\end{aligned} \tag{115}$$

where $x \geq 0$.

The detection probability can be obtained by integrating by parts

$$P_d(\beta, 1) = Q_1\left(\frac{\sqrt{2\lambda}}{\sigma_n}, \frac{2\sqrt{\beta}}{\sigma_n}\right) + \frac{1}{2}\exp\left(-\frac{2\beta + \lambda}{2\sigma_n^2}\right) - \frac{1}{2}\exp\left(\frac{2\beta - \lambda}{2\sigma_n^2}\right)Q_1\left(\frac{\sqrt{\lambda}}{\sigma_n}, \frac{2\sqrt{\beta}}{\sigma_n}\right) \tag{116}$$

The detection probability for a generic K does not admit an easy closed-form analytical expression, but it can be evaluated by using a numerical method for the inversion of the characteristic function (chf), which is reported in [50]. The corresponding chf is obtained by deriving Eq. (163) and by evaluating its Fourier transform. This computation leads to

$$Ch_d(t, 1) = \frac{\exp\left(-\frac{\lambda}{2\sigma_n^2}\right)}{1 + j\sigma_n^4 t^2}\left[j\sigma_n^2 t + \exp\left(\frac{\lambda}{\sigma_n^2}\right)\exp\left(\frac{j\lambda t}{1 - j\sigma_n^2 t}\right)\right] \tag{117}$$

The chf for a generic K is obtained by raising Eq. (117) to the power K:

$$Ch_d(t, K) = \frac{\exp\left(-\frac{K\lambda}{2\sigma_n^2}\right)}{(1 + j\sigma_n^4 t^2)^K}\left[j\sigma_n^2 t + \exp\left(\frac{\lambda}{\sigma_n^2}\right)\exp\left(\frac{j\lambda t}{1 - j\sigma_n^2 t}\right)\right]^K \tag{118}$$

The detection probability can be then evaluated by numerically inverting the chf in Eq. (118). It has to be remarked that the p.d.f. $f_{S_k(\bar{\tau}, \bar{f}_D)|H_1}(x|H_1)$ corresponding to Eq. (163) is not, in general, equal to zero in the origin. This corresponds to a discontinuity that would cause problems for the FFT based inversion algorithm. The problem can be solved by considering the regularized chf

$$\tilde{Ch}_d(t, K) = Ch_d(t, K) + Ch_d(-t, K) \tag{119}$$

The Fourier transform of Eq. (119) is given by

$$\tilde{f}_d(x, K) = f_d(x, K) + f_d(-x, K) \tag{120}$$

that is the sum of the p.d.f. $f_d(x, K)$ and of its symmetric $f_d(-x, K)$. $\tilde{f}_d(x, K)$ does not present discontinuities in the origin and thus it can be easily evaluated by means of FFT based techniques. $f_d(x, K)$ can be easily recovered from $\tilde{f}_d(x, K)$.

3.3. Two-step Differentially Coherent Signal Acquisition

In order to enhance the GNSS receiver sensitivity with the aim of reliable and robust signal acquisition in presence of bit sign transitions particularly at low C/N_0 values, the acquisition performance is often improved by post correlation integration techniques, such as coherent integration and non-coherent integration operations. Galileo will provide a navigation message at a higher bit rate resulting in a consequent possibility of a bit sign transition

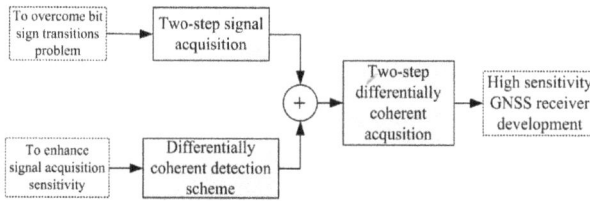

Figure 29. Two-step differentially coherent acquisition strategy for GNSS receivers.

in each primary spreading code period. In this case, if FFT's are exploited to perform the circular correlation, the bit sign transition occurring within an integration period may cause a splitting of the CAF main peak into two smaller lobes along the Doppler shift axis. In general this is a critical aspect for all the acquisition methods where the data are processed in blocks. The achievable maximum performance of extending the coherent integration time for improving the acquisition sensitivity has received significant attenuation in the presence of bit sign transitions. Similarly, the acquisition sensitivity could also be improved by increasing the non-coherent integration number, but the non-coherent integration approach is based on the sum of squared envelopes of correlator outputs, which presents the so called side effect of a relevant squaring loss.

To overcome this limitation, one possible way to achieve better acquisition sensitivity for a fixed SNR passes through the adoption of the differential detection scheme. Noise can be reduced by using differential detection since cross-correlated noise samples are assumed to be independent and their accumulation results in an equivalent noise term with asymptotically null power. The main disadvantage here is that traditional differential correlation for more than one bit duration is limited by the presence of unknown bit sign transitions that may collapse the accumulated signal peak energy.

In section 2.3, it has proved that the presence of bit sign transition in the GNSS signal does not destroy the information on the presence of the satellite in view, but introduces an erroneous Doppler frequency shift estimation. The two-step acquisition methodology can be adopted to mitigate the CAF main peak splitting effect and it can provide much improved performance. Toward the objective of a HS - GNSS receiver operated anywhere, a novel acquisition strategy, by introducing two-step acquisition to differentially coherent detection scheme, has been proposed in order to cope with the bit sign transitions problem for the received weak signals in indoor environments, which has been illustrated in Fig. 29. This combined acquisition strategy is therefore named two-step differentially coherent acquisition [48].

The performance of the proposed two-step differentially coherent signal acquisition technique has been investigated by means of ROC curves. Monte Carlo simulation campaigns have been performed on the simulated Galileo E1 OS signals to evaluate the performance of the proposed acquisition technique in order to consolidate theoretical analysis. In order to prove the advantage and effectiveness of the proposed acquisition technique, a comprehensive performance comparison has been deeply carried on among the traditional non-coherent integration approach, the existing single step differentially coherent detection and the two-step non-coherent integration strategies.

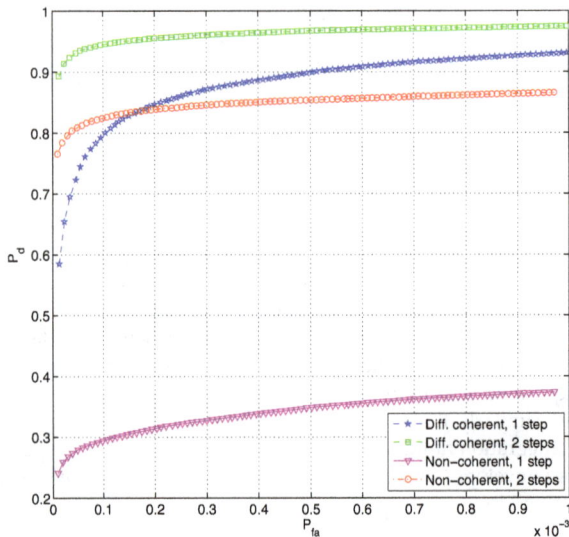

Figure 30. ROC curve comparison of all strategies: $C/N_0 = 32$ dB-Hz, $K = 2$. For the non-coherent integration scheme, the pre-detection integration time is 8 ms.

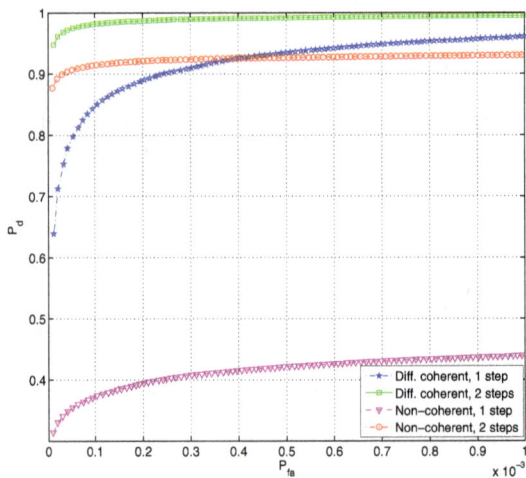

Figure 31. ROC curve comparison of all strategies: $C/N_0 = 34$ dB-Hz, $K = 1$. For the non-coherent integration scheme, the pre-detection integration time is 8 ms.

Fig. 30 and Fig. 31 prove that the performance of the two-step differentially coherent detection is always better than the others; and as expected, the non-coherent integration shows poor performance. Obviously, the proposed two-step differentially coherent acquisition technique outperforms the related two-step non-coherent integration approach requiring the same computational load.

In additional, it is able to know that the single step differentially coherent detection technique works slightly better than the two-step non-coherent integration scheme after a cross point between them is passed, so the single step differentially coherent detection is preferable for high value of P_{fa}.

The obtained results have revealed a significant performance improvement of the proposed acquisition technique over the aforementioned other acquisition approaches while false alarm and detection probabilities are used as measurement criteria. The developed technique overcomes the CAF peak splitting problem caused by the presence of bit sign transitions and also enhances the acquisition sensitivity particularly in weak signal environments, thus it can be well applied to the new generation GNSS signals where bit sign transition could change the relative polarity every primary code period.

The rationale behind this proposed acquisition technique is based on the hybrid combination between the two-step signal acquisition scheme for mitigating the CAF peak splitting effect due to the presence of bit sign transitions and the differentially coherent integration technique for improving the acquisition sensitivity. The improved performance with this proposed acquisition technique is achieved at the expense of increase of structural complexity and computation load of the acquisition process.

4. Channels Combining Strategies

4.1. Composite GNSS Signal Model

The signal at the input of a GNSS receiver, in a one-path additive Gaussian noise environment, can be written as

$$y_{RF}(t) = \sum_{i=1}^{N_s} r_{RF,i}(t) + \eta_{RF}(t) \tag{121}$$

that is the sum of N_s useful signals emitted by N_s different satellites and of a noise term $\eta_{RF}(t)$. $\eta_{RF}(t)$ is a stationary AWGN with PSD $N_0/2$. When considering composite GNSS signals with data and pilot components, such as Galileo E1 OS case, $r_{RF,i}(t)$ can be modeled as [51]:

$$r_{RF,i}(t) = \left\{ \sqrt{C_i} e_{D,i}(t - \tau_i) - \sqrt{C_i} e_{P,i}(t - \tau_i) \right\} \cos(2\pi (f_{RF} + f_{d,i})t + \varphi_{RF,i}) \tag{122}$$

where

- C_i is the received signal power at the output of the receiver antenna;
- $e_{D,i}(t)$ and $e_{P,i}(t)$ are the data and pilot components, respectively;

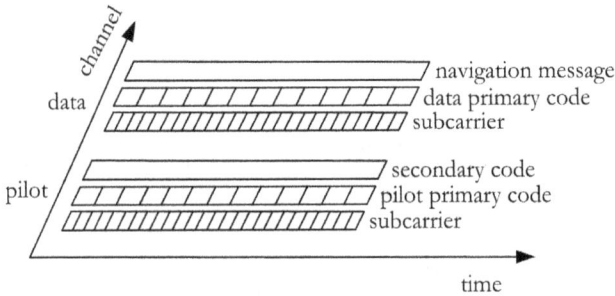

Figure 32. Data / pilot structure of a composite Galileo E1 OS signal. Each channel is given by different components: the periodic repetition of the primary spreading code, the subcarrier, and for the pilot channel, the secondary code.

- τ_i, $f_{d,i}$ and $\varphi_{RF,i}$ are the code delay, the Doppler frequency shift, and the carrier phase introduced by the transmission channel on the ith signal, respectively;

- f_{RF} is the carrier frequency, i.e. 1575.420 MHz for the Galileo E1 OS signal.

In Fig. 32 the structure of a composite Galileo E1 OS signal is depicted. In general the data and pilot components, $e_{D,i}(t)$ and $e_{P,i}(t)$ are given by the product of several terms

$$e_{D,i}(t) = d_i(t)c_{D,i}(t)s_{b,i}(t)$$
$$e_{P,i}(t) = s_i(t)c_{P,i}(t)s_{b,i}(t)$$

$$(123)$$

where $d_i(t)$ is the navigation data stream in the data channel; $s_{b,i}(t)$ is the signal obtained by periodically repeating the subcarrier; $s_i(t)$ is the secondary code adopted in the pilot channel; and $c_{D,i}(t)$ and $c_{P,i}(t)$ are the primary spreading code sequences for the data and pilot channels respectively.

The input signal in Eq. (121) is recovered by the receiver antenna, down-converted, and filtered by the receiver front-end. In this way the received signal, before the analog-to-digital (A/D) conversion is given by

$$y(t) = \sum_{i=1}^{N_s} \tilde{r}_i(t) + \eta(t)$$
$$= \sum_{i=1}^{N_s} \left\{ \sqrt{C_i}\tilde{e}_{D,i}(t-\tau_i) - \sqrt{C_i}\tilde{e}_{P,i}(t-\tau_i) \right\} \cos(2\pi(f_{IF} + f_{d,i})\,t + \varphi_i) + \eta(t)$$

$$(124)$$

where f_{IF} is the receiver intermediate frequency, $\tilde{e}_{D,i}$ and $\tilde{e}_{P,i}$ are the data and pilot components after filtering of the front-end and $\eta(t)$ is the down-converted and filtered noise component. Here the simplifying conditions

$$\tilde{e}_{D,i}(t) \approx e_{D,i}(t)$$
$$\tilde{e}_{P,i}(t) \approx e_{P,i}(t) \qquad (125)$$

are assumed and the impact of the front-end filter is neglected.

In a digital receiver the IF signal is sampled through an ADC. The ADC generates a sampled sequence $y(nT_s)$, obtained by sampling $y(t)$ at the sampling frequency $f_s = 1/T_s$. From now on the notation $x[n] = x[nT_s]$ will be adopted to indicate a generic sequence $x[n]$ to be processed in any digital platform. After the IF signal of Eq. (124) is sampled and digitized, by neglecting the quantization effect, the following signal model is obtained:

$$y[n] = \sum_{i=1}^{N_s} \left\{ \sqrt{C_i}\tilde{e}_{D,i}[n - \tau_i/T_s] - \sqrt{C_i}\tilde{e}_{P,i}[n - \tau_i/T_s] \right\} \cos(2\pi F_{D,i}\, n + \varphi_i) + \eta[n] \qquad (126)$$

where $F_{D,i} = (f_{IF} + f_{d,i})T_s$.

Due to the orthogonality property of the spreading code sequence, the different GNSS signals are analyzed separately by the receiver, and only a single satellite is considered in the following and the index i of a satellite is dropped. The resulting signal is written as

$$y[n] = \left\{ \sqrt{C}e_D[n - \tau/T_s] - \sqrt{C}e_P[n - \tau/T_s] \right\} \cos(2\pi F_D n + \varphi) + \eta[n] \qquad (127)$$

4.2. Single Channel Acquisition

Composite GNSS signals can be acquired by ignoring one of the two channels. In this case the input signal in Eq. (127) is correlated with either the data or the pilot primary spreading code sequence. Due to the orthogonality property of the spreading code sequences, one of the two channels is discarded and the acquisition process is equivalent to the conventional acquisition scheme for single channel component signals. The single channel acquisition scheme is shown in Fig. 33: the received input signal $y[n]$ is multiplied by two orthogonal reference sinusoids at the frequency $\bar{F}_D = (f_{IF} + \bar{f}_d)/f_s$, split at the in-phase (I) and quadrature (Q) branches, after the multiplication with a local code replica $c_{X,Loc}[n - \bar{\tau}/T_s]$, either of the data or of the pilot code sequence (i.e. $X = D, P$), delayed by $\bar{\tau}$, including the primary spreading code sequence and the subcarrier. The resulted signals on the I and Q branches are then coherently integrated, leading to the in-phase and quadrature components $Y_{X,I}(\bar{\tau}, \bar{F}_D)$ and $Y_{X,Q}(\bar{\tau}, \bar{F}_D)$, respectively. The correlator outputs of the I and Q branches are combined to form a complex correlation variable $Y_X(\bar{\tau}, \bar{F}_D)$:

$$Y_X(\bar{\tau}, \bar{F}_D) = Y_{X,I}(\bar{\tau}, \bar{F}_D) + jY_{X,Q}(\bar{\tau}, \bar{F}_D)$$

$$= \frac{1}{N} \sum_{n=0}^{N-1} \left\{ \left\{ \sqrt{C}\tilde{e}_D[n - \tau/T_s] - \sqrt{C}\tilde{e}_P[n - \tau/T_s] \right\} \cos(2\pi \bar{F}_D\, n + \varphi) + \eta[n] \right\}$$

$$\times \left\{ c_{X,Loc}[n - \bar{\tau}/T_s] \exp\left\{ -j2\pi \bar{F}_D n \right\} \right\} \qquad (128)$$

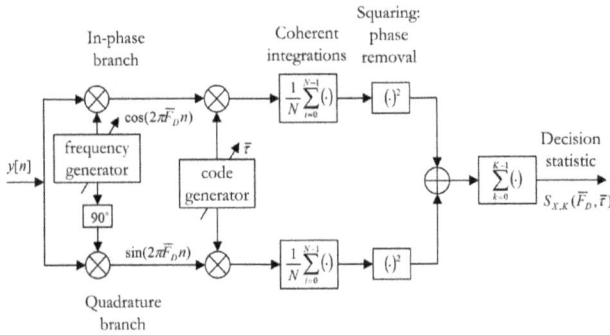

Figure 33. Single channel acquisition: the input signal is correlated with a delayed and modulated code replica, producing the final decision variable $S_X(\tilde{\tau}, \tilde{F}_D)$. The index X can be equal to D or P respectively referring to data and pilot channels.

where the index X can be either $X = D$ or $X = P$, which indicates quantities depending on the correlations with data and pilot local code replicas respectively; N is the number of samples used in the evaluation of the correlations between the received and local signals. The in-phase and quadrature components are then squared and summed, removing the dependence from the input signal phase φ. It is possible to obtain a two-dimensional decision statistic $S_X(\tilde{\tau}, \tilde{F}_D)$ for a coherent integration period, obtained as

$$S_X(\tilde{\tau}, \tilde{F}_D) = \left| Y_X(\tilde{\tau}, \tilde{F}_D) \right|^2 = Y_{X,I}^2(\tilde{\tau}, \tilde{F}_D) + Y_{X,Q}^2(\tilde{\tau}, \tilde{F}_D) \tag{129}$$

By considering Fig. 33, it is clear that all the operations before squaring blocks are linear. Their impact on the useful signal and on the noise can be studied separately. In particular, the in-phase and quadrature components $Y_{X,I}(\tilde{\tau}, \tilde{F}_D)$ and $Y_{X,Q}(\tilde{\tau}, \tilde{F}_D)$ are given by the following forms [52, 53]:

$$
\begin{aligned}
Y_{X,I}(\tilde{\tau}, \tilde{F}_D) &= Y_{X,I,0}(\tilde{\tau}, \tilde{F}_D) + \eta_{X,I} = \sqrt{\frac{C}{4}} d_X \frac{\sin(\pi N \Delta F)}{\pi N \Delta F} R(\Delta \tau) \cos(\Delta \varphi_X) + \eta_{X,I} \\
Y_{X,Q}(\tilde{\tau}, \tilde{F}_D) &= Y_{X,Q,0}(\tilde{\tau}, \tilde{F}_D) + \eta_{X,Q} = \sqrt{\frac{C}{4}} d_X \frac{\sin(\pi N \Delta F)}{\pi N \Delta F} R(\Delta \tau) \sin(\Delta \varphi_X) + \eta_{X,Q}
\end{aligned}
\tag{130}
$$

where:

- $R(\cdot)$ is the cross-correlation between the local code and the filtered incoming code;
- $\Delta F = F_D - \tilde{F}_D$ is the difference between the Doppler frequency of the local carrier and of the incoming signal;
- $\Delta \tau = \frac{\tau - \tilde{\tau}}{T_s}$ is the difference between the local code delay and the incoming code delay, normalized by the sampling interval;
- $\Delta \varphi$ is the difference between phases of received and local carriers;

- d is a value in the set $\{-1, 1\}$ that represents the effect of the navigation message or of the secondary code;

- η_I and η_Q are two independent centered Gaussian correlator output noise r.v.'s obtained by processing the noise term in Eq. (127).

It is clear to know that the I and Q components $Y_{X,I}(\bar\tau, \bar F_D)$ and $Y_{X,Q}(\bar\tau, \bar F_D)$ consist of signal and noise components. The signal components assume the following approximated expressions:

$$Y_{X,I,0}(\bar\tau, \bar F_D) = \begin{cases} \sqrt{\dfrac{C}{4}} \cos\varphi & \text{if } \bar F_D = F_D, \bar\tau = \tau \\ 0 & \text{otherwise} \end{cases}$$

$$Y_{X,Q,0}(\bar\tau, \bar F_D) = \begin{cases} \sqrt{\dfrac{C}{4}} \sin\varphi & \text{if } \bar F_D = F_D, \bar\tau = \tau \\ 0 & \text{otherwise} \end{cases} \tag{131}$$

The correlator noise outputs $\eta_{X,I}$ and $\eta_{X,Q}$ can be obtained in the following:

$$\eta_{X,I} = \frac{1}{N} \sum_{n=0}^{N-1} \eta[n] c_{X,Loc}[n - \bar\tau/T_s] \cos(2\pi \bar F_D n)$$

$$\eta_{X,Q} = -\frac{1}{N} \sum_{n=0}^{N-1} \eta[n] c_{X,Loc}[n - \bar\tau/T_s] \sin(2\pi \bar F_D n) \tag{132}$$

Since it has been assumed that the noise term in Eq. (127) is a white sequence and the considered blocks are linear, both $\eta_{X,I}$ and $\eta_{X,Q}$ are linear combinations of the samples of the Gaussian process $\eta[n]$, they are two Gaussian r.v.'s with zero mean and with equal variances, obtained in the following:

$$\begin{aligned} \text{Var}[\eta_{X,I}] &= E[\eta_{X,I}^2] - E^2[\eta_{X,I}] \\ &= \frac{1}{N^2} \sum_{n=0}^{N-1} \sum_{m=0}^{N-1} E\{\eta[n]\eta[m]\} c_{X,Loc}[n - \bar\tau/T_s] \cos(2\pi \bar F_D n) \cdot \\ &\quad c_{X,Loc}[m - \bar\tau/T_s] \cos(2\pi \bar F_D m) \\ &= \frac{1}{N^2} \sum_{n=0}^{N-1} \sigma_{IF}^2 \cos^2(2\pi \bar F_D n) \\ &\approx \frac{1}{2N} \sigma_{IF}^2 \end{aligned} \tag{133}$$

Similarly, $\text{Var}[\eta_{X,Q}] \approx \dfrac{1}{2N}\sigma_{IF}^2$. Thus, denote $\text{Var}[\eta_{X,I}] = \text{Var}[\eta_{X,Q}] = \sigma_n^2$.

Since the code multiplication and the subsequent integration act as a low-pass filter, it is possible to show that $\eta_{X,I}$ and $\eta_{X,Q}$ can be considered uncorrelated and thus independent.

In this way $Y_{X,I}(\bar{\tau}, \bar{F}_D)$ and $Y_{X,Q}(\bar{\tau}, \bar{F}_D)$ result in two independent Gaussian r.v.'s

$$Y_{X,I}(\bar{\tau}, \bar{F}_D) \sim \mathcal{N}(\sqrt{\frac{C}{4}} \cos \varphi, \sigma_n^2)$$
$$Y_{X,Q}(\bar{\tau}, \bar{F}_D) \sim \mathcal{N}(\sqrt{\frac{C}{4}} \sin \varphi, \sigma_n^2)$$

(134)

The quality of the GNSS signal is usually measured at this stage by the so called *coherent SNR*, defined as

$$\rho_c = \max_{\varphi_0} \frac{E[Y_{X,I}(\tau, F_D)]^2}{\text{Var}[Y_{X,I}(\bar{\tau}, \bar{F}_D)]} = \max_{\varphi_0} \frac{E[Y_{X,Q}(\tau, F_D)]^2}{\text{Var}[Y_{X,Q}(\bar{\tau}, \bar{F}_D)]}$$

(135)

By using Eq. (134), it results in

$$\rho_c = \frac{\frac{C}{4}}{\frac{N_0 f_s}{4N}} = \frac{C}{N_0} N T_s$$

(136)

that is the input C/N_0 multiplied by the coherent integration time NT_s. Eq. (136) has be obtained by assuming that both code phase delay and Doppler shift are perfectly matched.

If the local and received signals are not aligned or the useful signal is absent, that is under null hypothesis H_0, due to the quasi-orthogonality properties of the spreading codes, the decision variable $S_X(\bar{\tau}, \bar{F}_D)$ is a central χ^2 r.v. with 2 d.o.f.'s. When the local signal replica is aligned with the received signal, $\bar{F}_D \approx F_D$ and $\bar{\tau} \approx \tau$, that is under alternative hypothesis H_1, $S_X(\bar{\tau}, \bar{F}_D)$ is a non-central χ^2 r.v. with 2 d.o.f.'s and with non-centrality parameters λ equal to

$$\lambda = \frac{C}{4} \frac{\sin^2(\pi N \Delta F)}{(\pi N \Delta F)^2} R^2(\Delta \tau) \approx \frac{C}{4}$$

(137)

By using properties of central and non-central χ^2 r.v.'s, the false alarm and detection probabilities results can be obtained as follows

$$P_{fa}^{sc}(\beta, 1) = P(S_X(\bar{\tau}, \bar{F}_D) > \beta | H_0) = \exp\left(-\frac{\beta}{2\sigma_n^2}\right)$$

(138)

$$P_d^{sc}(\beta, 1) = P(S_X(\bar{\tau}, \bar{F}_D) > \beta | H_1) = Q_1\left(\frac{\sqrt{\lambda}}{\sigma_n}, \frac{\sqrt{\beta}}{\sigma_n}\right)$$

(139)

where $Q_1(a, b)$ is the generalized Marcum Q-function of order 1, defined as

$$Q_K(a, b) = \frac{1}{a^{K-1}} \int_b^{+\infty} x^K \exp\left\{-\frac{a^2 + x^2}{2}\right\} I_{K-1}(ax) \, dx \qquad (140)$$

in Eq. (67).

The r.v. $Y_X(\bar{\tau}, \bar{F}_D)$ represents the basic element for the decision variable that will be determined at the final acquisition stage. When non-coherent integrations are employed, the decision statistic is obtained by squaring $Y_{X,I}(\bar{\tau}, \bar{F}_D)$ and $Y_{X,Q}(\bar{\tau}, \bar{F}_D)$ and summing K different realizations of those r.v.'s, which is given as

$$\begin{aligned}
S_{X,K}(\bar{\tau}, \bar{F}_D) &= \sum_{k=0}^{K-1} \left| Y_{X,k}^2(\bar{\tau}, \bar{F}_D) \right|^2 \\
&= \sum_{k=0}^{K-1} \left[Y_{X,k,I}^2(\bar{\tau}, \bar{F}_D) + Y_{X,k,Q}^2(\bar{\tau}, \bar{F}_D) \right]
\end{aligned} \qquad (141)$$

where an index k is introduced to distinguish the different realizations of $Y_{X,k,I}(\bar{\tau}, \bar{F}_D)$ and $Y_{X,k,Q}(\bar{\tau}, \bar{F}_D)$, which are obtained by considering consecutive, non-overlapping portions of the input signal $y[n]$ and can be assumed statistically independent and identically distributed; K is the non-coherent integration number.

If the signal is not present or if it is not correctly aligned with the local code replica, $S_{X,K}(\bar{\tau}, \bar{F}_D)$ is a central χ^2 distributed r.v. with $2K$ d.o.f.'s; otherwise, when the code delay and the Doppler shift are properly aligned, $Y_{X,k,I}(\bar{\tau}, \bar{F}_D)$ and $Y_{X,k,Q}(\bar{\tau}, \bar{F}_D)$ are non-zero mean Gaussian r.v.'s, thus $S_{X,K}(\bar{\tau}, \bar{F}_D)$ is a non-central χ^2 distributed r.v. with $2K$ d.o.f.'s and with non-centrality parameter λ_K:

$$\lambda_K = \sum_{i=1}^{K} \lambda_i = K\lambda = K\frac{C}{4} \qquad (142)$$

Similarly, by using properties of non-central and central χ^2 r.v.'s, it is easy to obtain the false alarm and detection probabilities when adopting non-coherent integrations:

$$P_{fa}^{sc}(\beta, K) = P(S_{X,K}(\bar{\tau}, \bar{F}_D) > \beta | H_0) = \exp\left(-\frac{\beta}{2\sigma_n^2}\right) \sum_{i=0}^{K-1} \frac{1}{i!} \left(\frac{\beta}{2\sigma_n^2}\right)^i \qquad (143)$$

$$P_d^{sc}(\beta, K) = P(S_{X,K}(\bar{\tau}, \bar{F}_D) > \beta | H_1) = Q_K\left(\frac{\sqrt{\lambda_K}}{\sigma_n}, \frac{\sqrt{\beta}}{\sigma_n}\right) = Q_K\left(\frac{\sqrt{K\lambda}}{\sigma_n}, \frac{\sqrt{\beta}}{\sigma_n}\right) \qquad (144)$$

where $Q_K(a, b)$ is the Kth order generalized Marcum Q-function. The detection threshold β is usually chosen by fixing a false alarm probability and by inverting Eq. (143). This expression can be solved analytically only for $K = 1$, and numerical techniques have to be used for other cases.

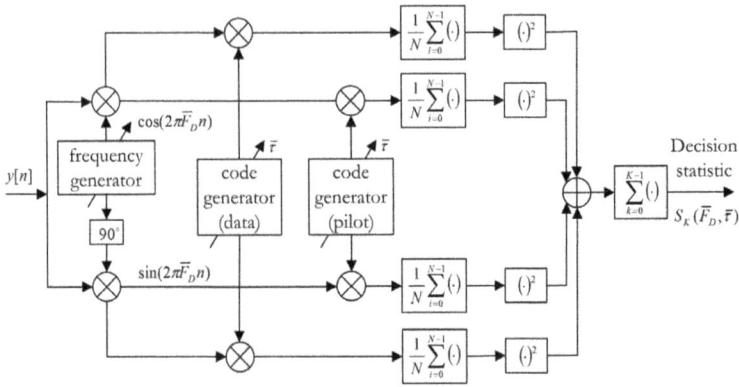

Figure 34. Non-coherent channels combining for composite GNSS signals. Each channel is given by different components: the periodic repetition of the primary spreading code, the subcarrier, and for the pilot channel, the secondary code.

4.3. Non-coherent Channels Combining

The non-coherent channels combining separately correlates the received composite GNSS signal defined in Eq. (127) with the data and pilot local replicas. The resulting signals are then coherently integrated, leading to four correlation outputs: $Y_{D,I}(\bar{\tau}, \hat{F}_D)$, $Y_{D,Q}(\bar{\tau}, \bar{F}_D)$, $Y_{P,I}(\bar{\tau}, \bar{F}_D)$ and $Y_{P,Q}(\bar{\tau}, \bar{F}_D)$. A scheme for the non-coherent channels combining strategy is depicted in Fig. 34.

When considering the quasi-orthogonality property of the spreading codes, the correlation outputs can be written in the following [54, 55]

$$
\begin{aligned}
Y_{D,I}(\bar{\tau}, \bar{F}_D) &= \sqrt{\frac{C}{4}} d_D \frac{\sin(\pi N \Delta F)}{\pi N \Delta F} R(\Delta \tau) \cos(\Delta \varphi_D) + \eta_{D,I} \\
Y_{D,Q}(\bar{\tau}, \bar{F}_D) &= \sqrt{\frac{C}{4}} d_D \frac{\sin(\pi N \Delta F)}{\pi N \Delta F} R(\Delta \tau) \sin(\Delta \varphi_D) + \eta_{D,Q} \\
Y_{P,I}(\bar{\tau}, \bar{F}_D) &= -\sqrt{\frac{C}{4}} d_P \frac{\sin(\pi N \Delta F)}{\pi N \Delta F} R(\Delta \tau) \cos(\Delta \varphi_P) + \eta_{P,I} \\
Y_{P,Q}(\bar{\tau}, \bar{F}_D) &= -\sqrt{\frac{C}{4}} d_P \frac{\sin(\pi N \Delta F)}{\pi N \Delta F} R(\Delta \tau) \sin(\Delta \varphi_P) + \eta_{P,Q}
\end{aligned}
\tag{145}
$$

where d_D and d_P are the signs of the data and pilot components; $\Delta \varphi_D = \Delta \varphi_P$; $\eta_{D,I}$, $\eta_{D,Q}$, $\eta_{P,I}$, and $\eta_{P,Q}$ are four independent zero mean Gaussian r.v.'s with the variance given by (133).

The non-coherent channels combining consists in simply non-coherently adding the data and pilot correlation components in (145), leading to the single coherent period decision variable:

$$
\begin{aligned}
S(\bar{\tau}, \bar{F}_D) &= |Y_D(\bar{\tau}, \bar{F}_D)|^2 + |Y_P(\bar{\tau}, \bar{F}_D)|^2 \\
&= Y_{D,I}^2(\bar{\tau}, \bar{F}_D) + Y_{D,Q}^2(\bar{\tau}, \bar{F}_D) + Y_{P,I}^2(\bar{\tau}, \bar{F}_D) + Y_{P,Q}^2(\bar{\tau}, \bar{F}_D)
\end{aligned}
\tag{146}
$$

In this case $S(\bar{\tau}, \bar{F}_D)$ is a χ^2 r.v. with 4 d.o.f.'s. When the received and the local signals are perfectly aligned, with respect to the code delay and the Doppler shift, $S(\bar{\tau}, \bar{F}_D)$ is non-central with non-centrality parameter equal to 2λ, where λ is defined by (137). Under null hypothesis H_0, $S(\bar{\tau}, \bar{F}_D)$ can be assumed to be a central χ^2 r.v.. From these considerations it is possible to evaluate the false alarm and detection probabilities for a single coherent period:

$$
P_{fa}^{nc}(\beta, 1) = P(S_X(\bar{\tau}, \bar{F}_D) > \beta | H_0) = \exp\left(-\frac{\beta}{2\sigma_n^2}\right)\left(1 + \frac{\beta}{2\sigma_n^2}\right)
\tag{147}
$$

$$
P_d^{nc}(\beta, 1) = P(S_X(\bar{\tau}, \bar{F}_D) > \beta | H_1) = Q_2\left(\frac{\sqrt{2\lambda}}{\sigma_n}, \frac{\sqrt{\beta}}{\sigma_n}\right)
\tag{148}
$$

In order to enhance the acquisition performance, different realizations of the decision variable $S(\bar{\tau}, \bar{F}_D)$ can be non-coherently combined in order to reduce the noise impact. In this way the final decision variable may be obtained

$$
\begin{aligned}
S_K(\bar{\tau}, \bar{F}_D) &= \sum_{i=1}^{K} S_k(\bar{\tau}, \bar{F}_D) \\
&= \sum_{i=1}^{K}\left\{|Y_{D,k}(\bar{\tau}, \bar{F}_D)|^2 + |Y_{P,k}(\bar{\tau}, \bar{F}_D)|^2\right\} \\
&= \sum_{i=1}^{K}\left\{Y_{D,k,I}^2(\bar{\tau}, \bar{F}_D) + Y_{D,k,Q}^2(\bar{\tau}, \bar{F}_D) + Y_{P,k,I}^2(\bar{\tau}, \bar{F}_D) + Y_{P,k,Q}^2(\bar{\tau}, \bar{F}_D)\right\}
\end{aligned}
\tag{149}
$$

where $S_K(\bar{\tau}, \bar{F}_D)$ is a χ^2 r.v. with 4K d.o.f.'s.

Under null hypothesis H_0, $S_K(\bar{\tau}, \bar{F}_D)$ is a central χ^2 r.v.; under alternative hypothesis H_1, $S_K(\bar{\tau}, \bar{F}_D)$ is non-central with non-centrality parameter equal to $2K\lambda$. Therefore, the expressions of the false alarm and detection probabilities are equal to [54]

$$
P_{fa}^{nc}(\beta, K) = P(S_{X,K}(\bar{\tau}, \bar{F}_D) > \beta | H_0) = \exp\left(-\frac{\beta}{2\sigma_n^2}\right)\sum_{i=0}^{2K-1}\frac{1}{i!}\left(\frac{\beta}{2\sigma_n^2}\right)^i
\tag{150}
$$

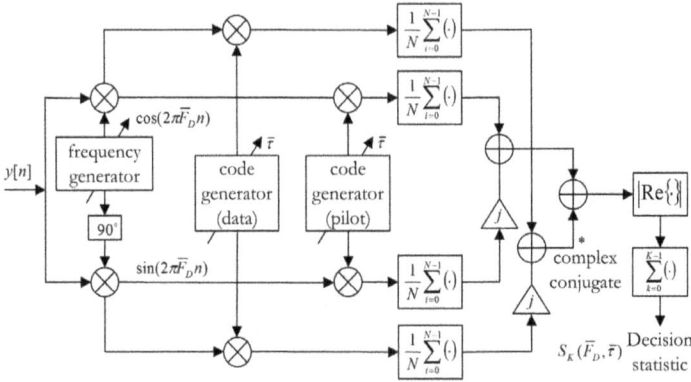

Figure 35. Acquisition scheme for differentially coherent channels combining: correlations with data and pilot local codes are performed separately and differentially coherently combined.

$$P_d^{nc}(\beta, K) = P(S_{X,K}(\bar{\tau}, \bar{F}_D) > \beta | H_1) = Q_{2K}\left(\frac{\sqrt{2K\lambda}}{\sigma_n}, \frac{\sqrt{\beta}}{\sigma_n}\right) \tag{151}$$

where $Q_{2K}(\cdot, \cdot)$ is the generalized Marcum-Q function of order $2K$.

4.4. Differentially Coherent Channels Combining

As described in section 3, in the traditional differentially coherent acquisition scheme the correlator outputs on two consecutive portions of the incoming signal are evaluated and the decision statistic is formed by taking the real part of the product of these two correlations. In this way the phase of the second correlation is employed to compensate the phase of the first one. Moreover, since the noise terms in the two correlations are independent, a lower noise amplification is expected with respect to non-coherent combining scheme. Differential combining is effective as long as the hypothesis of constant phase on the two subsequent correlations holds; degradations are expected in presence of a time-varying phase.

Considering composite GNSS signals, such as Galileo E1 OS signals defined in Eq. (122), the data and pilot channels experience the same transmission channel, and thus they are likely affected by the same code phase delay and Doppler frequency shift. Moreover, their phase difference strictly keeps 180°. In this way the differentially coherent acquisition scheme can be used in order to employ the data and pilot components instead of two subsequent portions of the same input signal. In Fig. 35 the acquisition scheme using differentially coherent channels combining strategy is illustrated. The input composite signal $y[n]$ is separately correlated with the data and the pilot local code replicas, and then two complex correlations are formed:

$$\begin{aligned} Y_{D,k}(\bar{\tau}, \bar{F}_D) &= Y_{D,k,I}(\bar{\tau}, \bar{F}_D) + jY_{D,k,Q}(\bar{\tau}, \bar{F}_D) \\ Y_{P,k}(\bar{\tau}, \bar{F}_D) &= Y_{P,k,I}(\bar{\tau}, \bar{F}_D) + jY_{P,k,Q}(\bar{\tau}, \bar{F}_D) \end{aligned} \tag{152}$$

where $Y_{D,k,I}(\bar{\tau}, \bar{F}_D)$, $Y_{D,k,Q}(\bar{\tau}, \bar{F}_D)$, $Y_{P,k,I}(\bar{\tau}, \bar{F}_D)$, and $Y_{P,k,Q}(\bar{\tau}, \bar{F}_D)$ are the in-phase and quadrature correlator outputs corresponding to the data and pilot channels, respectively.

The decision variable can be obtained as

$$
\begin{aligned}
S_k(\bar{\tau}, \bar{F}_D) &= \left| \mathcal{Re}\{Y_{D,k}(\bar{\tau}, \bar{F}_D) Y_{P,k}^*(\bar{\tau}, \bar{F}_D)\} \right| \\
&= \left| Y_{D,k,I}(\bar{\tau}, \bar{F}_D) Y_{P,k,I}(\bar{\tau}, \bar{F}_D) + Y_{D,k,Q}(\bar{\tau}, \bar{F}_D) Y_{P,k,Q}(\bar{\tau}, \bar{F}_D) \right|
\end{aligned}
\tag{153}
$$

In Eq. (153) the absolute value of the real part of the product $Y_{D,k}(\bar{\tau}, \bar{F}_D) Y_{P,k}^*(\bar{\tau}, \bar{F}_D)$ has been introduced in order to consider the phase difference of 180° between the data and pilot channels and also to remove the dependence on the product of the navigation message and secondary codes [54–56].

As proved in section 3.2, the real part of the product of the two independent Gaussian r.v.'s can be expanded as the difference of two independent χ^2 r.v.'s, here, concerning the decision statistic $S_k(\bar{\tau}, \bar{F}_D)$ in Eq. (153), we have

$$
\begin{aligned}
&\mathcal{Re}\{Y_{D,k} Y_{P,k}^*\} \\
&= \left| \frac{Y_{D,k} + Y_{P,k}}{2} \right|^2 - \left| \frac{Y_{D,k} - Y_{P,k}}{2} \right|^2 \\
&= \underbrace{\left[\left(\frac{Y_{D,k,I} + Y_{P,k,I}}{2} \right)^2 + \left(\frac{Y_{D,k,Q} + Y_{P,k,Q}}{2} \right)^2 \right]}_{\chi^2(2)} - \underbrace{\left[\left(\frac{Y_{D,k,I} - Y_{P,k,I}}{2} \right)^2 + \left(\frac{Y_{D,k,Q} - Y_{P,k,Q}}{2} \right)^2 \right]}_{\chi^2(2)}
\end{aligned}
\tag{154}
$$

In particular, when the useful signal is absent or not correctly aligned, i.e. under null hypothesis H_0, each element $\frac{Y_{D,k,I\{Q\}} \pm Y_{P,k,I\{Q\}}}{2}$ has a Gaussian distribution:

$$
\frac{Y_{D,k,I\{Q\}} \pm Y_{P,k,I\{Q\}}}{2} \sim \mathcal{N}\left(0, \frac{\sigma_n^2}{2}\right)
\tag{155}
$$

Thus, $\left| \frac{Y_{D,k} + Y_{P,k}}{2} \right|^2$ and $\left| \frac{Y_{D,k} - Y_{P,k}}{2} \right|^2$ are two independent central χ^2 r.v.'s with two d.o.f.'s. Therefore, $\mathcal{Re}\{Y_{D,k} Y_{P,k}^*\}$ is a r.v. which is equal to the difference of two independent χ^2 r.v.'s.

Let $f_{S_k(\bar{\tau}, \bar{F}_D)|H_0}(x|H_0)$ denote the c.p.d.f. of the decision variable $S_k(\bar{\tau}, \bar{F}_D)$ under null hypothesis H_0, as seen in section 3.2, its expression can be written in the following form

$$
f_{S_k(\bar{\tau}, \bar{F}_D)|H_0}(x|H_0) = \frac{1}{\sigma_n^2} \exp\left(-\frac{x}{\sigma_n^2}\right) \quad x \geq 0
\tag{156}
$$

In Eq. (156), it has clearly shown that the decision variable $S_k(\bar{\tau}, \bar{F}_D)$ is exponentially distributed under H_0, $S_k(\bar{\tau}, \bar{F}_D) \sim \text{Exp}(\frac{1}{\sigma_n^2})$, which is a special case of a Gamma distribution, i.e. $S_k(\bar{\tau}, \bar{F}_D) \sim \Gamma(1, \sigma_n^2)$. Thus the probability of false alarm $P_{fa}(\beta, 1)$ is obtained as

$$P_{fa}^{dc}(\beta, 1) = \int_{\beta}^{+\infty} f_{S(\bar{\tau}, \bar{F}_D)|H_0}(x|H_0)\, dx = \exp\left(-\frac{\beta}{\sigma_n^2}\right) \tag{157}$$

The acquisition over several periods can be performed by directly accumulating K independent realizations of $S_k(\bar{\tau}, \bar{F}_D)$:

$$S_K(\bar{\tau}, \bar{F}_D) = \sum_{i=0}^{K-1} S_k(\bar{\tau}, \bar{F}_D) = \sum_{k=0}^{K-1} \left| \mathcal{R}e\{Y_{D,k}(\bar{\tau}, \bar{F}_D) Y_{P,k}^*(\bar{\tau}, \bar{F}_D)\} \right| \tag{158}$$

$S_K(\bar{\tau}, \bar{F}_D)$ is the sum of K independent Gamma distributed r.v.'s, thus

$$S_K(\bar{\tau}, \bar{F}_D) = \sum_{i=0}^{K-1} S_k(\bar{\tau}, \bar{F}_D) \sim \Gamma(K, \sigma_n^2) \tag{159}$$

The false alarm probability for the differentially coherent channels combining assumes the following expression:

$$P_{fa}^{dc}(\beta, K) = \exp\left(-\frac{\beta}{\sigma_n^2}\right) \sum_{i=0}^{K-1} \frac{1}{i!} \left(\frac{\beta}{\sigma_n^2}\right)^i \tag{160}$$

Under alternative hypothesis H_1, it is able to prove that, $\mathcal{R}e\{Y_{D,k}Y_{P,k}^*\}$ in Eq. (154) can be seen as the difference of a non-central and a central χ^2 r.v.'s with two d.o.f.'s. In [49], the p.d.f. of $R_k(\bar{\tau}, \bar{F}_D)(= \mathcal{R}e\{Y_{D,k}Y_{P,k}^*\})$ under H_1 is obtained

$$p_{\mathcal{R}_k|H_1}(x) = \begin{cases} \dfrac{1}{2\sigma_n^2} e^{\frac{x}{\sigma_n^2}} e^{-\frac{\lambda}{2\sigma_n^2}} & x < 0 \\[3mm] \dfrac{1}{2\sigma_n^2} e^{\frac{x}{\sigma_n^2}} e^{-\frac{\lambda}{2\sigma_n^2}} Q_1\left(\sqrt{\frac{\lambda}{\sigma_n^2}}, \sqrt{\frac{4x}{\sigma_n^2}}\right) & x \geq 0 \end{cases} \tag{161}$$

Then the c.p.d.f. of the decision variable $S_k(\bar{\tau}, \bar{F}_D)(= |\mathcal{R}e\{Y_{D,k}Y_{P,k}^*\}|)$ can be derived in the following

$$\begin{aligned} f_{S_k(\bar{\tau}, \bar{F}_D)|H_1}(x|H_1) &= p_{\mathcal{R}_k|H_1}(x) + p_{\mathcal{R}_k|H_1}(-x) \\[2mm] &= \frac{1}{2\sigma_n^2} \exp\left(-\frac{\lambda}{2\sigma_n^2}\right)\left[\exp\left(-\frac{x}{\sigma_n^2}\right) + \exp\left(\frac{x}{\sigma_n^2}\right) Q_1\left(\sqrt{\frac{\lambda}{\sigma_n^2}}, \sqrt{\frac{4x}{\sigma_n^2}}\right)\right] \end{aligned} \tag{162}$$

where $x \geq 0$.

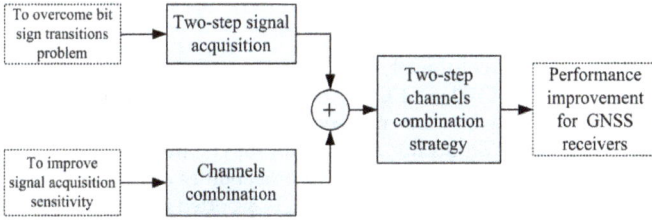

Figure 36. Two-step channels combination strategy for composite GNSS acquisition.

The detection probability can be obtained by integrating by parts

$$P_d^{dc}(\beta,1) = Q_1\left(\frac{\sqrt{2\lambda}}{\sigma_n},\frac{2\sqrt{\beta}}{\sigma_n}\right) + \frac{1}{2}\exp\left(-\frac{2\beta+\lambda}{2\sigma_n^2}\right) - \frac{1}{2}\exp\left(\frac{2\beta-\lambda}{2\sigma_n^2}\right)Q_1\left(\frac{\sqrt{\lambda}}{\sigma_n},\frac{2\sqrt{\beta}}{\sigma_n}\right) \quad (163)$$

The detection probability for a generic K does not admit an easy closed-form analytical expression, but it can be evaluated by using a numerical method for the inversion of the chf [57].

4.5. Two-step Channels Combining Acquisition Strategy

The presence of data message or a secondary code which modulates each primary code period reduces the possibility of increasing the integration time in a coherent way, since the data or the secondary code may lead to sign reversals in the correlation window. The achievable maximum performance of extending the coherent integration time for improving the acquisition sensitivity has received significant attenuation in the presence of bit sign transitions. In section 2.3, it has shown that the two-step acquisition methodology can be adopted to mitigate the CAF main peak splitting effect caused by bit sign transitions and it can provide much improved acquisition performance.

Meanwhile, due to the availability of data and pilot components separately broadcast in the new composite GNSS signals, the drawback of using only single channel independently is that half of the transmitted power is lost. When acquiring composite GNSS signals, such as the Galileo E1 OS modulation, if ignoring the pilot channel and processing only the data channel signal, only half of the useful signal is exploited and the GNSS receiver could not acquire signals that would be easily processed if all the useful signal power were used. This loss can be particularly troublesome at the acquisition stage especially in weak signal environment. In order to overcome the power loss problem, a series of channels combining techniques, such as non-coherent channels combination and differentially coherent channels combination in section 4.3 and section 4.4, can be used for the joint acquisition of data and pilot components of the new composite GNSS signals.

Therefore, in order to deal with bit sign transitions problem and also enhance acquisition sensitivity for GNSS receivers, the two-step acquisition scheme has been firstly applied to the channels combining techniques, thus forming novel two-step channels combination methodologies aiming at performance improvements for GNSS receivers [54, 55]. This proposed two-step channels combination methodology has been clearly illustrated in Fig. 36.

In order to support the theoretical analysis, Monte Carlo simulation campaigns have been performed on the simulated Galileo E1 OS signals in order to evaluate the performances of the proposed techniques. The described non-coherent channels combining and differentially coherent channels combining acquisition schemes for the composite GNSS signals have been deeply analyzed by characterizing the respective probabilities of detection and false alarm. Galileo E1 OS signals characterized by the parameters reported in Table 2 have been adopted in the simulation analysis. These signals have been acquired according to the different acquisition algorithms discussed in this section and false alarm and detection probabilities have been estimated by means of error counting techniques.

Parameter	Value
Sampling frequency, f_s	16.3676 MHz
Intermediate frequency, f_{IF}	4.1304 MHz
Code length	4092 chips
Pre-detection integration time	4 ms

Table 2. Simulation parameters

In order to enhance the acquisition sensitivity and to mitigate the CAF peak splitting effect due the bit sign transitions, the combination methodology between channels integration and two steps acquisition has been adopted specifically in weak signal scenario. In particular the different acquisition strategies are compared in terms of ROC curves and the performance of each strategy is analyzed by means of Monte Carlo simulations.

Moreover, in order to bring a whole view of the acquisition performance picture, the ROC curves for the single channel non-coherent integration, and the related single step channels integration schemes are also visualized, for comparison purposes. In Fig. 37 and Fig. 38, the simulated ROC curves for the different acquisition methods, and for $K = 1$, $C/N_0 = 33$ and 35 dB-Hz, respectively, have been compared. As expected, the traditional single channel acquisition always leads to the worst performance. This is due to the fact that only half of the available useful signal power is exploited. The advantage of this single channel signal acquisition is the relative simplicity of the algorithm, which requires only half of computational load needed by the related channels combining techniques.

From Fig. 37 and Fig. 38, it is clearly known that each two-step channels combining acquisition strategy much outperforms its counterpart - single-step channels combining acquisition approach. Concerning the single-step channels combining strategies, it emerges that non-coherent channels combining technique outperforms differentially coherent channels combining technique although differentially coherent channels combining tends to converge to the same curve for high C/N_0 value. When considering the two-step channels combining strategies, the differentially coherent combining technique works slightly better than the non-coherent channels combining one, which shows the highest detection probability.

In summary, the simulation results have revealed that the proposed two-step channels combining strategies provide much improved performance with respect to the conventional single channel non-coherent integration and the related single-step channels combining

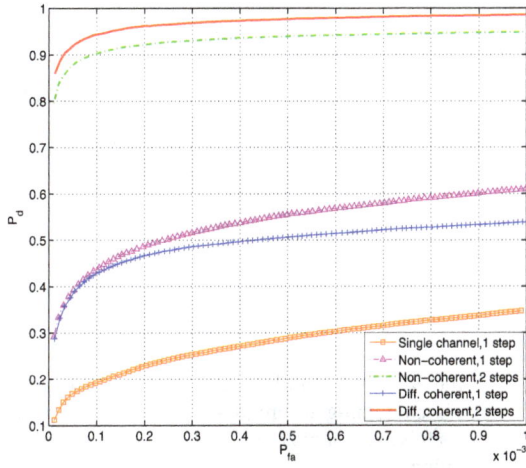

Figure 37. ROC comparison among different acquisition strategies: $C/N_0 = 33$ dB-Hz, K=1.

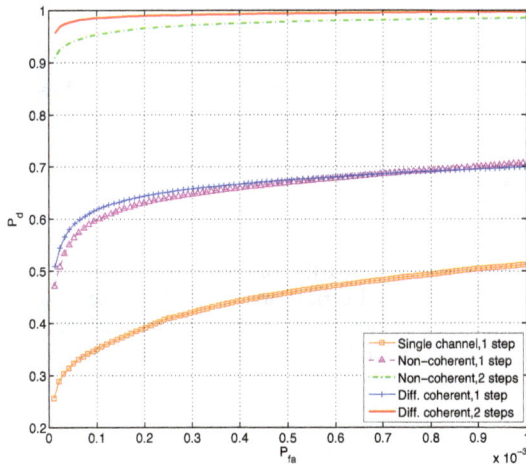

Figure 38. ROC comparison among different acquisition strategies: $C/N_0 = 35$ dB-Hz, K=1.

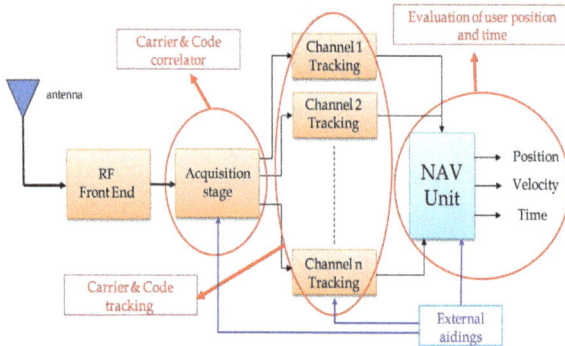

Figure 39. GNSS receiver functions scheme.

techniques, which prove the advantages and effectiveness of the developed theory. These proposed two-step channels combining techniques solve the CAF peak splitting problem present in the new composite GNSS signals acquisition and also enhance the acquisition sensitivity specifically adapting to weak signal environment.

In detail, from the developed analysis, it is clearly known that, when acquisition on a single primary code period is considered, among all the considered acquisition strategies, the two-step differentially coherent channels combing is the most effective acquisition strategy. The proposed innovative acquisition techniques improve the performance and provide more reliable signal detection even in weak signal environment, which can be applied to the new composite GNSS signals where the secondary codes could change the relative polarity every primary code period.

It is important to emphasize that a greater computational load is generally required to perform the acquisition process for each channels combining strategy when the two-step acquisition scheme is adopted.

5. GNSS Signal Tracking Techniques

5.1. Traditional GNSS Tracking Structure

Once the acquisition unit has detected the presence of a given satellite vehicle (SV) and estimated the offset on the residual carrier and the code phase delay with respect to the local replicas, a fine synchronization stage, named signal tracking, is activated to refine these values, keep track and demodulate the navigation data from the specific satellite. This fine synchronization is fundamental for measuring the pseudo range, based on code phase measurements, or also the carrier phase measurements.

The whole signal tracking process is a two-dimensional (code and carrier) signal replication process. It consists of two interoperating feedback loops, a Delay Lock Loop (DLL) for code tracking and a Phase Lock Loop (PLL) for carrier tracking (typically a Costas Loop). Fig. 39 shows a high-level block diagram of a typical GNSSS receiver tracking loops. Obviously, an n parallel channels receiver will have n sets of blocks corresponding to each independent

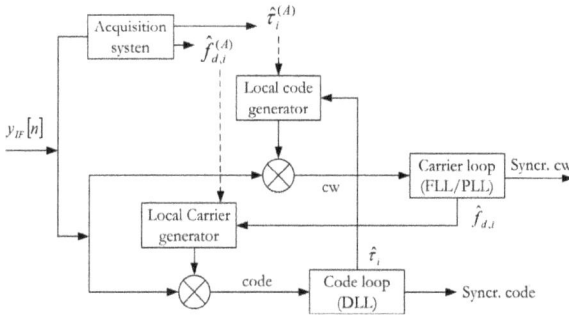

Figure 40. Generic tracking loop.

tracking loops. In a GNSS receiver, the digitized IF signal is input to each of these parallel channels.

The tracking is made of an iterative procedure during which the carrier tracking loop and the code tracking loop cooperate to provide the best estimates of the Doppler frequency shift ($\hat{f}_d$ and, in some cases, also of the phase $\hat{\varphi}$ of the incoming signal) and of the code phase delay ($\hat{\tau}$).

The outputs of the acquisition stage $\hat{p}_i^{(A)} = (\hat{f}_{d,i}^{(A)}, \hat{\tau}_i^{(A)})$ are used to initialize the two tracking loops. After the initialization they must operate together at the same time since:

- A good estimate of the code phase delay is not possible until the residual modulation due to the Doppler shift is removed;

- A good estimate of the residual Doppler shift is not possible until the code signal is canceled out, in order to allow the carrier loop to operate on a pure tone signal.

For these reasons the best estimate is obtained after several steps of approximation during which the output of the carrier tracking loop is used to remove the modulation (*carrier wipe-off*) for the code estimation, and the output of the code tracking loop is used to cancel out the code signal (*code wipe-off*) for the carrier estimation. This scheme is clearly illustrated in Fig. 40.

5.2. Carrier Tracking Loops

To demodulate the navigation data successfully an exact carrier wave replica has to be generated. To track a carrier wave signal, a carrier tracking loop is often adopted. The carrier tracking loop is a feedback loop able to finely estimate the frequency ($f_{IF} + f_d$) of a noisy sinusoidal wave and to track the frequency changes if the user and the satellite move. Most receivers also track the phase term φ present in the carrier wave signal.

Fig. 41 illustrates a block diagram of a GNSS receiver carrier tracking loop. The programmable designs of the carrier predetection integrators, the carrier loop discriminators, and the carrier loop filters characterize the receiver carrier tracking loop. These three functions determine the two most important performance characteristics of the receiver

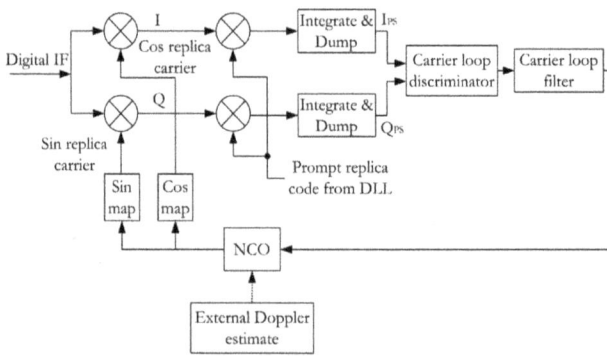

Figure 41. Generic GNSS receiver carrier tracking loop diagram.

carrier loop design: the carrier loop thermal noise error and the maximum line-of-sight (LOS) dynamic stress threshold. Since the carrier tracking loop is always one of the most sensitive blocks in a stand-alone GNSS receiver, its threshold characterizes the unaided GNSS receiver performance [22].

The purpose of the carrier tracking loop is to generate an estimate of the phase or frequency and Doppler shift of the received GNSS RF carrier. It does this by generating a replica of the IF carrier which is in phase with incoming signal. Carrier tracking loops which provide an estimate of phase are called *Phase Lock Loops* (PLLs).

The PLL provides an estimate of the true phase φ_t of the incoming GNSS signal seen at the receiver's antenna. Because of various errors, the actual observed phase at the antenna is different from the true phase that was broadcast by the satellite. This observed phase is the incoming signal phase and we denote it as φ_i. It is related to the true phase by the following equation [58]:

$$\varphi_i = \varphi_t + \delta\varphi_{sv} + \delta\varphi_a + \delta\varphi_i \tag{164}$$

Errors on the incoming signal caused by navigation satellite clock instabilities are represented by $\delta\varphi_{sv}$; errors on the incoming signal caused by propagation delays in the ionosphere and troposphere are represented by $\delta\varphi_a$; and wide band noise on the incoming signal is represented by $\delta\varphi_i$.

The PLL generates an estimate of φ_i (and, thus, indirectly an estimate of φ_t) by internally generating a replica of the measured signal at the antenna and synchronizing the phase of this replica with that of the measured signal. The phase of the internal replica is the output of the PLL and is denoted as φ_o. The phase of the replica or the observed output from the tracking loop is related to the measured or input phase by

$$\varphi_o = \varphi_i + \delta\varphi_{rx} + \delta\varphi_v + \delta\varphi_o + \delta\varphi_d \tag{165}$$

Errors on the output due to receiver clock (or oscillator) errors are represented by $\delta\varphi_{rx}$. Errors due to vibration (which induces phase error in the receiver's oscillator) are represented by $\delta\varphi_v$. Wide band noise on the output phase is represented by $\delta\varphi_o$. Transient errors due to abrupt platform motion are represented by $\delta\varphi_d$. Ideally, we would like a PLL to provide an accurate estimate of the true phase. That is, we would like to minimize the difference between φ_t and φ_o. This difference is called the phase error and it is difficult to measure it directly because we do not have access to φ_t. Therefore, the PLL tries to minimize the tracking error $\delta\varphi$ instead, which is the difference between the received signal phase and the phase of the replica. That is

$$\delta\varphi = \varphi_i - \varphi_o \tag{166}$$

In Fig. 41, the two first multiplications wipe off the carrier and the PRN code of the input signal. To wipe off the PRN code, the prompt output from the early-late code tracking loop is used. The carrier loop discriminator block is used to find the phase error on the local carrier wave replica. The output of the discriminator, which is the phase error (or a function of the phase error), is then filtered and used as a feedback to the numerically controlled oscillator (NCO), which adjusts the frequency of the local carrier wave. In this way the local carrier wave could be an almost precise replica of the input signal carrier wave.

If the GNSS receiver is tracking a data channel signal, it must be noticed that after the code wipe-off procedure the PLL receives a continuous wave signal still modulated by the navigation data. The problem with using an ordinary PLL is that it is sensitive to 180° phase shifts. Due to navigation bit transitions, a PLL used in a GNSS receiver has to be insensitive to 180° phase shifts.

Any carrier loop that is insensitive to the presence of data modulation is usually called a Costas loop. Typically a Costas loop implementation of the PLL model is utilized for carrier tracking and navigation data bit decoding. The Costas loop tolerates the presence of data modulation on the received signal (it is insensitive to 180° phase reversal due to data bit transitions) and then provides a carrier phase reference. Obviously, also the Costas loop requires prior PRN code despreading (code wipe-off) in order to correctly perform the carrier tracking.

Fig. 42 shows a Costas loop. One property of this loop is that it is insensitive for 180° phase shifts due to navigation bits. The carrier wipe-off process used in the generic receiver design requires only two two multiplications. Assuming that the carrier loop is in phase lock and that the replica cosine function is in phase with the incoming SV carrier signal (converted to IF), this results in a cosine squared product at the I output, which produces maximum I_{PS} amplitude (signal plus noise) following the code wipe-off and integrate and dump processes. The second multiplication is between a 90° phase-shifted carrier replica sine function and the incoming SV carrier. This results in a cosine × sine product at the Q output, which produces minimum Q_{PS} amplitude (noise only). For this reason, the Costas loop tries to keep all

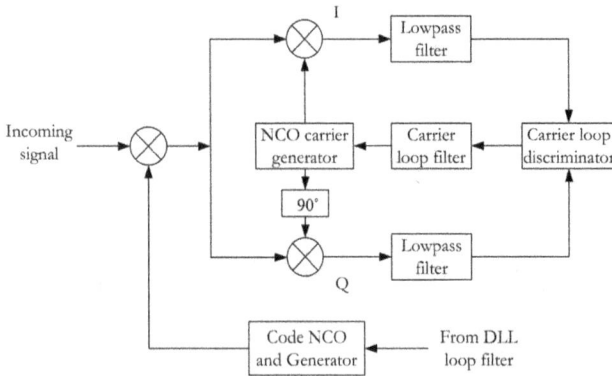

Figure 42. Costas loop used to track the carrier wave.

energy in the I arm. To keep the energy in the I arm, some kind of feedback to the oscillator is needed. If it is assumed that the code replica is perfectly aligned, the multiplication in the I arm yields the following sum:

$$
\begin{aligned}
& d(n) \cos(2\pi F_D n + \varphi_i) \cos(2\pi F_D n + \varphi_o) \\
&= \frac{1}{2} d(n) \cos(\varphi_i - \varphi_o) + \frac{1}{2} d(n) \cos(4\pi F_D n + \varphi_i + \varphi_o) \\
&= \frac{1}{2} d(n) \cos(\delta\varphi) + \frac{1}{2} d(n) \cos(4\pi F_D n + \varphi_i + \varphi_o)
\end{aligned}
\tag{167}
$$

where φ_i is the incoming signal carrier phase, φ_o is the phase of the local replica of the carrier phase, and $\delta\varphi$ is the phase difference between the phase of the input signal and the phase of the local replica carrier, called tracking error. The multiplication in the quadrature arm gives the following:

$$
\begin{aligned}
& d(n) \cos(2\pi F_D n + \varphi_i)[-\sin(2\pi F_D n + \varphi_o)] \\
&= \frac{1}{2} d(n) \sin(\varphi_i - \varphi_o) - \frac{1}{2} d(n) \sin(4\pi F_D n + \varphi_i + \varphi_o) \\
&= \frac{1}{2} d(n) \sin(\delta\varphi) - \frac{1}{2} d(n) \sin(4\pi F_D n + \varphi_i + \varphi_o)
\end{aligned}
\tag{168}
$$

If the two signals are lowpass filtered after the multiplication, the two terms with the double IF are eliminated and the following two signals remain:

$$
I_{PS} = \frac{1}{2} d(n) \cos(\delta\varphi)
\tag{169}
$$

$$Q_{PS} = \frac{1}{2}d(n)\sin(\delta\varphi) \tag{170}$$

To find a term to feed back to the carrier phase oscillator, it can be seen that the phase error of the local carrier phase replica can be found as

$$\frac{Q_{PS}}{I_{PS}} = \frac{\frac{1}{2}d(n)\sin(\delta\varphi)}{\frac{1}{2}d(n)\cos(\delta\varphi)} = \tan(\delta\varphi) \tag{171}$$

$$\delta\varphi = \tan^{-1}\left(\frac{Q_{PS}}{I_{PS}}\right) \tag{172}$$

From Eq. (172), it can be seen that the phase error is minimized when the correlation in the quadrature arm is zero and the correlation value in the in-phase arm is maximum. The arctan discriminator in Eq. (172) is the most precise of the Costas discriminators, but it is also the most time-consuming one. Table 3 also describes other possible Costas discriminators [22, 59].

Discr. Algorithm	Output phase error	Characteristics
$Q_{PS} \times I_{PS}$	$\sin 2\delta\varphi$	Classical Costas analog discriminator. Near optimal at low SNR. Slope proportional to signal amplitude A^2. Moderate computational burden.
$Q_{PS} \times \text{Sign}(I_{PS})$	$\sin \delta\varphi$	Decision directed Costas. Near optimal at high SNR. Slope proportional to signal amplitude A. Least computational burden.
$\tan^{-1}\frac{Q_{PS}}{I_{PS}}$	$\delta\varphi$	Two-quadrant arctangent. Optimal (maximum likelihood estimator) at high and low SNR. Slope not signal amplitude dependent. Highest computational burden.

Table 3. Common Costas Loop Discriminators.

The output of the phase discriminator goes through a loop filter to generate a signal which drives the NCO. The NCO generates a replica signal whose phase is synchronized to that of the incoming signal. Both the carrier tracking loop (Costas loop) and the following described code tracking loop have an analytical linear phase lock loop model that can be exploited to predict and analyze the performance.

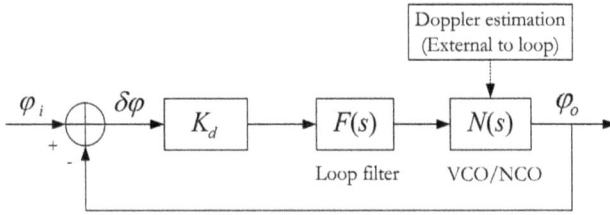

Figure 43. Linear model of PLL.

A linear model for the analog version of the PLL depicted in Fig. 42 is shown in Fig. 43. This linear PLL model is more suitable for analytical work, which could still be the basis of performance prediction. Thus, for simplicity, an analog model could be used at this stage. The second-order PLL system contains a first-order loop filter and a voltage controlled oscillator (VCO). The transfer functions of an analog loop filter and a VCO are

$$F(s) = \frac{1}{s} \frac{\tau_2 s + 1}{\tau_1} \tag{173}$$

$$N(s) = \frac{K_o}{s} \tag{174}$$

where $F(s)$ and $N(s)$ are the transfer functions of the loop filter and NCO, respectively; K_o is the NCO gain.

The transfer function of a linearized analog PLL is

$$H(s) = \frac{K_d F(s) N(s)}{1 + K_d F(s) N(s)} \tag{175}$$

where K_d is the gain of the phase discriminator.

Substituting Equations (173) and (174) into the closed-loop transfer function in Eq. (175) yields

$$H(s) = \frac{2\zeta \omega_n s + \omega_n^2}{s^2 + \zeta \omega_n s + \omega_n^2} \tag{176}$$

where the natural frequency $\omega_n = \sqrt{\dfrac{(K_o K_d)}{\tau_1}}$, and the damping ratio $\zeta = \dfrac{\tau_2 \omega_n}{2}$.

Another carrier tracking loop, which is able to track the frequency of the incoming signal's carrier ignoring its phase term φ_i, is known as *Frequency Lock Loop* (FLL). The FLL is often used to initialize the frequency wipe-off system, by providing a frequency estimate quite close to the correct one. Once the FLL has locked a frequency $\hat{f}_{d,FLL}$, the PLL can refine the frequency estimation working in a narrow band around $\hat{f}_{d,FLL}$.

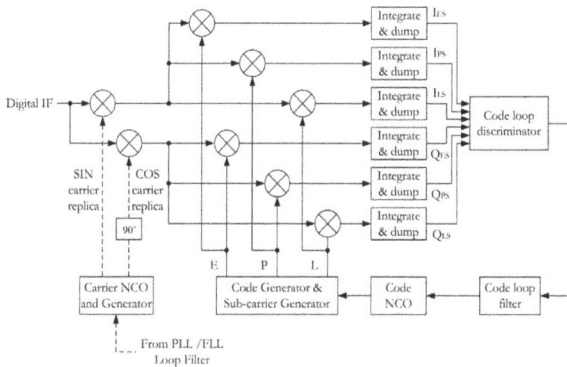

Figure 44. Generic GNSS receiver code tracking loop block diagram.

5.3. Code Tracking Loops

After the acquisition stage of a GNSS receiver has accomplished a rough alignment between the incoming and the local codes within a fraction of a chip interval, such an initial estimate of the code phase delay should be further refined. Furthermore, because of the relative motion between satellite and user receiver and the instability of clocks, correction must be made continuously. This process is performed by a *code tracking loop*, in order to keep track of the code phase of a specific code in the signal. The output of such a code tracking loop is a perfectly aligned replica of the code.

The code tracking loop is a feedback loop able to finely estimate the residual code delay $d_\tau = \tau - \hat{\tau}^{(A)}$ by means of a *Delay Lock Loop* (DLL) called an early-late tracking loop, which is a feedback system to control the behavior of the system itself. The main task of a DLL is to align a local replica code (called *prompt code*) to the code received from each detected satellite. The information about this relative delay between the incoming and the local codes is contained in the correlation peak, therefore the idea could be to finely estimate the correlation value. However, a search of the maximum of the correlation peak is not an effective approach, and it is not used in conventional GNSS receivers. It is then necessary to adopt a strategy insensitive to the absolute correlation peak value, based on a discrimination function that is null only when the incoming and the local codes are synchronized (null-seeker).

Fig. 44 illustrates a generic block diagram of a GNSS code tracking loop. The design of the programmable predetection integrators, the code loop discriminator, and the code loop filter characterizes the receiver code tracking loop. These three functions play a key role in determining the most important two performance characteristics of the receiver code tracking loop design: the code loop thermal noise error and the maximum LOS dynamic stress threshold [40].

Discr. Algorithm	Discr. Type	Characteristics
$I_{ES} - I_{LS}$	Coherent	Simplest of all discriminators. Does not require the Q branch but requires a good carrier tracking loop for optimal functionality.
$(I_{ES}^2 + Q_{ES}^2) - (I_{LS}^2 + Q_{LS}^2)$	*Non-coherent early minus late power*	The discriminator response is nearly the same as the coherent discriminator within $\pm\frac{1}{2}$ chip error.
$\dfrac{(I_{ES}^2 + Q_{ES}^2) - (I_{LS}^2 + Q_{LS}^2)}{(I_{ES}^2 + Q_{ES}^2) + (I_{LS}^2 + Q_{LS}^2)}$	*Normalized early minus late power*	Amplitude sensitivity removed, high computational load. The discriminator has a great property when the chip error is larger than a $\frac{1}{2}$ chip; this will help the DLL to keep track in noisy signals.
$\dfrac{1}{2}[I_{PS}(I_{ES} - I_{LS}) + Q_{PS}(Q_{ES} - Q_{LS})]$	*Quasi-coherent dot product power*	Use all six correlators. Low computational load.

Table 4. Common Delay Lock Loop Discriminators.

The idea behind the DLL is to correlate the input signal with three replicas of the code seen in Fig. 44. The first step is converting the PRN code to baseband, by multiplying the incoming signal with a perfectly aligned local replica of the carrier wave. Afterwards the signal is multiplied with three code replicas (early, prompt, and late). The three replicas are nominally generated with a spacing of $\pm\frac{1}{2}$ chip. After this second multiplication, the six outputs are integrated and dumped. The output of these integrations is a numerical value indicating how much the specific code replica correlates with the code in the incoming signal. The DLL design in Fig. 44 has the advantage that it is independent of the phase on the local carrier wave. If the local carrier wave is in phase with the input signal, all the energy will be in the in-phase arm. But if the local carrier phase drifts compared to the input signal, the energy will switch between the in-phase and the quadrature arms.

If the code tracking loop performance has to be independent of the performance of the phase lock loop, the code tracking loop has to use both the in-phase and quadrature arms to track the code. The DLL needs a feedback to the PRN code generators if the code phase has to be adjusted. The coherent discriminator and three non-coherent discriminators used for feedback are reported in Table 4. A complete analysis of the discriminator properties is illustrated in [22, 59].

The requirements of a DLL discriminator is dependent on the type of application and the noise in the signal. The space between the early, prompt, and late codes determines the noise bandwidth in the delay lock loop. If the discriminator spacing is larger than $\frac{1}{2}$ chip, the DLL would be able to handle wider dynamics and be more noise robust; on the other hand, a DLL with a smaller spacing would be more precise. In a modern GNSS receiver the discriminator spacing can be adjusted while the receiver is tracking the signal. The advantage from this is that if the SNR suddenly decreases, the receiver uses a wider spacing in the correlators to be able to handle a more noisy signal, and hereby a possible code lock loss could be avoided.

5.4. The problem of Data Present in the Code Tracking Loop

After the acquisition system has coarsely estimated the Doppler shift $\hat{f}_d^{(A)}$ and the code phase delay $\hat{\tau}^{(A)}$, the tracking stage is activated to refine the estimation of the two parameters f_d and τ, keep track and demodulate the navigation message from the specific satellite. As described in section 5.1, the tracking operation is an iterative procedure during which the carrier tracking loop and the code tracking loop cooperate to provide the best estimates of the Doppler frequency shift and the code phase delay. After the output of the carrier tracking loop is used to remove the modulation (carrier wipe-off) for the code phase estimation, the code tracking loop receives a spreading code still modulated by the navigation data, which is shown in Fig. 45. In the code tracking loop, it presents the similar CAF peak splitting impairment problem caused by the data sign transition, which has addressed in section 2.2. The DLL must consider the impact of the data bits and try to find a possible way to mitigate this effect; Assisted-GNSS (A-GNSS) approach can be exploited to overcome this problem.

GNSS was originally designed for military tasks, which was expected to work outside with a relatively clear view of the sky. Today GNSS is used for many more civilian than military purposes. In particular, the system demands of civilian applications far exceed those seen before. In very poor signal conditions, for example in a city, these signals may suffer multipath where signals bounce confusingly off buildings, or be weakened by passing through walls or tree cover. When first turned on in these conditions, some non-assisted GNSS navigation devices may not be able to work out a position due to the fragmentary signal, rendering them unable to function until a clear signal can be received continuously.

GNSS is expected to work almost anywhere, even, sometimes, indoors; push-to-fix applications have emerged where a single position is expected almost instantly; and all of this must be delivered in a way that adds little or no cost, size, or power consumption to the host device. These requirements are what drove the development of A-GNSS.

To calculate a position (or fix), a GNSS receiver must first find and acquire the signal from each satellite and then decode the data from the satellites. But before it can acquire each satellite signal, it must find the correct frequency for that satellite and the correct code delay. Each satellite appears on a different frequency, thanks to the Doppler shift induced by the high speeds at which the satellites move. The observed Doppler shift is a function of the location from which the receiver is acquiring the satellite signal. Before the receiver knows where it is, it cannot calculate the Doppler shift. Standard GNSS receivers with no a priori knowledge of these frequency variables would exhaustively search a large range of frequencies made up largely by the effects of satellite motion and receiver oscillator offset, and a small contribution from receiver velocity. However, even if the GNSS receiver has the

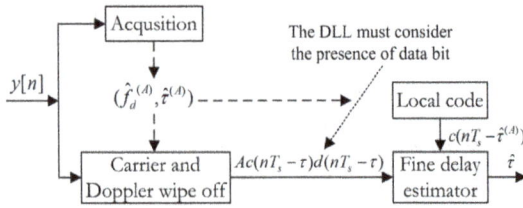

Figure 45. GNSS receiver code tracking loop in presence of data bit.

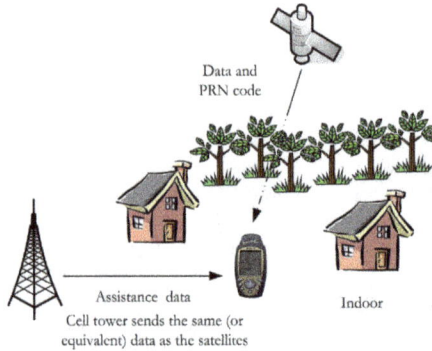

Figure 46. A-GNSS overview.

correct frequency, it must still find the correct code delay for the correlators to generate a correlation peak. The receiver without any a priori knowledge of code delay will also have to search all possible code delay bins. This gives the GNSS receiver a two-dimensional search space for each satellite. We call this the frequency / code delay search space. Having found a signal, it is then necessary to decode data to find the position of the satellite. Only after this satellite position data is decoded could a GNSS receiver compute the position [60].

A-GNSS improves on standard GNSS performance by providing information, through an alternative communication channel, that allows the GNSS receiver to know what frequencies to expect before it even tries, and then the assistance data provides the satellite positions for use in the position computation. Fig. 46 shows an overview of an A-GNSS system. Having acquired the satellite signals, all that is left to do is to take range measurements from the satellites, the A-GNSS receiver can do so more quickly, and with weaker signals, than an unassisted receiver, and then the A-GNSS receiver can compute the position. Furthermore, because the A-GNSS receiver is designed to know in advance what frequencies to search, the typical architecture of the receiver changes to allow longer dwell times, which increase the amount of energy received at each particular frequency. This increases the sensitivity of the A-GNSS receiver and allows it to acquire signals at much lower signal strengths.

Concerning the data bits problem present in the code tracking loop after the carrier wipe off procedure, A-GNSS scheme could be adopted to the data bits removal [61]. In this sense,

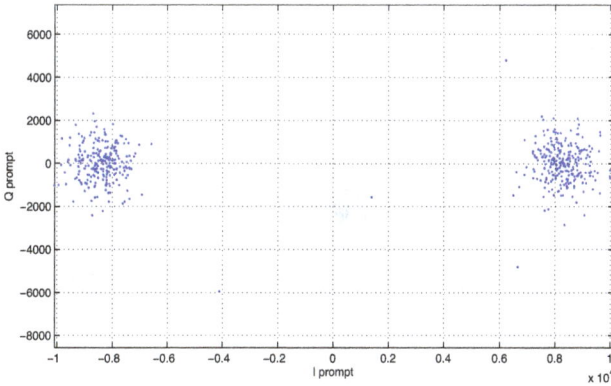

Figure 47. Scatter plot of the code tracking loop in presence of the data.

A-GNSS works by providing assistance data the same (or equivalent) as those present in the code tracking loop, in order to compensate the data bit transitions impact. The assistance data often comes from the cellular network.

In order to investigate the performance of the proposed A-GNSS based code tracking technique, simulations have been performed on the Galileo E1 OS BOC(1,1) signals. The Galileo E1 OS BOC(1,1) signals are simulated by N-FUELS signal generator which was developed by *Navigation Satellite Signal Analysis and Simulation* (NavSAS) research group of Politecnico di Torino, Italy, where the carrier-to-noise ratio (C/N_0) is 46 dB-Hz, the sampling frequency f_s is 16.3676 MHz, the intermediate frequency f_{IF} is 4.1304 MHz and the front-end quantization level is of 4 bits. Assistance data are provided in the A-GNSS based code tracking loop.

5.5. Simulation Analysis and Performance Evaluation

To clearly verify the impact of data bit transitions in a code tracking loop, the simulation campaigns are divided into two cases: the conventional code tracking and the data-assisted code tracking, aiming to have a whole view of the performance comparison among them. In the simulation campaigns, the predetection time is 8 ms in the *Integrate and Dump* block of a code tracking loop.

Here, the implemented code tracking loop discriminator is the normalized early minus late power. This discriminator is described as

$$D = \frac{(I_{ES}^2 + Q_{ES}^2) - (I_{LS}^2 + Q_{LS}^2)}{(I_{ES}^2 + Q_{ES}^2) + (I_{LS}^2 + Q_{LS}^2)} \tag{177}$$

where I_{ES}, Q_{ES}, I_{LS} and Q_{LS} are four outputs of the six correlators shown in Fig. 44. The normalized early minus late power discriminator is chosen because it is independent of the performance of the PLL as it uses both the in-phase and quadrature arms. The normalization

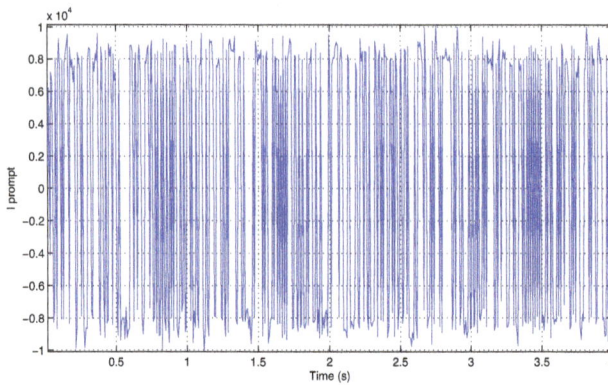

Figure 48. In-phase prompt accumulations of the code tracking loop in presence of the data.

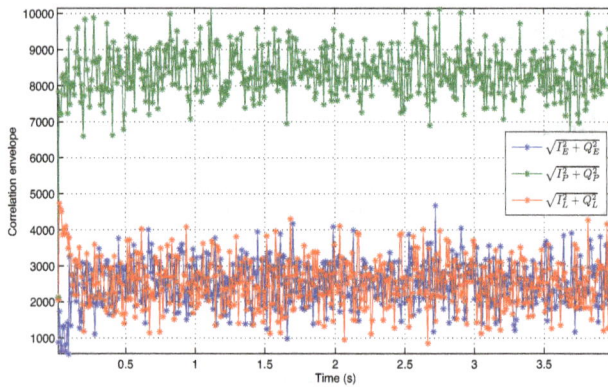

Figure 49. Early, prompt and late correlation envelope of the code tracking loop in presence of the data.

of the discriminator makes the discriminator adapt to be used with signals with different signal-to-noise ratios and different signal strengths.

Firstly, simulation test has been done for the conventional code tracking without data assistance. In Fig. 47, the scatter plot of the demodulated in-phase (I prompt) and quadrature (Q prompt) components is reported, where the navigation data bits are present in the correlator outputs of the code tracking loop. As expected, two bubbles appear due to the navigation bit sign transitions. Fig. 48 shows in-phase prompt accumulation I prompt, which indicates the navigation bits in the incoming signal. In Fig. 49, the early, prompt and late (E-P-L) correlation envelopes of the code tracking loop in presence of data are depicted.

Then, we consider the data-assisted code tracking case. Fig. 50 shows the scatter plot of the demodulated I prompt and Q prompt components, note that only one bubble appears, this is because the navigation data bits are wiped off by the assistance data. This point can be further verified in Fig. 51, where the values of the demodulated in-phase component

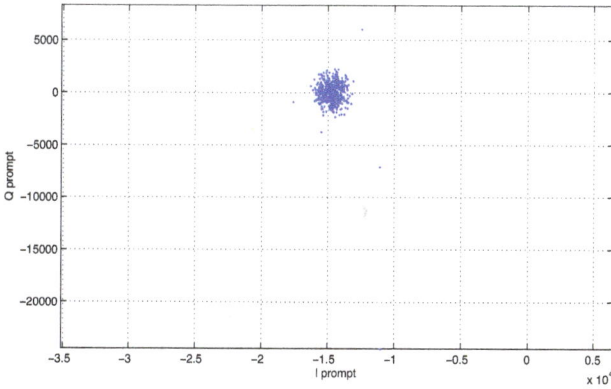

Figure 50. Scatter plot of the assisted code tracking loop.

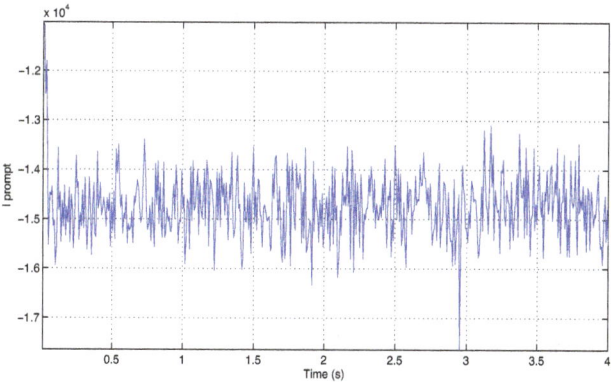

Figure 51. In-phase prompt accumulations of the assisted code tracking loop.

I prompt have always the same sign. In Fig. 52, the early, prompt and late correlation envelopes for the assisted code trackimng loop are reported, respectively; compared with the presented results in Fig. 49 for the conventional code tracking loop, much increased prompt correlation envelope values are obtained.

This novel A-GNSS based code tracking approach has been adopted to deal with the data bit transitions present in the code tracking loop of a GNSS receiver, which aims at enhancing the tracking performance. The performance evaluation of this proposed tracking technique has been provided by means of simulation analysis. From the analysis, it emerges that this technique mitigates the data bit transitions impact occurring in the code tracking loop by providing assistance data from cellular networks and enables signal tracking more robust in challenging environment.

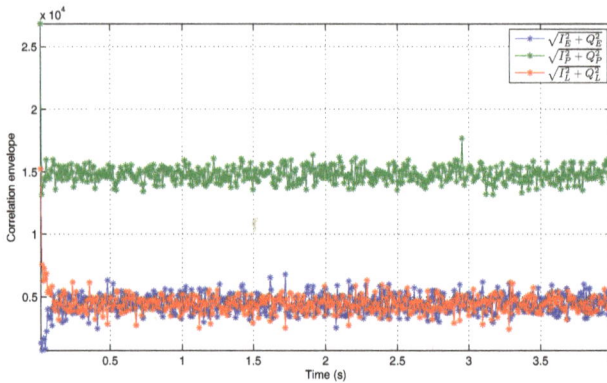

Figure 52. Early, prompt and late correlations of the assisted code tracking loop.

6. Neural Network based Adaptive GNSS Code Tracking Loop Architecture

In the code tracking loop for a GNSS receiver, a linear PLL model is usually adopted, which is an extremely powerful tool for the performance prediction and analysis. In general, a code tracking loop contains a first-order or second-order loop filter and a voltage controlled oscillator (VCO). Due to the nonlinear and time-delay property present in the practical code tracking loop, these systems generally have larger overshoot, longer adjusting time and are not stable. In classical control theory the Smith method [62, 63] can be used to construct controllers if the transfer function of the system has been known. But, the transfer function of a practical system is not easy to measure or to complete.

As is well known that, proportional-integral-derivative (PID) controlling is one of the most important control strategies [64–66]. The PID controller is widely used in the closed loop process control systems for its simplicity and robustness, especially in the systems for which can be established the precise mathematical model. Although PID controllers have strong abilities, they are not suitable for the control of long time delay and nonlinear complex systems, in which the P, I and D parameters are difficult to choose and can hardly adapt to time-varying characteristics in wide range; the conventional PID controller could not achieve the desired control effect because of the unsuitable parameters. Therefore, the application of conventional PID control is much more restricted and challenged.

The process of selecting the controller parameters to meet given performance specifications is known as controller tuning. PID Tuning for a control loop is the adjustment of its control parameters (proportional band / gain, integral gain / reset, derivative gain / rate) to the optimum values for the desired control response. PID tuning is a difficult problem, even though there are only three parameters and in principle is simple to describe, because it must satisfy complex criteria within the limitations of PID control. Stability (bounded oscillation) is a basic requirement, but beyond that, different systems have different behaviors, different applications have different requirements, and requirements may conflict with one another. One heuristic tuning method is formally known as the Ziegler-Nichols method, introduced by John G. Ziegler and Nathaniel B. Nichols in the 1940s. This tuning method proposes rules

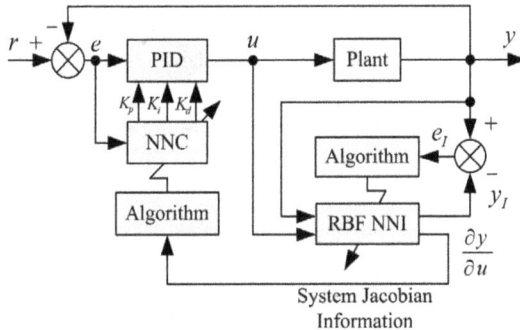

Figure 53. The block diagram of the adaptive PID controll system based on RBF neural network on-line identification process.

for determining values of the proportional gain, integral time and derivative time based on transient response characteristics of a given plant. Such determination of the parameters of PID controllers or tuning of PID controllers can be made by experiments on the plant, which will provide a stable operation of the control system [67]. However, this resulting system may exhibit a large overshoot in the step response, which is unacceptable. In this case, fine tunings are required until an acceptable result is obtained. Over the past years, many techniques are suggested for tuning of the PID parameters such as the refined Ziegler-Nichols method [68], the gain and phase margin method [69] and the internal model control (IMC) based method [70]. However, the limitations of PID control become evident when applied to more complicated systems such as those with a time delay, poorly damped, nonlinear and time-varying dynamics [71].

With the development of modern control theory, neural network has become a powerful computational tool that has been extensively used in the areas of adaptive control through learning process, therefore it could be adopted to regulate the parameters of the PID controller. Since neural network has arbitrarily approaching ability to nonlinear function, the control system based on neural network has strong self-adaptation ability, and best combined PID control effect can be achieved through self-learning process. In this way, good adjustments to proportion, integral and differential parameters of PID controller can be obtained to form cooperative and restrictive relationship in control quantity. In order to improve the GNSS code tracking performance, a novel adaptive PID controlling strategy based on Radial Basis Function (RBF) neural network online identification process is proposed in the loop filter design of the code tracking loop for GNSS receivers [72]. This proposed technique combines conventional PID control with neural network algorithm, and generates a new kind of PID controller with adaptability. Due to the self-learning ability of neural network, this proposed technique can self tune and automatically modify the robust PID parameters online by using gradient descent method.

The block diagram of the proposed adaptive PID control system in the code tracking loop for a GNSS receiver is shown in Fig. 53, which consists of three units:

• Conventional PID controller: it constitutes a closed-loop control and guarantees the stability of the whole control system;

- RBF neural network identifier (NNI): it employs RBF neural network to perform nonlinear system identification functionality, which is used for the prediction of the Jacobian information of the controlled plant in the system, i.e. $\partial y / \partial u$;
- Neural network controller (NNC): NNC uses a single neuron based adaptive PID controller which has self-tuning ability according to the system state in order to achieve optimal control effect.

6.1. Digital PID Controller

A PID controller is a generic control loop feedback mechanism (controller) widely used in practical control systems. It calculates an error value as the difference between a measured process variable and a desired setpoint, which is described in Fig. 54. The controller attempts to minimize the error by adjusting the process control inputs. The PID controller calculation (algorithm) involves three separate constant parameters, and is accordingly sometimes called three-term control: the proportional, the integral and derivative values, denoted as P, I and D, respectively. Heuristically, these values can be interpreted in terms of time: P depends on the present error, I on the accumulation of past errors, and D is a prediction of future errors, based on current rate of change. By tuning these three parameters in the PID controller algorithm, the controller can provide control action designed for specific process requirements.

The PID control scheme is named after its three correcting terms, and the proportional, integral and derivative terms are summed to constitute the output of the PID controller [73]. Defining $u(t)$ as the controller output, the equation for the output in the time domain is

$$
\begin{aligned}
u(t) &= K_p[e(t) + \frac{1}{T_i} \int_0^t e(\tau)d\tau + T_d \frac{d}{dt}e(t)] \\
&= K_p e(t) + K_i \int_0^t e(\tau)d\tau + K_d \frac{d}{dt}e(t)
\end{aligned}
\tag{178}
$$

where

- K_p: proportional gain;
- T_i: integral time;
- T_d: derivative time;
- $K_i = K_p / T_i$: integral gain;
- $K_d = K_p T_d$: derivative gain;
- $e(t) = r(t) - y(t)$: difference (error) between the measured process output and the reference input;
- t: time or instantaneous time (the present).

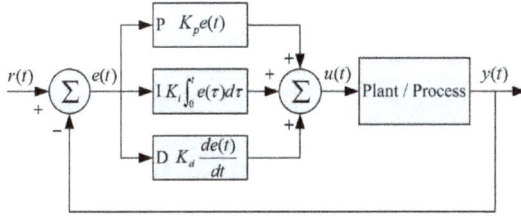

Figure 54. A block diagram of a PID controller.

The PID controller is modeled by the following transfer function

$$
\begin{aligned}
G(s) &= \frac{U(s)}{E(s)} \\
&= K_p(1 + \frac{1}{T_i s} + T_d s) \\
&= K_p + K_i \frac{1}{s} + K_d s
\end{aligned}
\tag{179}
$$

When the backward difference method is applied, the incremental digital PID algorithm can be obtained as follows

$$
\begin{aligned}
u[k] &= u[k-1] + K_p\{e[k] - e[k-1]\} + K_i e[k] \\
&\quad + K_d\{e[k] - 2e[k-1] + e[k-2]\}
\end{aligned}
\tag{180}
$$

6.2. RBF Neural Network Structure

The RBF neural network is a feed-forward network which consists of three layers: input layer, hidden layer and output layer. In the RBF neural network, the mapping from input to output is nonlinear, but it is linear from hidden layer to output layer; therefore the leaning speed with this type of neural network can be accelerated greatly. It has been proved that RBF neural network has the ability of approaching any continuous function with arbitrary accuracy. The structure of a typical RBF neural network is given in Fig. 55. In the RBF neural network, each input neuron corresponds to an element of the input vector and is fully connected to the neurons in the hidden layer. Similarly, each neuron of the hidden layer is also connected to the output layer neuron.

Suppose that $X = [x_1, x_2, \cdots, x_n]^T$ is the input vector of the neural network, neurons in the second layer (hidden layer) are activated by radial basis functions and the radial vector of the neural network is denoted as $H = [h_1, h_2, \cdots, h_m]^T$, where h_j $(j = 1, 2, \cdots, m)$ is supposed to be multivariate Gaussian function written as

$$
h_j = \exp\left(-\frac{\|X - C_j\|^2}{2b_j^2}\right)
\tag{181}
$$

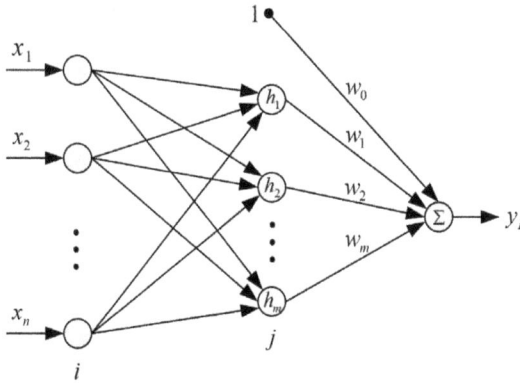

Figure 55. Structure of RBF neural network.

where the symbol $\| \cdot \|$ denotes the Euclidean norm; $C_j = [c_{j1}, c_{j2}, \cdots, c_{jn}]^T$ is the center vector for the jth node of the neural network; and b_j is the base width parameter of the node j in the hidden layer, whose value is greater than zero ($b_j > 0$), correspondingly, the base width vector $B = [b_1, b_2, \cdots, b_m]^T$ is formed in the hidden layer of the neural network.

The weight vector of the neural network is denoted as $W = [w_1, w_2, \cdots, w_m]^T$, therefore the output of the RBF neural network can be formed by a linearly weighted sum of the number of the radial basis functions in the hidden layer, which is written as follows

$$\begin{aligned} y_I &= w_0 + w_1 h_1 + \cdots + w_j h_j + \cdots + w_m h_m \\ &= w_0 + \sum_{j=1}^{m} w_j h_j \end{aligned} \tag{182}$$

where the subscript I means that the RBF neural network is used as an identifier in the control system, and w_j is the weight between the output neuron and the jth neuron in the hidden layer.

6.3. Identification Algorithm of RBF Neural Network

In this part, the identification algorithm of RBF neural network has been well explained. The performance function of the identification process can be defined as follows

$$J_I(k) = \frac{1}{2}[y(k) - y_I(k)]^2 = \frac{1}{2}e_I^2(k) \tag{183}$$

In order to minimize the error $e_I(k)$ between the RBF neural network output $y_I(k)$ and that of the real controlled plant output $y(k)$, gradient decent method can be adopted to adjust weights between output layer and hidden layer, node center and node radial width parameters for the hidden layer. Therefore, the corresponding iterative algorithm can be provided as follows

$$w_j(k) = w_j(k-1) + \eta[y(k) - y_I(k)]h_j + \\ \alpha[w_j(k-1) - w_j(k-2)] \tag{184}$$

$$b_j(k) = b_j(k-1) + \eta \triangle b_j + \alpha[b_j(k-1) - b_j(k-2)] \tag{185}$$

$$c_{ji}(k) = c_{ji}(k-1) + \eta \triangle c_{ji} + \alpha[c_{ji}(k-1) - c_{ji}(k-2)] \tag{186}$$

$$\triangle b_j = [y(k) - y_I(k)]w_j h_j \frac{\|X - C_j\|^2}{b_j^3} \tag{187}$$

$$\triangle c_{ji} = [y(k) - y_I(k)]w_j h_j \frac{x_i - c_{ji}}{b_j^2} \tag{188}$$

where η is the leaning rate and α is the momentum gene.

Through the RBF neural network online identification process, the Jacobian information, which is the sensitivity of the controlled plant output to the control input, can be obtained. The expression of the the Jacobian information is provided as follows

$$\frac{\partial y(k)}{\partial u(k)} \approx \frac{\partial y_I(k)}{\partial u(k)} = \sum_{j=1}^{m} w_j h_j \frac{c_{ji} - u(k)}{b_j^2} \tag{189}$$

The obtained Jacobian information will be sent to the following described single neuron based PID controller for optimal control effect.

6.4. Single Neuron based Self-adaptive PID Controller

Traditional PID controller could usually provide satisfactory performance if it is properly adjusted in the system which does not need very high requirement. However, traditional PID controller does not work well in the nonlinear and time-varying systems, since the PID controlling parameters are fixed after there are selected by appropriate tuning methods; traditional PID controller does not have self-learning ability and adaptability according to different real-time working conditions.

Here a single neuron is introduced in the self-adaptive PID controller design. Single neuron is a multiple input single output information processing unit which constitutes the basic

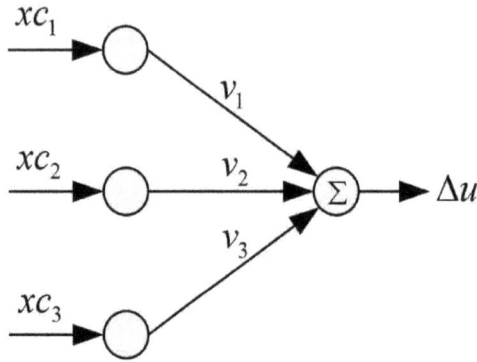

Figure 56. The structure of a single neuron based PID controller.

component of the neural network, with self-learning and adaptive abilities. The structure of a single neuron is shown in Fig. 56, which is simple and easy to be realized. Therefore, the combination between the single neuron and the traditional PID controller is made in order to form a single neuron based adaptive PID controller, which is simple in design and has strong adaptability and robustness. This proposed controller design uses the Jacobian information achieved from the RBF neural network identification unit for performing online adjustments to the control parameters, thus fulfills adaptive tuning of the PID controller in different real-time conditions.

In Fig. 56, $xc_i(i = 1,2,3)$ are the inputs, and $v_i(i = 1,2,3)$ are the corresponding weights of the single neuron. The output of the single neuron (that is the adaptive PID controller output) corresponds to the NCO output in the code tracking loop of the GNSS receiver, and it can be written as

$$\triangle u(k) = v_1(k)xc_1(k) + v_2(k)xc_2(k) + v_3(k)xc_3(k) \tag{190}$$

In comparison to Eq. (180), it is easy to get the following equations:

$$\begin{cases} xc_1(k) = e(k) - e(k-1) \\ xc_2(k) = e(k) \\ xc_3(k) = e(k) - 2e(k-1) + e(k-2) \end{cases} \tag{191}$$

$$\begin{cases} v_1(k) = K_p \\ v_2(k) = K_i \\ v_3(k) = K_d \end{cases} \tag{192}$$

In the neural network based PID controller, one of the fundamental tasks is to reduce the squared system error $J_c(k)$ to zero by adjusting the weights $v_i(k)(i = 1,2,3)$. $J_c(k)$ can be defined as follows

$$J_C(k) = \frac{1}{2}[r(k) - y(k)]^2 = \frac{1}{2}e^2(k) \tag{193}$$

where $e(k) = r(k) - y(k)$ is the error between the actual output of the control system and its desired reference input at present time.

In order to obtain optimal performance for the single neuron based adaptive PID controller, weights should be adjusted and updated on-line by adopting gradient decent method according to Hebb rule, which are written as follows

$$
\begin{aligned}
\triangle v_i(k) &= -\eta_i \frac{\partial J_C}{\partial v_i(k)} + \alpha_i \triangle v_i(k-1) \\
&= -\eta_i \frac{\partial J_C}{\partial y(k)} \frac{\partial y(k)}{\partial u(k)} \frac{\partial u(k)}{\partial v_i(k)} + \alpha_i \triangle v_i(k-1) \\
&= \eta_i \cdot e(k) \cdot \frac{\partial y(k)}{\partial u(k)} \cdot xc_i(k) + \alpha_i \triangle v_i(k-1)
\end{aligned}
\tag{194}
$$

Or, equivalently

$$
\begin{cases}
\triangle K_p(k) = \eta_1 \cdot e(k) \cdot \frac{\partial y(k)}{\partial u(k)} \cdot xc_1(k) + \alpha_1 \triangle K_p(k-1) \\
\triangle K_i(k) = \eta_2 \cdot e(k) \cdot \frac{\partial y(k)}{\partial u(k)} \cdot xc_2(k) + \alpha_2 \triangle K_i(k-1) \\
\triangle K_d(k) = \eta_3 \cdot e(k) \cdot \frac{\partial y(k)}{\partial u(k)} \cdot xc_3(k) + \alpha_3 \triangle K_d(k-1)
\end{cases}
\tag{195}
$$

where $\frac{\partial y(k)}{\partial u(k)}$ is the Jacobian information of the controlled plant in the system, which can be obtained from the RBF neural network identification device in Eq. (189).

6.5. Simulation Analysis and Performance Evaluation

To clearly verify the effectiveness and advantages of the proposed RBF neural network online identification based adaptive code tracking loop in a GNSS receiver, the simulation campaigns are performed on the Galileo E1 OS BOC(1,1) signals. The Galileo E1 OS BOC(1,1) signals are simulated by N-FUELS signal generator, where the C/N_0 is 46 dB-Hz, the sampling frequency f_s is 16.3676 MHz, the intermediate frequency f_{IF} is 4.1304 MHz and the front-end quantization level is of 4 bits. In the simulation campaigns, the predetection time is 8 ms in the *Integrate and Dump* block of a code tracking loop.

The quasi-coherent dot product power form of the discriminator has been adopted in the code tracking loop for a GNSS receiver, which uses all six correlator outputs. This type of discriminator can be written as

$$D = \frac{1}{2}[(I_{ES} - I_{LS})I_{PS} + (Q_{ES} - Q_{LS})Q_{PS}] \tag{196}$$

where I_{ES}, Q_{ES}, I_{PS}, Q_{PS}, I_{LS} and Q_{LS} are the six correlator outputs shown in Fig. 44.

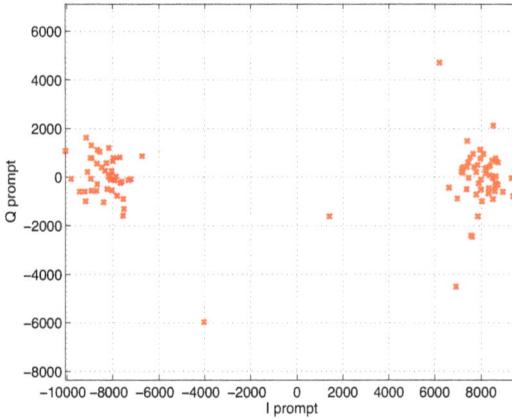

Figure 57. Scatter plot of the RBF neural network identification based adaptive code tracking loop in a GNSS receiver.

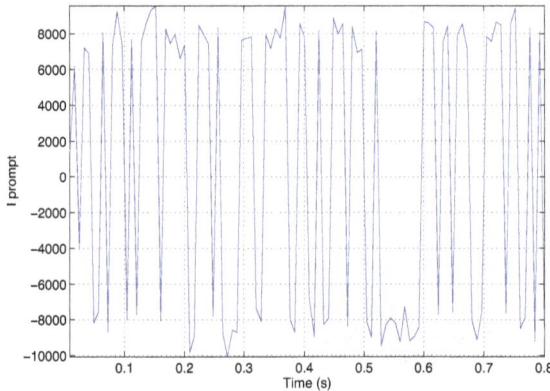

Figure 58. In-phase prompt accumulations of the RBF neural network identification based adaptive code tracking loop in a GNSS receiver.

The RBF neural network based online identification device adopted in the adaptive PID controller for the GNSS code tracking loop uses 4-8-1 topology structure. The RBF neural network has three layers: input layer, hidden layer and output layer. The input layer consists of 4 neurons, therefore the input vector of the RBF neural network can be selected as $X = [u(k), u(k-1), y(k), y(k-1)]^{\mathrm{T}}$, where u is the NCO output in the code tracking loop, and according to Eq. (196), the real output of the code tracking system y can be chosen as $1/2(I_{PS}^2 + Q_{PS}^2)$. In the hidden layer, 8 neurons are used. Considering Eq. (182), the output layer consists of only one neuron, whose output corresponds to y_I in Eq. (182).

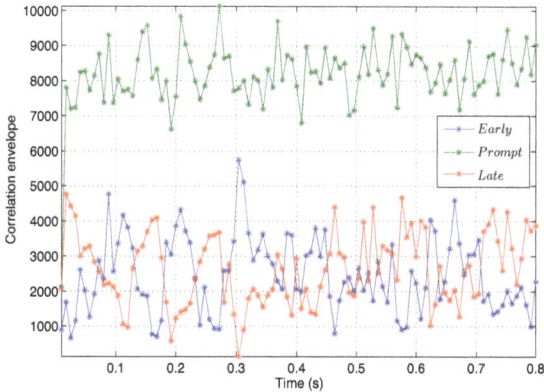

Figure 59. Early, prompt and late correlation envelopes of the RBF neural network identification based adaptive code tracking loop in a GNSS receiver.

In the RBF neural network based Jacobian information online identification device (that is, NNI), the learning rate η is 0.44 and the momentum gene α is 0.12 in the gradient decent algorithm; and in the single neuron based adaptive PID controller (that is, NNC), the learning rate settings are $\eta_1 = \eta_2 = \eta_3 = 0.2$, and the momentum gene settings are $\alpha_1 = \alpha_2 = \alpha_3 = 0.01$.

In Fig. 57, the scatter plot of the demodulated in-phase (I prompt) and quadrature (Q prompt) components is reported: two bubbles appear due to the navigation bit transitions. Fig. 58 shows I prompt accumulations, which indicate the navigation bits in the incoming signal. In Fig. 59, the early, prompt and late (E-P-L) correlation envelopes of the adaptive code tracking loop based on RBF neural network online system identification are depicted.

The Jacobian information of the RBF neural network identifier in the adaptive code tracking loop for the GNSS receivers is shown in Fig. 60, which will be sent to the NNC for weights adjustments in order to achieve optimal control effect. In the single neuron based adaptive PID controller, the PID tuning parameters K_p, K_i and K_d can be adaptively adjusted by using the Jacobian information obtained from the RBF neural network identification device in the code tracking system, which are shown in Fig. 61, respectively. From Fig. 61, it is clear that the adaptive PID controller adopted in the code tracking loop is able to self tune and automatically modify the robust PID parameters online, and these tuning parameters become almost constant values when the time instant arrives approximately in 0.48 s position.

The simulation results demonstrate that this proposed technique provides satisfactory performance in the code tracking loop for GNSS receivers, in particular, the neural network based PID controlling strategy shows robust performance to nonlinear and time-varying behaviors. The proposed adaptive code tracking technique makes the GNSS receiver have robustness, adaptability and self-learning properties and adapt to work in different signal conditions. This proposed novel adaptive code tracking loop design can be applied in the new generation intelligent GNSS receiver architecture.

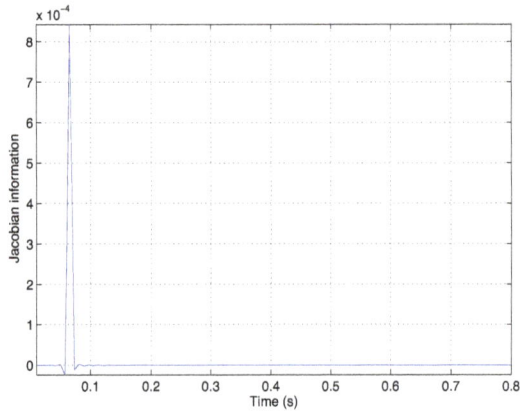

Figure 60. Jacobian information of RBF neural network online identification for the adaptive code tracking loop in a GNSS receiver.

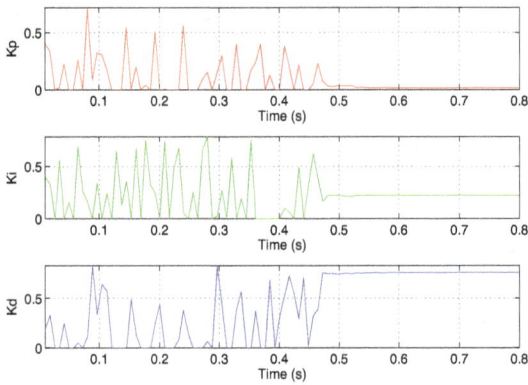

Figure 61. Parameters self-tuning of RBF neural network identification based adaptive code tracking loop in a GNSS receiver.

Author details

Kewen Sun

Hefei University of Technology, China

References

[1] *Signal In Space Interface Control Document (OS SIS ICD, Issue 1.1).* European Union / European GNSS Supervisory Authority, September 2010.

[2] *Interface specification NAVSTAR GPS space segment / navigation L5 user interfaces, Technical Report IS-GPS-705B.* http://www.gps.gov, September 2011.

[3] W. J. Hurd, J. I. Statman, and V. A. Vilnrotter, "High dynamic GPS receiver using maximum-likelihood estimation and frequency tracking," *IEEE Transactions on Aerospace and Electronics Systems*, vol. AES-23, no. 4, pp. 425–437, July 1987.

[4] U. Cheng, W. J. Hurd, and J. I. Statman, "Spread-spectrum code acquisition in the presence of doppler shift and data modulation," *IEEE Transactions On Communications*, vol. 38, no. 2, pp. 241–250, February 1990.

[5] H. Mathis and P. Flammant, "An analytic way to optimize the detector of a postcorrelation FFT acquisition algorithm," *Proc. ION/GNSS*, September 2003.

[6] L. Scharf, *Statistical Signal Processing: Detection, Estimation, and Time Series Analysis.* Prentice Hall, July 1991.

[7] C. O'Driscoll, *Performance Analysis of the Parallel Acquisition of Weak GPS Signals.* Ph.D. dissertation, National University of Ireland, Ireland, January 2007.

[8] C. W. Helstrom, *Elements of Signal Detection and Estimation.* Englewood Cliffs, NJ, USA: PTR Prentice Hall, 1995.

[9] H. L. V. Trees, *Detection, Estimation, and Modulation Theory. Part 1.* Wiley, 2001.

[10] L. L. Scharf, *Statistical Signal Processing: Detection, Estimation, and Time Series Analysis.* Addison-Wesley, 1991.

[11] A. Papoulis, *Probability, random variable and stochsastic processes*, 3rd ed. New York: McGraw Hill, 1991.

[12] J. Proakis, *Digital Communications*, forth ed. McGraw-Hill, August 2000.

[13] J. I. Marcum, "A statistical theory of target detection by pulsed radar," *IEEE Transaction on Information Theory*, pp. 59–267, 1 December 1947.

[14] C. Helstrom, "Computing the generalized marcum q-function," *IEEE Trans. Inform. Theory*, vol. 38, no. 4, pp. 1422–1428, July 1992.

[15] H. Urkowitz, "Energy detection of unknown deterministic signals," *Proceedings of the IEEE*, vol. 55, no. 4, pp. 523–531, April 1967.

[16] D. E. Cartier, "Partial correlation properties of pseudonoise (PN) code in noncoherent synchronization/detection schemes," *IEEE Transactions on Communications*, vol. 24, no. 8, pp. 898–903, August 1976.

[17] U. Cheng, "Performance of a class of parallel spread-spectrum code acquisition schemes in the presence of data modulation," *IEEE Transactions on Communications*, vol. 36, no. 5, pp. 596–604, May 1988.

[18] U. Cheng, W. J. Hurd, and J. I. Statman, "Spread-spectrum code acquisition in the presence of doppler shift and data modulation," *IEEE Transactions on Communications*, vol. 38, no. 2, pp. 241–250, February 1990.

[19] L. D. Davisson and P. G. Flikkema, "Fast single-element PN acquisition for the TDRSS MA system," *IEEE Transactions on Communications*, vol. 36, no. 11, pp. 1226–1235, November 1988.

[20] S. S. Rappaport and D. M. Grieco, "Spread-spectrum signal acquisition: Methods and technology," *IEEE Communications Magazine*, vol. 22, no. 6, pp. 6–21, June 1984.

[21] Y. T. Su, "Rapid code acquisition algorithms employing PN matched filters," *IEEE Transactions on Communications*, vol. 36, no. 6, pp. 724–733, June 1988.

[22] E. D. Kaplan, *Understanding GPS: Principles and Applications*, 2nd ed.   Norwood, MA 02062, Artech House, 2006.

[23] J. B.-Y. Tsui, *Fundamentals of Global Positioning System Receivers: A Software Approach*, 2nd ed.   New York: John Wiley and Sons, 2005.

[24] J. I. Marcum, "A statistical theory of target detection by pulsed radar," *IRE Transactions on Information Theory*, vol. 6, pp. 59–267, 1 December 1949.

[25] P. Swerling, "Probability of detection for fluctuating targets," *IEEE Trans. Inform. Theory*, vol. 6, no. 2, pp. 269–308, April 1960.

[26] D. A. Shnidman, "Radar detection in clutter," *IEEE Trans. Aerosp. Electron. Syst.*, vol. 41, no. 3, pp. 1056–1067, July 2005.

[27] D. Borio and D. Akos, "Noncoherent integrations for GNSS detection: Analysis and comparisons," *IEEE Trans. on Aerospace and Electronics Systems*, vol. 45, no. 1, pp. 360–375, January 2009.

[28] J. Betz, *Systems, Signals and Receiver Signal Processing*.   Navtech GPS, September 2006, vol. 3.

[29] R. Pulikkoonattu and M. Antweiler, "Analysis of differential non coherent detection scheme for CDMA pseudo random (PN) code acquisition," *Proc. of IEEE ISSSTA*, pp. 212–217, August 1984.

[30] R. N. McDonough and A. D. Whalen, *Detection of Signals in Noise*, 2nd ed.   Academic Press, 1995.

[31] J. V. DiFranco and W. L. Rubin, *Radar Detection*.   Artech House, 1980.

[32] J. B.-Y. Tsui, *Fundamentals of Global Positioning System Receivers: A Software Approach*, 2nd ed.   New York, USA: Wiley, 2005.

[33] E. Kaplan and C. Hegarty, *Understanding GPS: Principles and Applications*, 2nd ed. Manassas, VA, USA: Artech House, 2005.

[34] P. Misra and P. Enge, *Global Positioning System:Signals, Measurements and Performance*, 2nd ed. Lincoln, MA, USA: Ganga-Jamuna Press, 2006.

[35] C. W. Helstrom, *Statistical Theory of Signal Detection*, 2nd ed. Oxford, New York, USA: Pergamon Press, 1968.

[36] K. Sun, L. L. Presti, and M. Fantino, "GNSS signal acquisition in presence of sign transitions," *Proc. of the European Navigation Conference*, 3-6 May 2009.

[37] K. Sun and L. L. Presti, "A two steps GNSS acquisition algorithm," *22nd International Meeting of the Satellite Division of the Institute of Navigation*, 22-25 September 2009.

[38] ——, "Bit sign transition cancellation method for gnss signal acquisition," *The Journal of Navigation*, vol. 65, no. 01, pp. 73–97, January 2012.

[39] K. Sun, *GNSS Signal Acquisition and Tracking in Presence of Data*. Ph.D. dissertation, Politecnico di Torino, Italy, March 2010.

[40] E. D. Kaplan, *Understanding GPS: Principles and Applications*. Norwood, MA Artech House, 1996.

[41] L. L. Presti, X. Zhu, M. Fantino, and P. Mulassano, "GNSS signal acquisition in the presence of sign transitions," *IEEE Journal of Selected Topics in Signal Processing.*, vol. 3, no. 4, pp. 557–570, August 2009.

[42] D. J. R. V. Nee and A. J. R. M. Coenen, "New fast GPS code-acquisition technique using FFT," *Electronics Letters*, vol. 27, no. 2, 17 January 1991.

[43] J. Marcum, "A statistical theory of target detection by pulsed radar," *IRE Transactions on Information Theory*, vol. 6, no. 2, pp. 59–267, April 1960.

[44] M. H. Zarrabizadeh and E. S. Sousa, "A differentially coherent PN code acquisition receiver for CDMA systems," *IEEE Trans. Commun.*, vol. 45, no. 11, pp. 1456–1465, November 1997.

[45] C. Roberto and P. d Silvano, "Performance of CDMA with differential detection in the presence of phase noise and multiuser interference," *IEEE Transactions on Communications*, vol. 52, no. 3, pp. 498–506, March 2004.

[46] R. Tapani and J. Jyrki, "Code timing acquisition for DS-CDMA in fading channels by differential correlations," *IEEE Transactions on Communications*, vol. 49, no. 5, pp. 899–910, May 2001.

[47] L. Le and F. Adachi, "Joint frequency-domain differential detection and equalization for DS-CDMA signal transmissions in a frequency-selective fading channel," *IEEE Journal on Selected Areas in Communications*, vol. 24, no. 3, pp. 649–658, March 2006.

[48] K. Sun and L. L. Presti, "A differential post detection technique for two steps GNSS signal acquisition," *Proc. of IEEE / ION PLANS 2010*, May 4-6 2010.

[49] M. K. Simon, *Probability Distributions Involving Gaussian Random Variables: A Handbook for Engineers and Scientists*, 1st ed.   Springer, May 2002.

[50] D. Borio, C. O'Driscoll, and G. Lachapelle, "Composite GNSS signal acquisition over multiple code periods," *IEEE Trans. on Aerospace and Electronics Systems*, September 2008.

[51] *Galileo Open Service Signal In Space Interface Control Document*.   European Space Agency / Galileo Joint Undertaking, GAL OS SIS ICD/Draft 0, May 2006.

[52] C. Hegarty, M. Tran, and A. J. V. Dierendonck, "Acquisition algorithms for the GPS L5 signal," *Proc. of ION GPS/GNSS 2003*, 9-12 September 2003.

[53] F. Bastide, O. Julien, C. Macabiau, and B. Roturier, "Analysis of L5/E5 acquisition, tracking and data demodulation thresholds," *Proc. of ION GPS*, 24-27 September 2002.

[54] K. Sun and L. L. Presti, "Channels combining techniques for a novel two steps acquisition of new composite GNSS signals in presence of bit sign transitions," *Proc. of IEEE / ION PLANS 2010*, May 4-6 2010.

[55] K. Sun, "Composite GNSS signal acquisition in presence of data sign transition," *Proc. of 2010 International Conference on Indoor Positioning and Indoor Navigation (IPIN)*, September 15-17 2010.

[56] ——, "Differential channels combining strategies for composite GNSS signal acquisition," *Proc. of 2011 ION International Technical Meeting (ITM)*, January 24-26 2011.

[57] A. Requicha, "Direct computation of distribution functions from characteristic functions using the fast Fourier transform," *Proceedings of IEEE*, vol. 58, no. 7, pp. 1154–1155, July 1970.

[58] A. Razavi, D. Gebre-Egziabher, and D. Akos, "Carrier loop architectures for tracking weak GPS signals," *IEEE Trans. on Aerospace and Electronics Systems*, vol. 44, no. 2, pp. 697–710, April 2008.

[59] K. Borre, D. M. Akos, N. Bertelsen, P. Rinder, and S. H. Jensen, *A Software-Defined GPS and Galileo Receiver: A Single-Frequency Approach*, 1st ed.   Birkhäuser, 2007.

[60] F. van Diggelen, *A-GPS: Assisted GPS, GNSS, and SBAS*.   Artech House, March 2009.

[61] K. Sun, "GNSS code tracking in presence of data," *Proc. of 7th International Conference on Wireless Communications, Networking and Mobile Computing (WiCOM)*, September 23-25 2011.

[62] K. Warwick and D. Rees, *Industrial Digital Control Systems*, 2nd ed.   Institution Of Engineering And Technology, 1988.

[63] J. E. Normey-Rico, J. G. mez Ortega, and E. F. Camacho, "A smith-predictor-based generalised predictive controller for mobile robot path-tracking," *Control Engineering Practice*, vol. 7, no. 6, pp. 729–740, June 1999.

[64] S. Bennett, "A history of control engineering," *Institution Of Engineering And Technology*, pp. 142–148, June 1986.

[65] K. Ang, G. Chong, and Y. Li, "PID control system analysis, design, and technology," *IEEE Trans. on Control Systems Technology*, vol. 13, no. 4, pp. 559–576, July 2005.

[66] S. Bennett, "Nicholas minorsky and the automatic steering of ships," *IEEE Control Systems Magazine*, vol. 4, no. 4, pp. 10–15, November 1984.

[67] K. Ogata, *Modern Control Engineering*, 5th ed. Prentice Hall PTR Upper Saddle River, NJ, USA, 2009.

[68] C. Hang, K. Astrom, and W. Ho, "Refinements of the Ziegler-Nichols tuning formula," *IEE Proceedings-D, Control Theory and Applications*, vol. 138, no. 2, pp. 111–118, March 1991.

[69] W. Ho, C. Hang, and J. Zhou, "Performance and gain and phase margins of well-known PI tuning formulas," *IEEE Transactions on Control Systems Technology*, vol. 3, no. 2, pp. 245–248, June 1995.

[70] Q. Wang, C. Hang, and X. Yang, "Single-loop controller design via IMC principles," *Automatica*, vol. 37, no. 12, pp. 2041–2048, December 2001.

[71] S. Huang, K. Tan, and T. Lee, "A combined PID / adaptive controller for a class of nonlinear systems," *Automatica*, vol. 37, no. 4, pp. 611–618, April 2001.

[72] K. Sun, "Adaptive code tracking loop design for GNSS receivers," *Proc. of IEEE / ION PLANS 2012*, April 24-26 2012.

[73] R. C. Dorf, *Modern Control Systems*, 12th ed. Prentice Hall PTR Upper Saddle River, NJ, USA, 2010.

Chapter 3

Multipath Propagation, Characterization and Modeling in GNSS

Marios Smyrnaios, Steffen Schön and
Marcos Liso Nicolás

Additional information is available at the end of the chapter

1. Introduction

GNSS signals may arrive at the receiving antenna not only through the direct path, i.e. the line-of-sight (LOS) path, but also on multiple indirect paths, due to different electromagnetic effects as signal reflection or diffraction. These signal components arrive with a certain delay, phase, and amplitude difference relative to the LOS component. We will call these signal components multipath components (MPCs) and the phenomena multipath propagation.

Multipath propagation degrades the positioning accuracy. Moreover, in precise applications, multipath errors dominate the total error budget. Despite the different approaches developed, several aspects of multipath propagation are still not fully understood. The generally unknown number of MPCs and their path geometry, the signal characteristics, the diffraction and reflection effects as well as their changing nature together with a complex antenna and receiver design make multipath mitigation very challenging. Furthermore, the site-dependent characteristics of multipath decorrelate the errors caused by multipath propagation at different antenna locations and thus, differential techniques, like e.g. double differences (DD), cannot mitigate it.

The superposition of the MPCs and the LOS signals yields a compound signal at the receiving antenna. Depending on the relative phase between the MPCs and the LOS signal, constructive or destructive interference appears. As a result, during signal tracking the correlation output between the received signal and the local pseudorandom noise (PRN) code replica generated by the receiver is deformed. Since MPCs arrive generally at the receiving antenna with small extra paths, up to 20 m, relative to the LOS signal, the correlation output is biased and the receiver is not able to discriminate between MPC and the

LOS signals. This correlation output is the fundamental input for the next iteration of the code and phase tracking loops of the receiver as well as for C/N_0 estimation algorithms. As a result, the three GNSS observables code-phase, carrier-phase, and C/N_0 are biased by multipath propagation. In this text, errors in code-phase and carrier-phase observations caused by multipath propagation are referred to as code multipath and carrier-phase multipath, respectively, and in general as multipath errors.

In the observation domain, multipath errors are not constant in time. They show a sinusoidal behavior which can be noticed in carrier-phase residuals from Precise Point Positioning (PPP), double differences (DD) or C/N_0 time series. This behavior is due to the change of the relative phase between the direct and indirect signals as the satellite vehicle moves above the local horizon of the antenna. The magnitude of these oscillations depends on the relative amplitude of the MPC which varies as geometry changes. The C/N_0 observable is the only GNSS observation type in which multipath propagation effects are directly visible without any sophisticated data pre-processing. In contrast, in the phase or code domain, residuals should be analyzed or differences should be formed in order to eliminate all other errors sources. This is one of the main reasons why signal strength measures have attracted much attention in GNSS multipath studies. Since the relative signal amplitude between the LOS and MPC signals plays a key role for the understanding of multipath propagation and also for the magnitude of the multipath error in the GNSS observables, the following contribution focuses on an extended description and proposes an analytical model for modeling GNSS signal amplitudes.

This chapter is structured as follows. A compact overview on different approaches for multipath mitigation or characterization will be presented next. The approaches are categorized into techniques in the observation domain, receiver-internal as well as antenna-related techniques and further methods. Cornerstone methods of each category will be highlighted. In the third section, the multipath phenomenon and its impact on GNSS code, carrier phase and C/N_0 will be summarized. Special emphasize is given to the reflection process including signal polarization. An analytical model for GNSS signal amplitude is proposed. The equations for phase and code errors due to multipath propagation are extended so that the signal amplitude can be analytically calculated for each epoch. Finally, results from a dedicated experiment are shown in order to highlight the key features of multipath propagation.

2. Overview of multipath related studies in GNSS

In the beginning of the 1970s multipath effects on L1 frequency were first studied by [1] and the fundamental relationships between code error due to multipath and the driving parameters were derived. In [2], it was shown that the presence of multipath can be identified by using double differenced phase observations. Since then and during the last 4 decades many researchers have been involved in the characterization and modeling of this propagation phenomenon. A large number of scientific papers have been published on this topic, where different approaches and aspects of the problem have been investigated under certain predefined assumptions. Consequently, the scientific literature on multipath propagation for GNSS positioning can appear very rich for scientists. Despite the large number of different approaches developed, a universal solution of this problem is not achieved until the time of writing. Nevertheless, different promising approaches are under consideration and development.

In almost all textbooks on geodesy or navigation, with very few exceptions (e.g. [3]), multipath propagation is presented in short texts of a couple of pages (e.g. [4, 5, 6]). Most of the time, the phenomenon is explained geometrically, while other physical or electromagnetic properties of the reflected signals are not discussed. The progress in multipath-related studies is documented in various PhD thesis, we cite here exemplarily [7, 8, 9, 10, 11, 12, 13, 14, 15, 16].

A prominent example for the successful reduction of *code multipath* is the smoothing of the code observations by the about two orders of magnitude more precise carrier-phase observations [17]. Some manufacturers apply code smoothing as a default setting in the receiver. Longer smoothing periods give better performance in general [18, 9]. According to [9], the benefits of such an approach can yield a significant reduction of multipath impact given a sufficient large smoothing time constant, e.g. in aviation typically 100 sec are used. However, the variability of the ionospheric conditions may create additional range biases when smoothing. Van Nee in [19] showed that due to the non-zero mean of code-phase multipath, multipath effects cannot be eliminated by simply averaging over longer periods.

One of the most popular methods to characterize and quantify code multipath, are the so-called multipath linear combinations [20]. The original code and carrier-phase observations from dual frequency receivers are combined in such a way, that the code multipath can be isolated. Due to its computational simplicity this approach is often used to assess the overall multipath contamination at continuously operating reference stations (CORS), for example, the IGS network, or to characterize the performance of new receivers or new satellite signals, like e.g., the upcoming Galileo or GPS L5. However, it should be noted that the characterization is only valid if no code smoothing was applied.

Contrary to the code observations, the *multipath error on the carrier phase observations* is restricted in magnitude, since it is smaller than a quarter of the respective wavelength, i.e. about 5 cm maximum for the GPS L1 frequency. However, this is still large compared to the precision that carrier-phase observation could reach. Wanninger and May [21] proposed a method for carrier-phase multipath characterization of GPS reference stations. They analyze the double difference residuals in GPS networks. Consequently, it may sometimes be challenging to assign exactly the multipath signature to a specific site or satellite.

In post-processing, sidereal filtering or sidereal differencing is often applied. Taking advantage of the sidereal repetition of the GPS orbits, observation or coordinate time series of subsequent days can be subtracted in order to reduce the impact of multipath. Genrich and Bock in [22] showed that a reduction of about 80% can be obtained in this way. However, strictly speaking, each GPS satellite has its own, time-varying orbital period differing up to 10 sec with respect to the nominal sidereal period. Different approaches have been proposed to find the correct individual repeat times cf. [23, 24]. In addition, due to the nodal precession, the satellite ground track deviates over time, so that different reflectors will be illuminated, cf. for exemplary ground track variations [21, 16, 24]. Finally, changing weather conditions like rain or snow will influence the reflection properties of the antenna vicinity, so that the similarity of multipath errors is reduced. All three effects restrict the power of multipath mitigation by sidereal differencing, especially over longer time intervals.

Finally, the analysis of un-differenced carrier-phase residuals from PPP is a useful tool for accessing the impact of multipath effects [25]. For this purpose, the residuals are color-coded and depicted in a sky-plot. The variations of the residuals translate into a sequence of concentric rings in the sky-plot. But, since PPP residuals contain further remaining systematic effects, like e.g., varying tropospheric refraction, averaging strategies may be necessary [26].

Observations of the *signal strength*, like SNR or C/N_0, have attracted much attention in multipath related studies although most of the approaches are found in post-processing applications [27]. Compared to code or carrier-phase observations, the C/N_0 values are usable without sophisticated pre-processing steps and attributed to one satellite-receiver propagation channel, i.e. no double differences are formed. First results of this type of investigation were presented in [28] while newer ones can be found in [29, 27, 10]. The basic idea is that the C/N_0 values follow a nominal curve with respect to the satellite elevation that is mainly determined by the antenna gain pattern. Thus, deviations from this pattern can be easily identified and attributed to reflected and/or diffracted signals. Although, the qualitative analysis is straight-forward, a quantitative analysis of multipath by C/N_0 is still limited. The major restrictions are (i) proprietary algorithms and definitions of the C/N_0 values given by the commercial receivers. Different manufactures use different algorithms for the calculation of this type of observations. (ii) Different receivers (especially older ones) use different quantization levels (e.g. 1 dB-Hz instead of 0.1 dB-Hz) which can be very coarse for certain type of applications. (iii), the gain pattern of the receiving antenna is often unknown. Then empirical methods are proposed to determine the form of the C/N_0 curves [29, 27, 10]. Recently and very encouraging, some antenna manufacturers have published receiver antenna gain patterns for right-hand and left- hand circular polarization [30]. It would be very useful if in future more manufacturers could follow this example. Finally, C/N_0-based observation weighting has been found to be very efficient to reduce the impact of reflection and/or diffraction effects on the observation level [31, 32, 33].

Receiver-internal multipath mitigation/detection techniques incorporate different signal processing strategies for the reduction of this type of errors. The cornerstone approach for this category of approaches is the narrow-correlator technology [34]. It was demonstrated that by reducing the spacing of the early and late correlators from 1 chip to 0.1 chips, a significant reduction of multipath error could be achieved. In this way, MPCs with large extra path delay could be filtered out. Since then, several other approaches were developed, most of them for the mitigation of code multipath and much fewer for phase multipath. The majority of the internal approaches incorporate the use of several early and late correlators with different spacing between them. The correlator outputs are then used for the formation of different multipath mitigation discriminators or for the detection of the deformed slopes of the correlation peak.

One of the most characteristic approaches is the High Resolution Correlator [35], where the code discriminator is formed by two pairs of early-late correlators. Strobe Correlator [36, 37] and Vision Correlator [38] are other receiver internal techniques. A major breakthrough in the receiver internal approaches happened in 1995, when Novatel introduced the Multipath Estimating Delay-Lock Loop (MEDLL) [39, 40]. According to [41], the MEDLL is a maximum likelihood estimation technique pioneered by Van Nee [11] at Delft University of Technology and it improves the C/A-code narrow correlator performance by configuring the residual pseudorange error to a smaller region of secondary path relative delay. Since then, different approaches have been developed incorporating multiple correlators and estimation theory. Despite the evolvement of the receiver internal mitigation approaches, MPCs of relative short extra paths (less that 30 m) still cause errors in all types of GNSS observables. An overview work on receiver internal approaches with very interesting references can be found in [9, 42]. In addition, aspects on multipath propagation and the impact on the receiver's signal processing modules can be found in text books about software-defined GNSS receivers, like [43, 44, 45].

A lot of effort was given on the *antenna design*. In [46], basic aspects of a GNSS antenna are presented. A first approach consists in using antenna elements with a large ground-plane which increases the directivity of the antenna for the upper hemisphere and reduces reflections from below the antenna horizon. However, diffraction at the edges occurs and cause severe problems e.g. [47]. Subsequently, choke ring antennas were developed [48] and are widely used now, especially for reference station applications. Even though they attenuate MPCs coming from negative elevation angles with respect to the antenna horizon, their design cannot mitigate MPCs coming from positive elevation angles. Modern designs use variable choke-ring depths [49]. Assuming a change of polarization of the RHCP GNSS signals upon reflection, a basic principle applied in all GNSS antennas is to increase the sensitivity for RHCP and to simultaneously decrease the sensitivity of LHCP signals. Finally, it should be stated that different attempts to reduce the multipath reflection by applying micro-wave absorbing material are reported in literature, e.g. [50].

Closely-spaced antennas [12, 51] are also developed for the reduction of multipath errors. They form a type of small antenna arrays. Based on a least-squares adjustment, the multipath relative amplitude α, the multipath relative phase ($\Delta\Phi$) as well as azimuth and elevation of the assumed reflector can be estimated [9]. Further developments leads to beam steering and adaptive beam forming antennas or antenna-receiver combination, like e.g., DLR's GALANT receiver [52]. However, this interesting technology seems to be not mature enough to be installed at GNSS reference stations.

In [53] the concept of *station calibration* for multipath mitigation by a parabolic antenna was presented, however the concept is not in operational use, today. Further approaches of station calibration try to randomize the multipath effect by shifting the antenna in a controlled manner using a robot [15, 54, 55]. This method is very successful for in-situ calibration; however the efforts due to the operation of the robot are very large. The validity of the corrections is again restricted by the individual repeat times of the satellite orbits. In [16, 24] a separation of the so-called near- and far-field multipath is proposed, by calibrating the antenna with and without its mount, like e.g., tripods or special metal adaptations. The difference in the determined phase center variations is attributed to the near-field multipath.

New GNSS signals with different signal structures (e.g. AltBOC) have a better performance against multipath; cf. e.g. [56, 57, 58]. Nevertheless, short delay MPCs will still cause problems in all types of GNSS signal.

In order to get a better understanding of the physical processes involved in the multipath phenomenon, models from *wave propagation* are very useful. First results were given in [59]. More elaborated models are based on ray-tracing tools that use as a fundamental input the physical environment in which the antenna is placed, the receiving antenna position and the transmitting antenna position. Based on these input parameters, all possible signal paths are estimated and the geometric and electromagnetic properties of them are calculated. Ray-tracing tools are widely used for the simulation of wireless networks. In GNSS-related studies they were presented in [8] and [60]. Another ray-tracing approach [14] was used for the characterization and modeling of P-code multipath in different environments. Based on digital terrain models, more complex scenarios like urban canyons are analyzed, cf. [61]. Besides deterministic channel models, stochastic modeling is applied in complex scenarios, like e.g., DLR's land mobile and aeronautic channel model [62] and subsequent work.

In recent years, the concept of GNSS reflectometry and scatterometry has strongly evolved, cf. e.g., [63, 64, 65, 66, 67, 68, 69]. Here multipath propagation is not considered as bias but as basic information. In the context of GNSS-R, much progress has been made, especially in the mathematical and physical modeling of the wave propagation. Hence, these publications can be a valuable source for mitigating multipath in positioning.

3. Multipath characterization

3.1. Basic considerations

After reaching the receiving antenna, the GNSS signals are down-converted from radio-frequency (RF) to an intermediate frequency (IF) by the RF front-end of the receiver. Afterwards, the signal is digitized, down-converted to baseband and correlated with locally-generated replicas of the PRN codes. Then, the result of the correlation is accumulated for a certain time interval. In a typical GNSS receiver architecture, three replicas for each PRN code are generated, the so-called prompt, early and late replicas. The resulting correlation outputs of the prompt replica of the PRN code in the presence of a number of k MPCs are written as in [34, 17]:

$$I_P = AD(\tau)R(\tau)\cos\left(\theta_{dir}\right) + A\sum_{k=1}^{n} \alpha_k D(\tau)R(\tau - \Delta\tau_k)\cos\left(\theta_{dir} + \Delta\Phi_k\right), \tag{1}$$

$$Q_P = AD(\tau)R(\tau)\sin\left(\theta_{dir}\right) + A\sum_{k=1}^{n} \alpha_k D(\tau)R(\tau - \Delta\tau_k)\sin\left(\theta_{dir} + \Delta\Phi_k\right), \tag{2}$$

where A is the amplitude of the direct signal, D is the navigation bit, R is the correlation function, τ is the code tracking error and θ_{dir} is the phase of the LOS component. The multipath components are characterized by the relative amplitude of the k-th MPC α_k the relative delay $\Delta\tau_k$ and the relative phase $\Delta\Phi_k$ with respect to the LOS signal. Equations for the early and late replicas of the signal can be written in a similar way, although in these cases the chip spacing should also be considered.

Based on the correlation outputs (I_P and Q_P) of the prompt correlator, a GNSS receiver is able to calculate the amplitude (A_c) and phase (θ_c) of the compound signal (see Figure 1). However, the characteristics of the direct and the MPC signals cannot be estimated since the geometry is not known and the receiver cannot discriminate between them, especially in the case of short delay MPCs. In Figure 1, a vectorial representation for the case of the direct component and one multipath component is presented. The direct signal component is characterized by a certain phase (θ_d) and amplitude (A_d), while the MPC is characterized by a relative phase ($\Delta\Phi$) and a relative amplitude (α) with respect to the direct component. It should be mentioned that due to the motion of the satellite and maybe of the receiving antenna, none of the above parameters is constant in time. Directly from this vector diagram, an expression for the phase error and the amplitude of the compound signal can be derived as in [2, 3]:

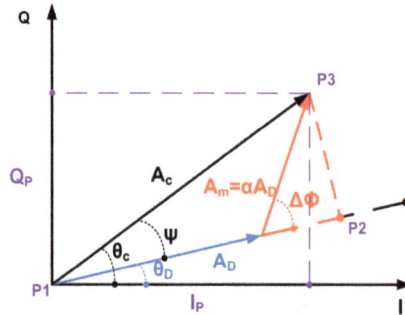

Figure 1. Phase error of PLL due to one multipath component

Assuming a perfect tracking (i.e. R=1) and the relationship $A_m = \alpha A_d$, the phase error (ψ) can be expressed as a function of α and $\Delta\Phi$:

$$\tan\psi = \frac{A_m\sin\Delta\Phi}{A_d + A_m\cos\Delta\Phi} => \psi = \arctan\left(\frac{\alpha\sin\Delta\Phi}{1 + \alpha\cos\Delta\Phi}\right) \tag{3}$$

The compound signal amplitude can be derived as a function of α, $\Delta\Phi$ and A_d from the orthogonal triangle (P1 P2 P3):

$$A_c^2 = \left(A_d + A_m\cos\Delta\Phi\right)^2 + A_m^2\sin^2\Delta\Phi$$

By using trigonometric identities, the expression for the compound signal amplitude is rearranged as:

$$A_c^2 = A_d^2 + 2\alpha A_d^2 \cos\Delta\Phi + \alpha^2 A_d^2. \tag{4}$$

An equation for pseudo-range error (ϱ) is given in [13]:

$$\varrho = \frac{\alpha\delta\cos\Delta\Phi}{1 + \alpha\cos\Delta\Phi}, \tag{5}$$

with δ being the extra path length of the MPC expressed in meters.

From equations (3), (4) and (5) the 90° phase shift between errors in code, amplitude domains and phase domain can be noticed. When $\Delta\Phi = 0°/180°$ and $\psi = 0$, then ϱ, A_c are maximum/minimum respectively, while when $\Delta\Phi = 90°$, then ψ is maximum and $\varrho = 0$. Furthermore, for a number of k multipath components, equations (3), (4) and (5) can be written as:

$$\Psi = \arctan\left(\frac{\sum\limits_{k=1}^{n} \alpha_k \sin\Delta\Phi_k}{1 + \sum\limits_{k=1}^{n} \alpha_k \cos\Delta\Phi_k}\right), \tag{6}$$

$$A_c = A_d\sqrt{\left(1 + \sum\limits_{k=1}^{n} \alpha_k \cos\Delta\Phi_k\right)^2 + \left(\sum\limits_{k=1}^{n} \alpha_k \sin\Delta\Phi_k\right)^2}, \tag{7}$$

$$\varrho = \frac{\sum\limits_{k=1}^{n} \alpha_k \delta_k \cos\Delta\Phi_k}{1 + \sum\limits_{k=1}^{n} \alpha_k \cos\Delta\Phi_k}. \tag{8}$$

The relative phase of each multipath component can be expressed as a function of the extra path length (δ) and the wavelength of the carrier signal: $\Delta\Phi = 2\pi\delta/\lambda$, where δ is expressed in meters. Moreover, δ is a function of station height and reflection angle, which in turns depends on the satellite elevation and the orientation of the reflector in space. In the special case of ground multipath, which we will investigate in the following parts of this chapter, the extra path delay is a function of the station height (h) and the satellite elevation (el):

$$\delta = 2h\sin(el). \tag{9}$$

In conclusion, it can be stated that some of the introduced equations can be calculated based on the geometry of the scenario. The signal amplitudes for both LOS and MPC are not directly accessible. However, this information or at least the relative amplitude is a crucial part of all introduced formulas. Therefore, in the following an analytical model of the relative amplitude α is developed.

3.2. Polarization state of the GNSS Signals

In order to understand the amplitude relation between the LOS and the MPCs, in a first step, the polarization state of a signal is introduced and linked to the right-hand circular polarization (RHCP) of the GNSS signal. According to [70], there are many ways to represent the polarization state of a signal or of an antenna. Some are graphic in nature (e.g. Poincaré sphere) and can be easily visualized, while others can be more appropriate for specific applications, like e.g., investigation of antenna wave interaction (e.g. Stokes parameters and/or complex vector representation).

In the case of a circularly polarized (CP) planar wave, the electric field vector is propagating in a helical way (see Figure 2a), where the projection of the tip of this vector forms a circle in a fixed plane normal to the direction of propagation (z). The clockwise or counter clockwise sense of rotation of the electric field vector looking towards the direction of propagation defines whether the signal is left or right-hand circularly polarized, respectively. At each instant in time, the electric field vector is a combination of two components (see Figure 2 b, c):

$$\vec{E}(t) = E_1 \cos(\omega t)\vec{x} + E_2 \cos(\omega t + \zeta)\vec{y}, \tag{10}$$

where E_1 and E_2 are the amplitudes of the instantaneous electric field [V/m] in x and y directions, respectively, ω is the angular frequency in [radians/sec] and ζ is the relative phase shift [radians] by which the y component leads the x component.

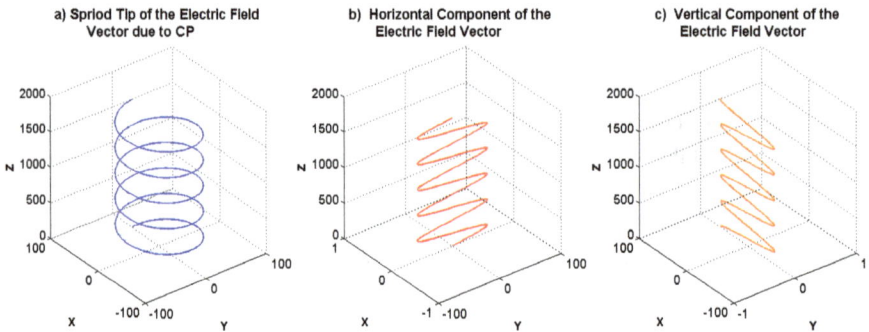

Figure 2. The combination of the two electric field components, in horizontal and vertical direction, results in the CP electric field vector, propagating in z direction in a spiroidal way

Each of the two components represents a linearly polarized wave. When $E_2 = 0$, the wave has a linearly polarization along the x-axis and when $E_1 = 0$, the wave is linearly polarized along the y-axis. Furthermore, if $E_1 = E_2$ and $\zeta = 0°$, then the wave is linearly polarized with a 45° tilt. In general, it can be stated that the polarization of a signal can be completely described by the relative amplitudes and phase of these two components. For circular polarization, there are two requirements on the linear components [71]: i) time quadrature, i.e. $\zeta = \pi/2$ for LH or $\zeta = -\pi/2$ for RH polarization, and ii) equal amplitude, i.e. ($E_1 = E_2$).

Another decomposition of the electric field vector yields orthogonal circular polarization states (LHCP and RHCP). Furthermore, the electric field phasor expressed in terms of CP phasors can be written as [70]:

$$\vec{E}(t) = \vec{E}_L(t) + \vec{E}_R(t) = \frac{\left[(E_{L0} + E_{R0}\, e^{j\zeta})\vec{x} + j(E_{L0} - E_{R0}\, e^{j\zeta})\vec{y} \right]}{\sqrt{2}}, \tag{11}$$

where $E_{L\,0}$ and E_{R0} are the amplitude of the two orthogonal components.

Finally, the Jones vector is a further way for the representation of signal polarization. According to [72], the Jones vector can be used only for completely polarized signals, e.g. GNSS signals. Equation (10) can be rewritten in complex vector notation as:

$$\vec{E}(t) = \left(|E_1| e^{j\zeta_x}\vec{x} + |E_2| e^{j\zeta_y}\vec{y} \right) e^{j\omega t}, \tag{12}$$

where the relative phase shift between the two electric field components is $\zeta = \zeta_x - \zeta_y$ (see Figure 3). According to [72], this can be further factorized as:

$$\vec{e} = A\vec{x} + Be^{j\zeta}\vec{y}, \tag{13}$$

which is the so-called Jones vector ($\vec{e}$) and the coefficients A and B are:

$$A = \frac{|E_1|}{\sqrt{|E_1|^2 + |E_2|^2}}, \quad B = \frac{|E_2|}{\sqrt{|E_1|^2 + |E_2|^2}},$$

where $A^2 + B^2 = 1$ and the E_{eff} represents the strength of an effective linearly polarized wave that would give the same intensity that is described by eq. (10) [72].

$$E_{eff} = \sqrt{|E_1|^2 + |E_2|^2}\, e^{j\zeta_x}.$$

Eq. (13) can be written in a column vector form as:

$$\vec{e} = \begin{bmatrix} A \\ Be^{j\zeta} \end{bmatrix}. \tag{14}$$

Eq. (14) completely describes the polarization state of GNSS signals since information of the amplitude of both electric field components as well as their relative phase can be extracted. Particular cases for the Jones vector can be seen in Table 1:

Polarization State	
Linearly polarized in x direction	$\begin{bmatrix} 1 \\ 0 \end{bmatrix}$
Linearly polarized in y direction	$\begin{bmatrix} 0 \\ 1 \end{bmatrix}$
Right-hand circularly polarized (RHCP)	$\frac{1}{\sqrt{2}}\begin{bmatrix} 1 \\ -j \end{bmatrix}$, with $\zeta = -90°$
Left-hand circularly polarized (LHCP)	$\frac{1}{\sqrt{2}}\begin{bmatrix} 1 \\ j \end{bmatrix}$, with $\zeta = 90°$

Table 1. Jones vector representation for the polarization state of characteristic cases.

3.3. Reflection process

In multipath propagation, one or multiple reflections as well as diffraction of the transmitted signal may occur. In the following, we focus on the description of the reflection process. In the case of reflection of the incident field, the reflection coefficients will indicate how much the reflected field will be attenuated and how the polarization state of the incident field will be deformed. The reflection coefficients used in this investigation are as in [7]:

$$R_{\perp}(\theta_{ref}) = \frac{\cos\theta_{ref} - \sqrt{Y - \sin^2\theta_{ref}}}{\cos\theta_{ref} + \sqrt{Y - \sin^2\theta_{ref}}}, \tag{15}$$

$$R_{\parallel}(\theta_{ref}) = \frac{Y\cos\theta_{ref} - \sqrt{Y - \sin^2\theta_{ref}}}{Y\cos\theta_{ref} + \sqrt{Y - \sin^2\theta_{ref}}}, \tag{16}$$

With

$$Y = \varepsilon - i*60*\lambda*\sigma, \tag{17}$$

where θ_{ref} is the reflection angle and λ is the carrier wavelength. The relative permittivity (ε) and conductivity (σ) depend on the material properties of the reflector. Typical values for the material properties for GNSS frequencies can be found in [7]. The exemplary material properties used in this text can be seen in Table 2.

Reflector Material	Relative Permittivity (ε)	Conductivity (σ) [S/m]
Concrete	3	$2*10^{-5}$
Sea Water	20	4
Wet Ground	30	$2*10^{-1}$

Table 2. Material properties of the exemplary reflectors used in this investigation

The reflection coefficients can also be expressed as circular components of the two orthogonal polarizations (R_{co}, R_{cross}):

$$R_{co} = \frac{R_{\perp}(\theta_{ref}) + R_{\parallel}(\theta_{ref})}{2},$$ (18)

$$R_{cross} = \frac{R_{\perp}(\theta_{ref}) - R_{\parallel}(\theta_{ref})}{2}.$$ (19)

For the RHCP satellite signals, R_{co} can be associated with the RHCP component, while R_{cross} represents the orthogonal LHCP component.

In Figure 3a), the circular reflection coefficients are plotted for the L1 carrier. The co-polarized components are plotted with solid lines and the cross-polarized with dashed. The material properties of the exemplary materials (see Table 2) were used for the calculation of the reflection coefficients. The reflection coefficients for a concrete reflector are plotted in black, for a sea water reflector in blue, and for a wet ground reflector in red. They indicate how the electrical field components of the incident field will be changed upon reflection, in terms of magnitude and relative phase.

Looking at reflection coefficients for the different reflection angles in Figure 3a) and for a concrete reflector, it can be seen that when the reflection angle approaches 0°, then the cross-polarized components (LHCP component) tends to 0 and the co-polarized components (RHCP component) is 1, i.e. no loss. When the reflection angle is between 0° and 30°, both components exist but still the co-polarized one is larger in magnitude and the reflected signal is right-hand elliptically polarized (RHEP), with the eccentricity of the polarization ellipse getting bigger. At around 30°, the magnitude of the reflection coefficients is equal. In geodetic literature, this angle is referred to as the Brewster angle [7, 9] for a concrete reflector, although this definition deviates from the common use of the Brewster angle in electromagnetic theory. For these material properties (concrete) and for reflection angle of 30°, the reflected signal is linear polarized. When the reflection angle is between 30° and 90°, the cross polarized component has a higher magnitude and this results in a change of the initial polarization from RH to LH. The eccentricity of the ellipse is then getting smaller as the reflection angle increases. When the reflection angle approaches 90°, the reflected signal is LHCP since the co-polarized components is zero. When the material properties change, then the changes on the reflected field are different. For example, the angle for which the magnitude of the circular reflection coefficients is the same, may vary significantly for different reflectors. Furthermore, small magnitude changes between the two different frequencies (L1 and L2) may occur. These differences may also vary between different reflectors.

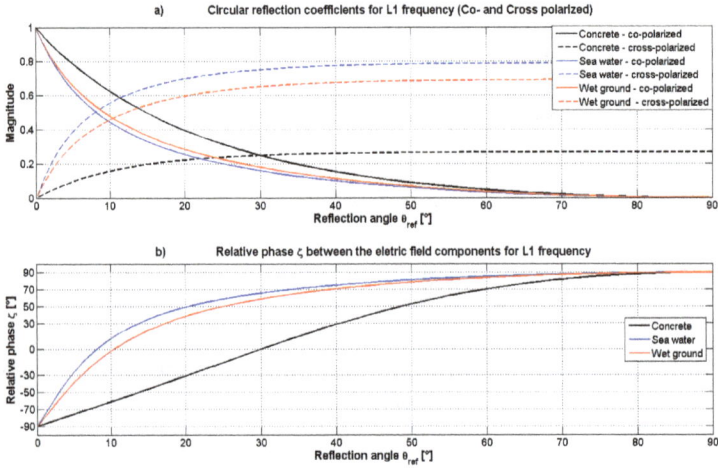

Figure 3. Co- and cross-polarized circular reflection coefficients for L1 (a); and relative phase change between the electric field components due to reflection process for L1 (b). For the exemplary material properties of concrete, sea water and wet ground.

In Figure 3b), the relative phase (ζ) between the field components is illustrated for L1 for the three exemplary material properties. The relative phase between the two components is plotted for all possible reflection angles of the incident field. Without reflection (reflection angle = 0°), the relative phase (ζ) is -90°, which is the relative phase shift for RHCP signals. As the reflection angle increases, ζ is getting smaller and smaller. When the reflection angle is equal to the Brewster angle (for this particular reflector), then ζ=0°, which yields linear polarization. Finally, for reflection angles larger that the Brewster angle, ζ has an opposite sign which indicates that the reflected signal has changed polarization.

3.4. Signal amplitude modeling

According to [73, 74], the signal after the receiving antenna can be written as:

$$S_R = \vec{e}_{rec}^{H} H \vec{e}_{tran} S_T \; , \tag{20}$$

where S_R is the received signal, $\vec{e}_{rec}^{H}$ is the Hermitian conjugate of the Jones vector of the receiving antenna, and $\vec{e}_{tran}$, is the Jones vector of the transmitting antenna. They are derived from the antenna gain patterns (for both RH and LH circular polarizations) and thus depend on of the angle of arrival of the signals to the antenna:

$$\vec{e}_{rec_{(el)}} = \frac{1}{\sqrt{2}}\begin{bmatrix} E_{\theta(el)} \\ -jE_{\varphi(el)} \end{bmatrix} = \frac{1}{\sqrt{2}}\begin{bmatrix} \left(E_{rec_{RHCP(el)}} + E_{rec_{LHCP(el)}}\right) \\ -j\left(E_{rec_{RHCP(el)}} - E_{rec_{LHCP(el)}}\right) \end{bmatrix},$$ (21)

More information on the RHCP gain pattern of the satellite antennas can be found in [75]. The signal ellipticity should not be worst that 1.2 dB for L1 and 3.2 dB for L2, [76].

$$\vec{e}_{tran_{(el)}} = \frac{1}{\sqrt{2}}\begin{bmatrix} E_{\theta(el)} \\ -jE_{\varphi(el)} \end{bmatrix} = \frac{1}{\sqrt{2}}\begin{bmatrix} \left(E_{tran_{RHCP(el)}} + E_{tran_{LHCP(el)}}\right) \\ -j\left(E_{tran_{RHCP(el)}} - E_{tran_{LHCP(el)}}\right) \end{bmatrix}.$$ (22)

The matrix H is defined for the MPC components as:

$$\begin{aligned} H_{MPC} &= A_{MPC}e^{-j\beta}H_{ref}, & \text{a} \\ H_{LOS} &= A_{LOS}e^{-j\alpha}, & \text{b} \end{aligned}$$ (23)

and for the LOS components as (23b)

where A_{LOS} and A_{MPC} account for the free-space loss attenuation, which for this investigation may be considered equal. $e^{-j\beta}$ and $e^{-j\alpha}$ introduce the phase changes for the LOS and MPC components respectively caused by the free-space propagation. Finally H_{ref} is the channel polarization matrix for one single reflection:

$$H_{ref} = \begin{bmatrix} \cos\Psi_{rec} & \sin\Psi_{rec} \\ -\sin\Psi_{rec} & \cos\Psi_{rec} \end{bmatrix} * \begin{bmatrix} R_{\perp}(\theta_{ref}) & 0 \\ 0 & R_{\parallel}(\theta_{ref}) \end{bmatrix} * \begin{bmatrix} \cos\Psi_{tran} & \sin\Psi_{tran} \\ -\sin\Psi_{tran} & \cos\Psi_{tran} \end{bmatrix},$$ (24)

where Ψ_{rec} and Ψ_{tran} are the rotation angles between the normal of the incident plane and E_{θ} component of the electric field vector (see Figure 4), $R_{\perp}$ and $R_{\parallel}$ are given by eq. (18) and (19).

This expression is divided into three subsequent matrix multiplications. First, the selected base of the **E** field of the transmitting signal has to be aligned with the coordinate system of the so-called incident plane. This plane is defined by the position of the transmitter, the receiver and the reflection point. Next, the signal attenuation and the polarization change are computed taking the electromagnetic properties of the reflecting material into account (e.g. eq. (15) and (16)). Finally, the resulting **E** vector is rotated into the coordinate system associated with the receiving antenna. In this way, both the variation of the orientation of the antennas and the impact of the reflection on the polarization state are taken into account. The geometric situation is depicted in Figure 4.

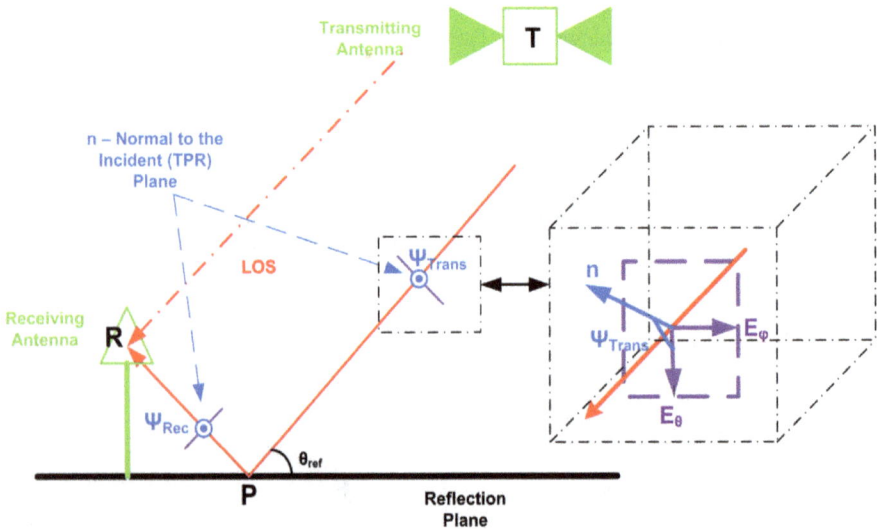

Figure 4. Reflection geometry of ground MPC. The electric field components are expressed in spherical coordinates.

The signal amplitudes for the ground-reflected MPC and for the LOS component are calculated as the absolute values of eq. (25) and (26), where the transmitted signal S_T in this investigation is considered equal to 1. Practically speaking, S_{LOS} reflects the antenna gain pattern value for each specific angle of arrival of the LOS component.

$$S_{MPC} = \vec{e}^H_{rec(el+90°)} H_{MPC} \vec{e}_{tran_{(el)}} S_T , \quad (25) \tag{25}$$

$$S_{LOS} = \vec{e}^H_{rec(el)} H_{LOS} \vec{e}_{tran_{(el)}} S_T , \quad (26) \tag{26}$$

yielding the final formulas for the phase error, for the code error and for the compound signal amplitude (for the case of one ground MPC):

$$\psi = \arctan\left(\frac{\left(\frac{|S_{MPC}|}{|S_{LOS}|}\right)\sin(\Delta\Phi)}{1 + \left(\frac{|S_{MPC}|}{|S_{LOS}|}\right)\cos(\Delta\Phi)}\right), \tag{27}$$

$$\varrho = \frac{\left(\frac{|S_{MPC}|}{|S_{LOS}|}\right)\delta\cos(\Delta\Phi)}{1 + \left(\frac{|S_{MPC}|}{|S_{LOS}|}\right)\cos(\Delta\Phi)}, \tag{28}$$

$$A_c = |S_{LOS}| \cdot \sqrt{\left(1 + 2\left(\frac{|S_{MPC}|}{|S_{LOS}|}\right)\cos(\Delta\Phi) + \left(\frac{|S_{MPC}|}{|S_{LOS}|}\right)^2\right)}. \tag{29}$$

When multiple multipath components are received, the above equations can be rewritten in a similar way as:

$$\Psi = \arctan\left(\frac{\sum\limits_{k=1}^{n}\left(\frac{|S_{MPC_k}|}{|S_{LOS}|}\right)\sin(\Delta\Phi_k)}{1 + \sum\limits_{k=1}^{n}\left(\frac{|S_{MPC_k}|}{|S_{LOS}|}\right)\cos(\Delta\Phi_k)}\right), \tag{30}$$

$$Q = \frac{\sum\limits_{k=1}^{n}\left(\frac{|S_{MPC_k}|}{|S_{LOS}|}\right)\zeta_k\cos(\Delta\Phi_k)}{1 + \sum\limits_{k=1}^{n}\left(\frac{|S_{MPC_k}|}{|S_{LOS}|}\right)\cos(\Delta\Phi_k)}, \tag{31}$$

$$A_c = |S_{LOS}| \cdot \sqrt{\left(1 + \sum\limits_{k=1}^{n}\left(\frac{|S_{MPC_k}|}{|S_{LOS}|}\right)\cos(\Delta\Phi_k)\right)^2 + \left(\sum\limits_{k=1}^{n}\left(\frac{|S_{MPC_k}|}{|S_{LOS}|}\right)\sin(\Delta\Phi_k)\right)^2}. \tag{32}$$

In a similar sense, eq. (1) and (2) can be also rewritten in order to model analytically the relative amplitude factor α which is present also in these equations.

4. Example

4.1. Measurement set up

In order to illustrate the above derived formulas, various experiments were carried out. In the followings we describe one experiment in details. A dedicated experiment was conducted in a controlled environment at the PTB antenna test facility in Braunschweig, i.e. an environment where only one ground reflection would occur (Figure 5). Two antennas were set up with different heights above the reflector. The antennas were spaced about 21.3 m. The observation period lasted for about 7 hours and a cut-off angle of 0° was applied. The data rate was 1 Hz. One pair of AX1202GG LEICA antennas were used together with LEICA GRX1200+GNSS receivers.

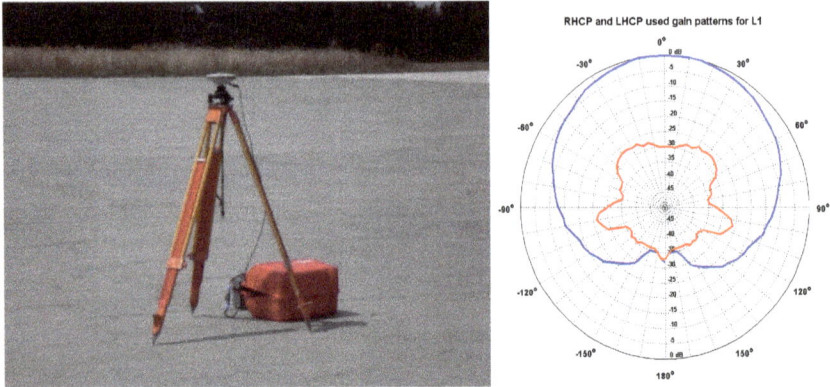

Figure 5. Left: Experimental set up of the Leica AX1202GG antenna. Right: The NOV702GG gain patterns [dB] for both orthogonal polarizations (RHCP in blue and LHCP in red) were used to model the antenna. Adopted from [77]

4.2. Simulations

The following assumptions were used for the simulations: The receiving antenna gain patterns are taken from a NOV702GG antenna assumed to be similar to the AX1202 (see Figure 5). The normalized gain patterns were assumed symmetrical in azimuth. In this way the Jones vector of the receiving antenna for all possible azimuths and elevation angles of the incoming LOS and MPCs GNSS signals can be computed as in eq. (21). The satellite antenna was assumed here perfect RHCP and modeled as:

$$\vec{e}_{trans} = \frac{1}{\sqrt{2}} \begin{bmatrix} 1 \\ -j \end{bmatrix} \tag{33}$$

Finally, for simplicity the angles ψ_{rec} and ψ_{trans} were assumed to be equal and constant over time. Absolute and individual antennas phase center corrections from IfE Hannover were also taken into account during data processing. Furthermore, it is assumed that the ground reflector is a perfect horizontal reflector. Using eq. (27) and (29), the simulated time series for the phase error and the compound signal amplitude (dB) were computed. Under the assumption the values are independent of the satellite azimuth; one exemplary satellite (PRN 12) is plotted covering the whole elevation range. Two different antenna heights above the reflector were used according to the measuring set up. The heights of the two antennas A1, A2 were 1.358 m and 2.053 m, respectively.

In Figure 6, simulated normalized amplitudes [dB] of the MPC, the direct and compound signals are plotted for PRN12 versus satellite elevation for both antennas of the observed baseline. The direct signal's amplitude is plotted in red, the amplitude of the compound signal in blue, and in black the difference between the direct signal and the compound one is given, thus the impact of the multipath.

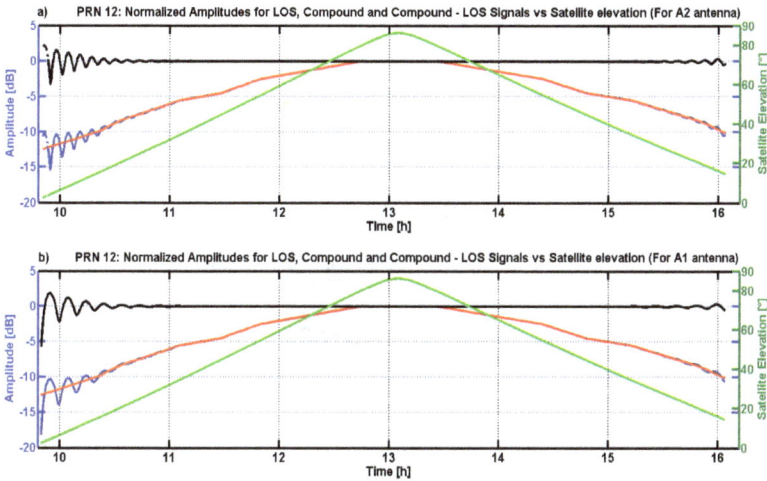

Figure 6. Simulated and normalized amplitude values versus satellite elevation for PRN 12 and reflector made of concrete. a) LOS, ground MPC and compound signal amplitudes for antenna position A2. b) LOS, ground MPC and compound signal amplitudes for the antenna position A1.

The normalized amplitude of the LOS is a smooth curve mainly determined by the gain pattern of the receiving antenna. The resulting multipath effect (black) shows the typical amplitude variations of few dB. Finally, the compound amplitude shows a superposition of both features. Furthermore, a frequency difference of the oscillations can be noticed between the antennas, depending on the height difference. For the maximum elevation angle of 90° PRN 12 has a minimum attenuation of 0 dB. For the same satellite elevation, the amplitude of the ground MPC is almost -32 dB. The reason is that the reflected MPC, at a reflection angle of about 90° and with angle of arrival near -90° with respect to the antenna horizon, is almost LHCP (see Figure 3a) and it is strongly attenuated due to the LHCP gain pattern of the antenna for such angles of arrival (see Figure 5, right side). Thus, the LOS signal is dominating. At elevation angles near 0°, the MPC is still having most of the signal energy in the RHCP component (see Figure 3) and only a small part of it in the LHCP component. The relative amplitude α of the MPC w.r.t. LOS is about 0.6. So in this case, the MPC reflection loss is minimum, the MPC RHCP component is dominating and the antenna applies a very similar gain for both LOS and ground MPC signal components.

In a next step, the impact of different reflector material properties will be investigated. We chose exemplary material properties of perfect, concrete, sea water and wet ground reflectors (see Table 2). Since the impact will be much stronger at low elevation angles, only a part of the arc of PRN 12 is plotted. Figure 7 shows the resulting curves for the compound signal amplitude for two different antenna heights.

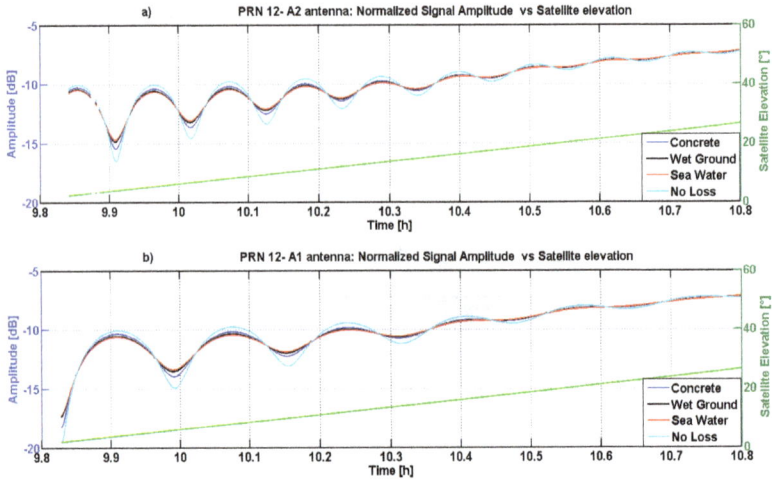

Figure 7. Normalized amplitude in [dB] of the compound signal for different material properties. Zoom at low elevation angles. (a): antenna A2, 2.053 m above the reflector, (b): antenna A1, 1.358 m above the reflector.

The change in the material properties creates mainly a change in the amplitude of the variations. A perfect reflector creates the strongest oscillation. The concrete reflector is also creating MPC with higher amplitude than water reflector and the wet ground reflector. This is because for small reflection angles the material properties of concrete are causing a smoother change of the signal polarizations (from RH to LH) in contrast to the reflections coefficients of the other two reflectors that are creating a more abrupt change (see Figures 3).

4.3. Comparison of simulations and real data

The C/N$_0$ values obtained from the RINEX observation files can be compared with the simulated values. In Figure 8 for two exemplary satellites, PRN 12 and PRN 14, the corresponding time series are compared. Here, only the data for the antenna A1 at a height of 1.358 m above the reflector is considered. The reflection properties are modeled for a concrete reflector (see Table 2). Both curves show that the simulations can explain the main features present in the observations. The scale on both axes is the same. However, some disturbances occur which are due to our simplifications, e.g. neglecting a small tilting of the reflector, the approximated satellite and receiver antenna patterns.

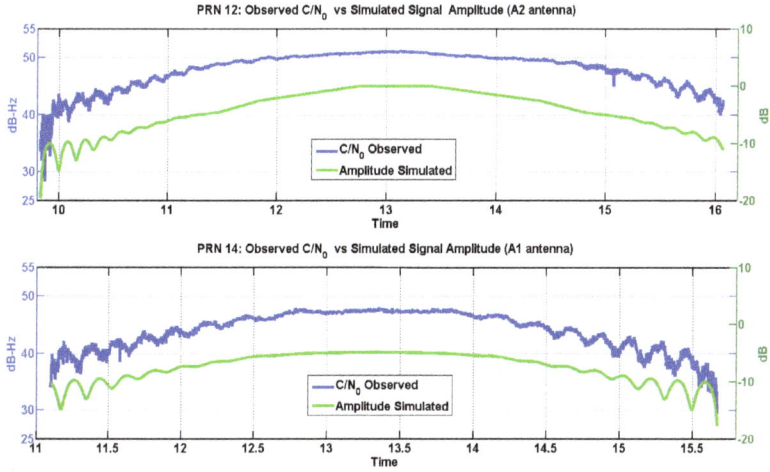

Figure 8. Observed C/N_0 values versus simulated signal amplitude for PRN 12 and 14 for antenna A1

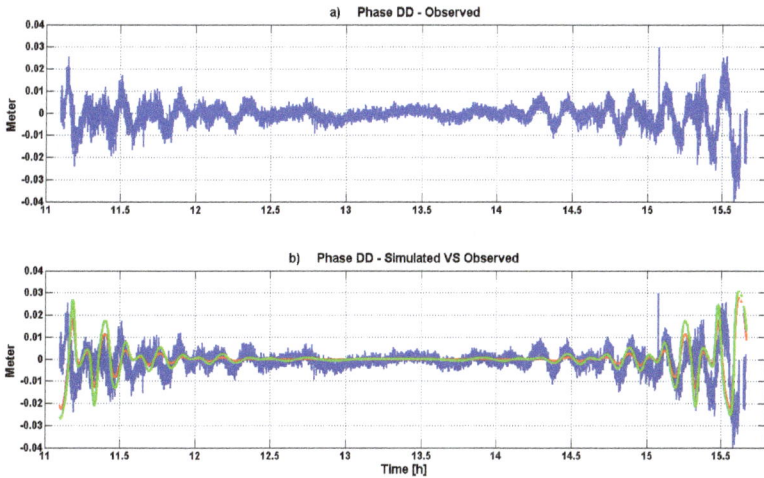

Figure 9. Simulated phase error - DD values versus observed phase DD (in blue) formed from PRN 12 and 14. Simulated DD for a concrete reflector are indicated in red and for a perfect reflector in green.

Finally, the carrier phase observations are investigated. To this end, double-differences (DD) are formed between the two antennas on the short baseline. Thus, most distance-dependent systematics as well as the receiver clock errors will cancel out. Again we use PRN 12 and 14. In Figure 9, the phase DD formed by PRN 12 and 14 of the short baseline are plotted together with double-differenced simulated phase error as in eq. (27). The time series of the observed values for an observational period of about 4.5 hours are plotted in blue in Figure 9a). In Figure 9b) the time series is overlaid by the simulated DD phase errors, where the relative amplitude of each multipath component is calculated analytical. In green no reflection loss is considered and in red the material properties of a concrete reflector are used.

5. Conclusion

Multipath propagation is still limiting the accuracy of precise GNSS applications despite the four decades of intensive research in this field. Signal amplitude, both for direct and indirect GNSS signal components that arrive at the receiving antenna with small relative delays, are crucial for the resulting phase error magnitude due to multipath propagation.

After a short introduction and a discussion on the different approaches that can be found in literature, aspects of ray propagation and signal polarization modeling are introduced. Then, an extensive description of the reflection process and its impact on the polarization state of the reflected signals is presented and all involved parameters are characterized. An analytical model for GNSS signal amplitudes is presented after. The model is then integrated into the equations for the phase error and the compound signal amplitude. Simulated and real data from a dedicated experiment are used to highlight the main properties of multipath propagation on the signal amplitude and carrier-phase domains.

Acknowledgements

This work has been realized within the BERTA project (50NA1012), funded by the Federal Ministry of Economics and Technology, and based on a resolution by the German Bundestag. The authors thank the BERTA partner Prof. Thomas Kürner (Institut für Nachrichtentechnik, TU Braunschweig) for the fruitful discussions. The authors would like to thank Dr. Thorsten Schrader (PTB Braunschweig) for supporting the experiments at the antenna reference open area test site at PTB.

Author details

Marios Smyrnaios[1*], Steffen Schön[1] and Marcos Liso Nicolás[2]

*Address all correspondence to: smyrnaios@ife.uni-hannover.de

1 Institut für Erdmessung, Leibniz-Universität Hannover, Hannover, Germany

2 Institut für Nachrichtentechnik Technische Universität Braunschweig, Braunschweig, Germany

References

[1] L. Hagerman, Effects on multipath of coherent and non coherent PRN ranging receiver, The Aerospace Corporation, 1973.

[2] Y. Georgiadou and A. Kleusberg, On carrier signal multipath effects in relative GPS positioning, *Manuscripta Geodaetica, 13,* pp. 172-179, 1988.

[3] M. Braasch, Multipath effects, in *GPS Positioning System Theory and Applications Vol. 1,* vol. Progress in Astronautics and Aeronautics Vol. 163., 1996, pp. 547-568.

[4] G. Seeber, Satellite Geodesy, Walter de Gruyter, 2003 2nd edition, pp. 316 - 320.

[5] A. Leick, GPS Satellite Surveying, Hoboken, New Jersey: John Wiley & Sons, Inc., 2004, pp. 237 - 243.

[6] B. Hofmann-Wellenhof, H. Lichtenegger and E. Wasle, GNSS Global Navigation Satellite Systems, Springer Wien New York, 2008, pp. 154 - 160.

[7] B. Hannah, Modeling and Simulation of GPS Multipath Propagation, Queensland: Queensland University of Technology, 2001, PhD thesis.

[8] L. Lau, Phase Multipath Modeling and Mitigation in Multiple Frequency GPS and Galileo Positioning, London: University College London, University of London, 2005, PhD thesis.

[9] M. Irsigler, Multipath propagation, mitigation and monitoring in the light of Galileo and the modernized GPS, München, 2008, PhD thesis.

[10] C. Rost, Phasenmehrwegereduzierung basierend auf Signalqualitätsmessungen geodätischer GNSS-Empfänger, Bayerische Akademie der Wissenschaften, 2011, Nr 665, PhD thesis.

[11] R. Van Nee, Multipath and Multi-Transmitter Interference in Spread-Spectrum Communication and Navigation Systems, Delf: Delf University of Technology, 1995, PhD thesis.

[12] J. Ray, Mitigation of GPS Code and Carrier Phase Multipath Effects Using a Multi-Antenna System, University of Calgary, Calgary, Alberta, Canada, 2000, PhD thesis.

[13] A. Bilich , Improving the Precision and Accuracy of Geodetic GPS: Application to Multipath and Seismology, Colorado: University of Colorado, 2006, PhD thesis.

[14] J. Weiss, Modeling and Characterization of Multipath in Global Navigation Satellite System Ranging Signals, Colorado: University of Colorado, 2007, PhD thesis.

[15] V. Böder, Zur hochpräzisen GPS-Positions- und Lagebestimmung unter besonderer Berücksichtigung mariner Anwendungen, Hannover: Wissenschaftliche Arbeiten der Fachrichtung Geodäsie und Geoinformatik der Universität Hannover, Nr. 245, 2002, PhD thesis.

[16] F. Dilssner, Zum Einfluss des Antennenumfeldes auf die hochpräzise GNSS-Position-sbestimmung, Hannover: Wissenschaftliche Arbeiten des Fachrichtung Geodäsie und Geoinformatik der Leibniz Universität Hannover, Nr. 253, 2008, PhD thesis.

[17] P. Misra and P. Enge, Global Positioning System: Signals, Measurements, and Per-formance, Jamuna Press, 2006, 2nd edition.

[18] R. Hatch , Dynamic advanced GPS at the centimeter level, in *Proceedings of the 4th international geodetic symposium on satellite positioning*, Austin, Texas, 1986.

[19] R. van Nee , Multipath effects on GPS code phase measurements, in *Proceedings of the 4th International Technical Meeting of the Satellite Division of The Institute of Navigation (ION GPS 1991)*, Washington, DC, September 11 - 13, 1991.

[20] C. Rocken, C. Meertens, B. Stephens, J. Braun, T. Van Hove, S. Perry, O. Ruud, M. McCallum and J. Richardson, UNAVCO Academic Reaserch Infrastructure (ARI) Receiver and Antenna Test Report, 1996.

[21] L. Wanninger and M. May, Carrier Phase Multipath Calibration of GPS Reference Stations, in *Proceedings of the 13th International Technical Meeting of the Satellite Division of The Institute of Navigation (ION GPS 2000)*, Salt Lake City, Utah, September 19 - 22, 2000.

[22] J. Genrich and Y. Bock , Rapid resolution of crustal motion at short ranges with the global positioning system., in *Geophys Res Lett 97:3261-3269*, 1992.

[23] D. Agnew and K. Larson, Finding the repeat times of the GPS constallation., *GPS Solution*, vol. 11, pp. 71-76, 2007.

[24] F. Dilssner, G. Seeber, G. Wübbena and M. Schmitz, Impact of Near-Field Effects on the GNSS Position Solution, in *Proceedings of the 21st International Technical Meeting of the Satellite Division of The Institute of Navigation (ION GNSS 2008), pp. 612-624* , Savannah, GA , September 16 - 19, 2008.

[25] C. Granström and J. Johansson , Site-Dependent Effects in Hight-Accuracy Applica-tions of GNSS, In Report on the Symposium of the IAG Subcommission for Europe (EUREF)., London, 2007.

[26] T. Iwabuchi, Y. Shoji, S. Shimada and H. Nakamura, Tsukuba GPS Dense Net Cam-paign Observations: Comparison of the Stacking Maps of Post-fit Phase Residuals Estimated from Three Software Packages, *Journal of the Meteorological Society*, vol. 82(1B), pp. 315-330, 2004.

[27] C. Rost and L. Wanninger, Carrier phase multipath mitigation based on GNSS signal quality measurements, *Journal of Applied Geodesy, 3(2), 81- 87.103*, 2009.

[28] P. Axelrad, C. Comp and P. Macdoran , SNR-based multipath error corection for GPS differential phase., in *IEEE Transaction on Aerospace and Electronic Systems*, 1996, 32(2), 650-660.99.

[29] A. Bilich , K. Larson and P. Axelrad, Modeling GPS phase multipath with SNR: Case study from the Salar de Uyuni, Boliva, *Journal of Geophysical Research*, vol. 113, 2008.

[30] Leica, http://www.surveyequipment.com/PDFs/AR25_Brochure.pdf, Leica AR25. [Online].

[31] L. Lau and P. Cross, A new signal-to-noise-ratio based stochastic model for GNSS highprecision carrier phase data processing algorithms in the presence of multipath errors, in *Proceedings of the 19th International Technical Meeting of the Satellite Division of The Institute of Navigation (ION GNSS 2006)*, Fort Worth, Texas, September 26 - 29, 2006.

[32] H. Hartinger and F. Brunner, Variances of GPS phase observations: the SIGMA-ε model, *GPS Solutions*, pp. Volume 2, Issue 4 , pp 35-43, 1999.

[33] A. Wieser, Robust and fuzzy techniques for parameter estimation and quality assessment in GPS, Graz, Technische Universität Graz, 2001, PhD thesis.

[34] A. Van Dierendonck, P. Fenton and T. Ford, Theory and performance of narrow correlator spacing in a GPS receiver, *Journal of the Institute of Navigation*, Vols. 39(3), 265-283.47, 90, 1992.

[35] G. McGraw and M. Braasch, GNSS multipath mitigation using Gated and High-Resolution correlator concepts, in *Proceedings of the 1999 National Technical Meeting of The Institute of Navigation*, San Diego, California, January 25 - 27, 1999, 333-342.

[36] L. Garin, F. van Diggelen and J. Rousseau, Storbe and Edge Correlator multipath mitigation for code, in *Proceedings of the 9th International Technical Meeting of the Satellite Division of The Institute of Navigation (ION GPS 1996)*, Kansas City, Missouri, September 17 - 20, 1996, 657-664.

[37] L. Garin and J. Rousseau, Enhanced Storbe Correlator Multipath Rejection for Code and Carrier, in *Proceedings of the 10th International Technical Meeting of the Satellite Division of The Institute of Navigation (ION GPS 1997)*, Kansas City, Missouri, September 16 - 19, 1997, 559-568.

[38] P. Fenton and J. Jones, Theory and Performance of Novatel Inc.'s Vision Correlator, in *Proceedings of the 18th International Technical Meeting of the Satellite Division of The Institute of Navigation (ION GNSS 2005)*, Long Beach, California, September 13 - 16, 2005, 2178 - 2186.

[39] B. Townsend, R. Van Nee, P. Fenton and K. Van Dierendonck, Performance Evaluation of the Multipath Estimating Delay Lock Loop, *Navigation*, Vols. 42, No. 3, pp. 503-514, 1995.

[40] B. Townsend , P. Fenton , K. Van Dierendonck and R. Van Nee, L1 Carrier Phase Multipath Error Reduction Using MEDLL Technology, in *Proceedings of the 8th International Technical Meeting of the Satellite Division of the Institute of Navigation, ION - GPS 1995*, Palm Springs, California, 1995, 1539-1544.

[41] R. Lawrence, How Good Can It Get with New Signals? Multipath Mitigation, *GPS World*, 2003.

[42] A. Van Dierendonck, Evaluation of GNSS Receiver Correlation Processing Techniques for Multipath and Noise Mitigation, in *Proceedings of the 1997 National Technical Meeting of The Institute of Navigation*, Santa Monica, CA, January 14 - 16, 1997, 207-215.

[43] K. Borre, D. Akos, N. Bertelsen, P. Rinder and S. Jensen, A Software-Defined GPS and Galileo Receiver. A Single Frequency Approach, Boston: Birkhäuser, 2007.

[44] J. Tsui, Fundamentals of Global Positioning System Receivers. A Software Approach, New York: John Wiley & Sons, INC, 2000.

[45] T. Pany, Navigation Signal Processing for GNSS Software Receivers, Artech House, 2010.

[46] G. Moernaut and D. Orban, GNSS Antennas An Introduction to Bandwidth, Gain Pattern, Polarization, and All That, *GPS World*, pp. 42 - 48, February 2009.

[47] J. Tranquilla, J. Carr and H. Al-Rizzo, Analysis of a choke ring groundplane for multipath control in Global Positioning Systems (GPS) applications, in *IEEE Transactions on Antennas and Propagation, 42(7), 905-911*, 1994.

[48] T. Blakney, D. Connell, B. Lamberty and J. Lee, Broad-band antenna sructure having frequency-independent, low loss ground plane (Patent Nr. 4608572), Seattle, WA, 1986.

[49] V. Filippov, I. Sutiadin and J. Ashjaee, Measured Characteristics of Dual Depth Dual Frequency Choke Ring for Multipath Rejection in GPS Receivers, in *In Proceedings of the 12th International Technical Meeting of the Satellite Division of the Institute of Navigation (ION GPS 1999), 793 – 796*, Nashville, 1999.

[50] T. Ning, G. Elgered and J. Johansson, The impact of microwave absorber and radome geometries on GNSS measurements of station coordinates and atmospheric water vapour, *Adv Space Sci, 47(2), 186*, 2011.

[51] J. Ray, M. Cannon and P. Fenton , Mitigation of Static Carrier Phase Multipath Effects Using Multiple Closely-Spaced Antennas, in *Proceedings of the 1th International Technical Meeting of the Satellite Division of the Institute of Navigation, ION - GPS98*, Nashville, Tennessee, September 15-18, 1998, 1025-1034.

[52] M. Cuntz, H. Denks, A. Konovaltsev, A. Hornbostel, A. Dreher and M. Meurer, GALANT - Architecture Design and First Results of A Novel Galileo Navigation Receiver Demonstrator With Array Antennas, in *Proceedings of the 21st International Technical Meeting of the Satellite Division of the Institute of Navigation* , Savannah, september 16-19, 2008, pp. 1470-1476.

[53] K. Park, P. Elosegui, J. Davis, P. Jarlemark, B. Corey, A. Niell, J. Normandeau, C. Meertens and V. Andreatta, Development of an Antenna and Multipath Calibration System for Global Positioning System Sites, *RADIO SCIENCE*, vol. 39, 2004.

[54] V. Böder , F. Menge , G. Seeber, G. Wübbena and M. Schmitz, How to Deal With Station Dependent Errors - New Developments of Absolute Field Calibration of PCV and Phase-Multipath With a Precise Robot, in *ION 2001*, Salt Lake City, Utah, 2001.

[55] G. Wübbena, M. Schmitz and G. Boettcher, Near-field Effects on GNSS Sites: Analysis using Absolute Robot Calibration and Procedures to Determine Corrections, in *IGS Workshop* , Drmstadt, Germany, 2006.

[56] P. Shetty, A. Kakkar, U. Weinbach and S. Schön, Experimental Analysis of Multipath Linear Combination of GPS and Galileo Signals, Poster, in *Geodätische Woche*, Köln, Oktober 5-7, 2010.

[57] E. Schönemann , A. Hauschild, P. Steigenberger, T. Springer, J. Dow, O. Montenbruck, U. Hugentobler and M. Becker, New Results from GIOVE: The CONGO-Network and the Potential of Tracking Networks with Multiple Receiver and Antenna Types, in *EGU General Assembly*, Vienna, Austria, 2010.

[58] O. Montenbruck, A. Hauschild and U. Hessels, Characterization of GPS/GIOVE Sensor Stations in CONGO Network, *GPS Solution, 15(3), 193-205*, 2011.

[59] P. Elósegui, J. Davis, R. Jaldehag, J. Johansson, A. Niell and I. Shapiro, Geodesy using the Global Positioning System: The effects of signal scattering on estimates of site position, *Journal of Geophysical Research, 100(B7), 9921-9934*, 1995.

[60] M. Liso , M. Smyrnaios, S. Schön and T. Kürner, Basic Concepts for the Modeling and Correction of GNSS Multipath Effects Using Ray-Tracing and Software Receivers, in *IEEE APS Topical Conference on Antennas and Propagation in Wireless Communication*, Torino, Italy, 2011.

[61] J. Bradbury, M. Ziebart, P. Cross, P. Boulton and A. Read, Code multipath modelling in the urban environment using large virtual reality city models: Determining the local environment, *Journal of Navigation, 60(1), 95-105*, 2007.

[62] A. Steingass and A. Lehner, Measuring the Navigation Multipath Channel ... A Statistical Analysis, *Proceedings of the 17th International Technical Meeting of the Satellite Division of The Institute of Navigation (ION GNSS 2004)*, September 21 - 24, 2004, pp. 1157 - 1164.

[63] R. Treuhaft, S. Lowe, C. Zuffada and Y. Chao, 2-cm GPS altimetry over Crater Lake, *Geophysical Research Letters, 22(23), 4343-4346*, 2001.

[64] A. Helm, Ground-based GPS Altimetry with the L1 Open GPS receiver using carrier phase-delay observations of reflected GPS signals, Scientific Technical report STR08/10,GFZ Potsdam, Potsdam, 2008, PhD thesis.

[65] J. Garrison, S. Katzberg and C. Howell, Detection of Ocean Reflected GPS Signals: Theory and Experiment, in *Proceedings of the IEEE Southeast*, Blacksburg, USA, April 12-14, 1997, pp. 290-294.

[66] K. Larson, E. Small, E. Gutmann, A. Bilich, P. Axelrad and J. Braun, Using GPS multipath to measure soil moisture fluctuations: initial results, *GPS Solution, Vol 12(3)*, pp. 173-177, 2008.

[67] M. Martin-Neira, A Passive Reflectometry and Interferometry System (PARIS): Application to Ocean Altimetry, *ESA Journal*, vol. 17(4), pp. 331-355, 1993.

[68] V. Zavorotny, K. Larson, J. Braun, E. Small, E. Gutmann and A. Bilich, A physical model of GPS multipath caused by land reflections: toward bare soil moisture retrievals, in *IEEE J-STARS*, 2010.

[69] A. Rius, J. Aparicio, E. Cardellach and M. Martin-Neira, Sea Surface State Measured Using GPS Reflected Signals, *Geophysical Reaserch Letters*, vol. 29(23), 2002.

[70] W. Stutzman, Polarization in Electromagnetic Systems, Artech House, Boston - London, 1993.

[71] C. Balanis, Antenna Theory, New Jersey: John Wiley & Sons, Inc., 2005.

[72] J. Peatross and M. Ware, Physics of Light and Optics, Brigham Young University, 2008.

[73] A. Maltsev, V. Erceg, E. Perahia, C. Hansen, R. Maslennikov, A. Lomayev, A. Sevastyanov and A. Khoryaev, Polarization model for 60 GHz WLAN Systems, 2009.

[74] A. Maltsev, R. Maslennikov, A. Lomayev, A. Sevastyanov and A. Khoryaev, Statistical Channel Model for 60 GHz WLAN Systems in Conference Room Enviroment, *Radioengineering*, vol. 20, pp. 409-422, June 2011.

[75] W. Marquis and D. Reigh, On-Orbit Performance of the Improved GPS Block IIR Antenna Panel, in *Proceedings of the 18th International Technical Meeting of the Satellite Division of The Institute of Navigation (ION GNSS 2005)*, Long Beach, CA, September 13 - 16, 2005.

[76] "ICD-GPS-200c 10-10-1993 to 12-4-2000".

[77] "GPS-701-GG and GPS-702-GG Antennas," Novatel, [Online]. Available: http://webone.novatel.ca/assets/Documents/Papers/GPS701_702GG.pdf.

[78] F. Czopek and S. Shollenberger, Description and Performance of the GPS Block I and II L-Band Antenna and Link Budget, in *Proceedings of the 6th International Technical Meeting of the Satellite Division of The Institute of Navigation (ION GPS 1993)*, Salt Lake City, UT, 37-43, September 22 - 24, 1993.

[79] S. Ericson, K. Shallberg and C. Edgar, Characterization and Simulation of SVN49 (PRN01) Elevation Dependent Measurement Biases, in *Proceedings of the 2010 International Technical Meeting of The Institute of Navigation*, San Diego, CA, January 25 - 27, 2010.

Chapter 4

Satellite Gravimetry: Mass Transport and Redistribution in the Earth System

Shuanggen Jin

Additional information is available at the end of the chapter

1. Introduction

The Earth's gravity field is a basic physical parameter, which reflects mass transport and re-distribution in the Earth System. It not only contributes to study the Earth's interior physical state and the dynamic mechanism in geophysics, but also provides an important way to re-search the Earth's interior mass distribution and characteristics. The gravity field and its changes with time is of great significance for studying various geodynamics and physical processes, especially for the dynamic mechanism of the lithosphere, mantle convection and lithospheric drift, glacial isostatic adjustment (GIA), sea level change, hydrologic cycle, mass balance of ice sheets and glaciers, rotation of the Earth and mass displacement [33; 37; 7; 39; 17 and 18]. For Geodesy, the gravity field is an important parameter to study the size and shape of the Earth. Meanwhile the Earth's gravity field is very important to determine the trajectory of carrier rocket, long-range weapons, artificial Earth's satellites and spacecrafts. In addition, the gravity field could provide some signals of pre-, co-, and post-earthquake with mass transport following earthquakes [25; 14]. Therefore, precisely determining Earth's gravity field and its time-varying information are very important in geodesy, seismology, oceanography, space science and national defense as well as geohazards.

The global Earth's gravity field is described by spherical harmonics. The non-rotating part of the potential is mathematically described as [15]:

$$V(\theta, \phi) = \frac{GM}{r}[1 + \sum_{n=2}^{\infty} \sum_{m=0}^{n} (\frac{R}{r})^n \tilde{P}_{nm}(\sin \theta)(C_{nm} \cos m\phi + S_{nm} \sin m\phi)] \tag{1}$$

where θ and ϕ are geocentric (spherical) latitude and longitude respectively, $\tilde{P}_{nm}$ are the fully normalized associated Legendre polynomials of degree n and order m, and C_{nm}, S_{nm} are the numerical coefficients of the model. For the Earth's gravity field model, the potential co-efficient of the Earth (C_{nm}, S_{nm}) should be determined.

Traditional measurements of Earth's gravity field mainly use three techniques. The first one is the terrestrial gravimeter, while the cost is high and the labor work is hard, and further-more the temporal-spatial resolution is low. The second one is satellite altimetry, which can estimate the gravity field and geoid over the ocean. However, it is still subject to various er-rors and temporal-spatial resolutions. The third one is to use the laser ranging of artificial Earth's satellites. Because the satellite orbital motion is largely affected by gravitational force and other non-conservation forces, orbit solutions based on precise satellite tracking obser-vations can estimate the gravity field. While, it only provided long-wavelength gravity field information as such satellite orbits are very high. Combination of these three kinds of techni-ques can give comprehensive gravity field models, however, the accuracy of the model based on satellite orbit tracking data sharply decrease with the increase of the gravity coeffi-cients' degree. Furthermore, due to the sparse surface gravimetric data, uncertain weighting of various measurements and truncation of the spherical harmonic coefficients, these obser-vations are very difficult to obtain a more precise gravity field model.

With the recent development of the low-earth orbit (LEO) satellite gravimetry, it has greatly increased the Earth's gravity field model's precision and temporal-spatial resolution, partic-ularly recent Gravity Recovery and Climate Experiment (GRACE). Satellite gravimetry is a successful innovation and breakthrough in the field of geodesy, following the Global Posi-tioning System (GPS). Unlike the traditional gravity measurements, such as satellite altime-try and high-altitude orbital perturbation analysis, the most advanced SST (Satellite-to-Satellite Tracking) and SGG (Satellite Gravity Gradiometry) techniques are used to estimate the global high-precision gravity field and its variations. Satellite-to-Satellite Tracking tech-nique includes the so-called high-low satellite-to-satellite tracking (hl-SST) [1] and low-low satellite-to-satellite tracking (ll-SST) [43], which can precisely determine the variation rate of the distance between two satellites. The satellite gravity gradiometric (SGG) technique uses a gradiometer carried on the low-orbit satellite to determine directly the second order deriv-atives of gravity potential (gradiometric tensor), which can recover the Earth's gravity field precisely. Therefore, the satellite gravimetry has greatly improved the gravity field precision and its applications in geodesy, oceanography, hydrology and geophysics.

2. Gravity field from satellite gravimetry

Since 2000, three gravity satellites missions have been launched and dedicated to gravity field recovery, i.e., CHAMP (Challenging Mini-Satellite Payload for Geophysical Research and Application), GRACE (Gravity Recovery and Climate Experiment) and GOCE (Gravity Field and Steady-state Ocean Circulation Explorer).

2.1. High-low satellite to satellite tracking (hl-SST)

CHAMP satellite has been successfully launched on July 15, 2000 using the hl-SST technical mode, and the high orbit satellites were GPS satellites [29]. CHAMP was a German small satellite mission for geoscientific and atmospheric research and applications. The three primary scientific objectives of the CHAMP mission were to obtain highly precise global long-wavelength features of the static Earth's gravity field and its temporal variation with unprecedented accuracy, crustal magnetic field of the Earth and atmospheric and ionospheric products from GPS radio occultation, including temperature, pressure, water vapour and electron content. The GPS receiver on-board CHAMP and ground-based satellite laser ranging were used to determine the CHAMP's orbit. The three-axes STAR accelerometer measured the non-gravitational accelerations of perturbing CHAMP's orbit. Therefore, the long- to mid-scale Earth's gravity field can be recovered from the above data with an unprecedented accuracy.

2.2. High-low/low-low satellite to satellite tracking (hl-SST/ll-SST)

The Gravity Recovery and Climate Experiment (GRACE), a joint mission of NASA and the German Aerospace Center (DLR), has been launched in March 2002 to recover detailed Earth's gravity field [42; 38]. GRACE has twin satellites with distance of about 220 kilometers and used the typical high-low/low-low satellite-to-satellite tracking (hl-SST/ll-SST) techniques. The primary objective is to obtain extremely high-resolution global Earth's gravity field and its changes with time. The k-band ranging system is used to measure the precise distance change rate between twin satellites. With the accelerometer, the GRACE could determine the gravity field and its change with time. These estimates provide a comprehensive understanding of how mass is distributed globally and how that distribution varies over time in the Earth system.

2.3. High-low satellite to satellite tracking/satellite gravity gradient mode

The Gravity Field and Steady-State Ocean Circulation Explorer (GOCE) mission has been launched on March 17, 2009 with taking high-low satellite-to-satellite tracking and satellite gravity gradiometer (hl-SST/SGG), which is the first satellite mission to employ the concept of gradiometry [8]. The mission objectives are to determine gravity-field anomalies with an accuracy of 10^{-5} ms^{-2} (1 mGal) and the geoid with an accuracy of 1-2 cm, and to achieve a spatial resolution better than 100 km. Unlike the previous two modes, GOCE was equipped with three pairs of ultra-sensitive accelerometers and onboard GPS/GLONASS receiver to determine the exact position of the satellite with high-low satellite-to-satellite tracking mode (hl-SST). The non-conservative forces on the gradiometer such as the linear and angular inertia acceleration produced by the atmosphere drag and the solar radiation pressure can be accurately balanced by a non-conservation control system (Drag-free) Therefore, GOCE could recover the global earth gravity field with higher resolution and higher accuracy.

These satellite gravimetric techniques greatly improved the knowledge about the Earth's gravity field, which could provide more abundant information on mass transport and redistribution in the Earth system. These products will make an important contribution to some key scientific issues of global change, such as global sea level changes, ocean circulation, ice sheets and glaciers mass balance and hydrologic cycle This chapter focuses on the mass transport and redistribution in the Earth system with monthly resolution are derived from approximate 10 years of monthly GRACE measurements (2002 August-2011 December).

3. Mass transport and redistribution

3.1. Terrestrial water storage from GRACE

The GRACE mission was launched in March 2002 and began operating nearly continuously since August 2002 [37]. One of the scientific objectives of the GRACE mission is to produce high-quality terrestrial water storage and ocean mass estimates. GRACE delivers monthly averages of the spherical harmonic coefficients, which are sensitive to fluctuations in continental water storage and the polar ice sheets, as well as changes in atmospheric and oceanic mass distribution [40; 17]. At this point, the terrestrial water storage anomalies over the land can be directly estimated by gravity coefficient anomalies for each month ($\Delta C_{lm}, \Delta S_{lm}$) (40):

$$\Delta \eta_{land}(\theta,\phi,t) = \frac{a\rho_{ave}}{3\rho_w}\sum_{l=0}^{\infty}\sum_{m=0}^{l}\tilde{P}_{lm}(\sin\theta)\frac{2l+1}{1+k_l}(\Delta C_{lm}\cos(m\phi)+\Delta S_{lm}\sin(m\phi)) \qquad (2)$$

where ρ_{ave} is the average density of the Earth, ρ_w is the density of fresh water, a is the equatorial radius of the Earth, $\tilde{P}_{lm}$ is the fully-normalized Associated Legendre Polynomials of degree l and order m, k_l is Love number of degree l[13], θ is the geographic latitude and ϕ is the longitude. The precise terrestrial water storages (TWS) are estimated using monthly GRACE solutions (Release-04) from the Center for Space Research (CSR) at the University of Texas, Austin from August 2002 until December 2011, except for June 2003, January 2011 and June 2011 without data. The degree 2 and order 0 (C_{20}) coefficients are replaced from Satellite Laser Ranging (SLR) due to large uncertainties in GRACE coefficients [5]. The degree 1 spherical harmonics coefficients (C_{11}, S_{11}, and C_{10}) are used from 34) and the postglacial rebound (PGR) influences is removed with 23). In addition, since GRACE solutions have larger noise and strips [40; 36], the 500km width of Gaussian filter and de-striping filter are used to mitigate these effects [36]. Thus, about 10 years of global terrestrial water storages (TWS) are estimated from GRACE.

3.2 Ocean bottom pressure from GRACE

Monthly GRACE gravity changes over oceanic regions can be transformed to ocean mass or ocean bottom pressure (OBP) at latitude θ, longitude ϕ as described by 40):

$$\Delta\eta_{ocean}(\theta,\phi,t) = \frac{a\rho_{ave}}{3\rho_w}\sum_{l=0}^{\infty}\sum_{m=0}^{l}\frac{(2l+1)}{(1+k_l)}W_l\tilde{P}_{lm}(\sin\theta)\begin{Bmatrix}(\Delta C_{lm}(t)+\Delta C_{lm}^{GAD}(t))\cos(m\phi)+\\(\Delta S_{lm}(t)+\Delta S_{lm}^{GAD}(t))\sin(m\phi)\end{Bmatrix} \tag{3}$$

where the variables have the same meaning with equation (2), and W_l is the Gaussian averaging function with increasing l. As the coefficients from the GRACE are deviations from a background model, we have to add back the monthly OBP modeled in the GRACE processing. A new OBP product (GAD) is now available by [9]. 4] demonstrated that GRACE could measure the variation in the global mean ocean mass (and hence OBP) quite accurately. 2] found that the seasonal mode of OBP variation in the North Pacific extracted from GRACE data agreed qualitatively with that of an ocean model. Therefore, the reliable monthly OBP time series could be precisely estimated from the most recent GRACE gravity field solutions (Release-04) from the Center for Space Research (CSR) at the University of Texas, Austin [4]. Here monthly grid OBPs are used with a 500-km Gaussian smooth from August 2002 until December 2011, except for June 2003, January 2011 and June 2011 when no solutions exist.

4. Results and Discussions

4.1. Global hydrological cycle

4.1.1. Seasonal changes of Terrestrial water storage

The TWS time series have significant seasonal variations. Amplitude and phase of annual and semi-annual variations at grid points are estimated from GRACE TWS time series (August 2002-December 2011) through the method of least squares fit to a bias, trend, and seasonal period sinusoids as:

$$TWS(t) = A_a\sin(\omega_a t - \varphi_a) + A_{sa}\sin(\omega_{sa}t - \varphi_{sa}) + B + C(t-t_0) + \varepsilon(t) \tag{4}$$

where B is the constant, t_0 is on January 1st 2002, φ is the phase and A is the amplitude of period p as 1 and 0.5 years. The GRACE results are further compared with the Global Land Data Assimilation System (GLDAS) model. GLDAS model is a hydrological model, which is jointly developed by the National Aeronautics and Space Administration (NASA) Goddard Space Flight Center (GSFC) and the National Oceanic and Atmospheric Administration (NOAA) National Centers for Environmental Prediction (NCEP) [31]. Figure 1 shows the annual amplitude and phase of global terrestrial water storages from GRACE and GLDAS model. It has clearly shown that annual amplitude of GRACE-derived terrestrial water storage is up to 20 cm in South America's Amazon River Basin and about 10 cm in the Niger, Lake Chad and Zambezi River Basins in the African continent, the Ganges and the Yangtze River region in Southeast Asia, and in other areas the annual variations of terrestrial water storage are not significant. The annual amplitudes from GRACE have similar patterns with the GLDAS, but a little larger than GLDAS results as the GLDAS model does not include groundwater. In addition, for the most parts of the world, the terrestrial water storage reaches the maximum in September-October each year, and the minimum in March-April. The semi-annual signals in most regions of the world are not significant, so here we don't discuss.

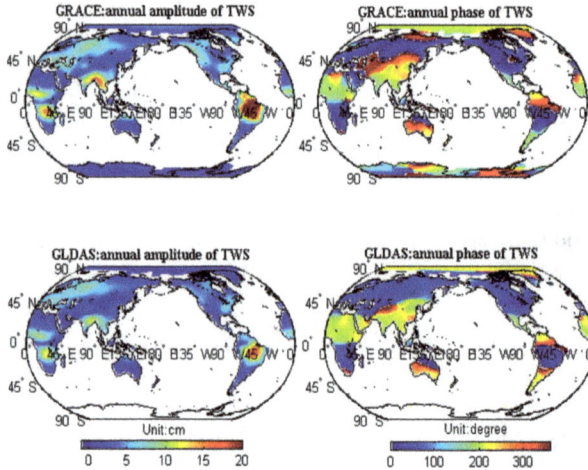

Figure 1. Annual amplitude and phase of TWS based on GRACE and GLDAS model.

4.1.2. Long-term trend of terrestrial water storage

The long-term trends of global terrestrial water storage are further analyzed. Figure 2 shows the long-term trend of global terrestrial water storage from GLDAS and GRACE data. For some parts, they agreed each other, but the GLDAS model cannot capture the detailed extreme climate and human groundwater depletion signals in terrestrial water storage, e.g., great groundwater depletion in Northwest India. While GRACE results in Figure 2(b) have clearly shown that the terrestrial water storage is decreasing at about -15.5 mm/y in Northwest India, which have been proved that over groundwater depletion lead to decrease in TWS [32]. The terrestrial water storage in North China Plain is reducing at -4.8mm/yr, mainly due to the sparse vegetation of the region, the larger evaporation and huge groundwater depletion. While in Antarctica, Greenland and Canadian Archipelago, Alaska, Patagonia glaciers as well as the Himalayan glaciers, the TWS is significantly decreasing due to rapid glacier melting. In addition, the flood in Amazon River Basin of South America, results in increase of terrestrial water storage at about 20.5mm/yr. In La Plata region, the terrestrial water storage is reducing at about -9.8mm/y due to recent drought. Our results almost confirmed the early results based on short-time GRACE data. For example, 39) found that the mass of the Antarctic ice sheet in decreased significantly during 2002–2005, at a rate of 152 ± 80 cubic kilometers of ice per year, which is equivalent to 0.4 ± 0.2 millimeters of global sea-level rise per year, Luthcke's studies show that during 2002-2005, the Greenland ice sheet lost at the speed of (239 ± 23) km³ /year [21].

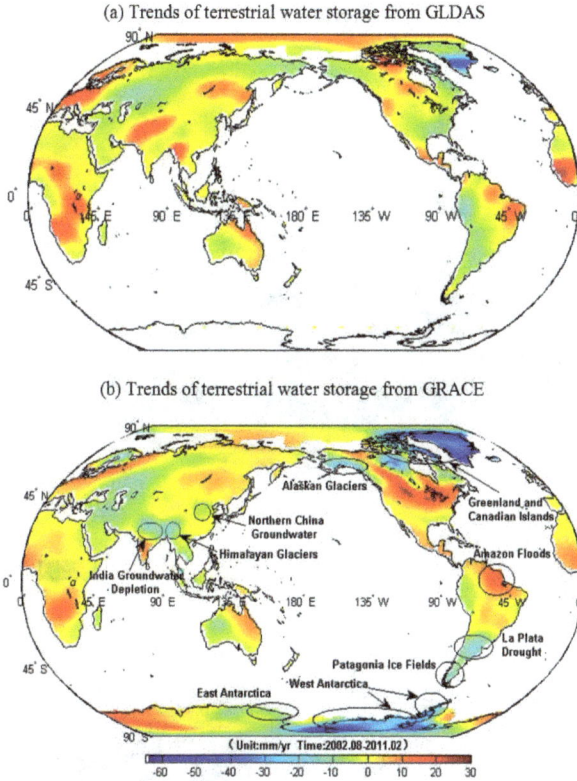

(a) Trends of terrestrial water storage from GLDAS

(b) Trends of terrestrial water storage from GRACE

Figure 2. The long-term trend of terrestrial water storage from GLDAS and GRACE.

4.2 Global Ocean Bottom Pressure variations

4.2.1 Seasonal OBP variation

The OBP time series also have significant seasonal variations. Figure 3 shows the amplitude distributions of annual OBP variations from GRACE and ECCO. Larger amplitudes of annual OBP variations from GRACE are found in the Pacific and Indian oceans with up to 3.5±0.4 cm, particularly in the west of Australia, Pacific sector of the Southern Ocean, and the northwest corner of the North Pacific as well, while the lower annual amplitudes are in Atlantic at less than 1.0±0.3 cm. However, ECCO estimates for all oceans are generally less than 1.5±0.3 cm, much weaker than GRACE. The phase patterns of annual OBP variations are both closer from GRACE and ECCO. For example, the phase of annual OBP variations both shows an asymmetry in middle north Pacific and south Pacific (Figure 4). The semi-annual OBP variations from GRACE and ECCO are relatively weaker and most semi-annual amplitudes are less than 1.0±0.3 cm.

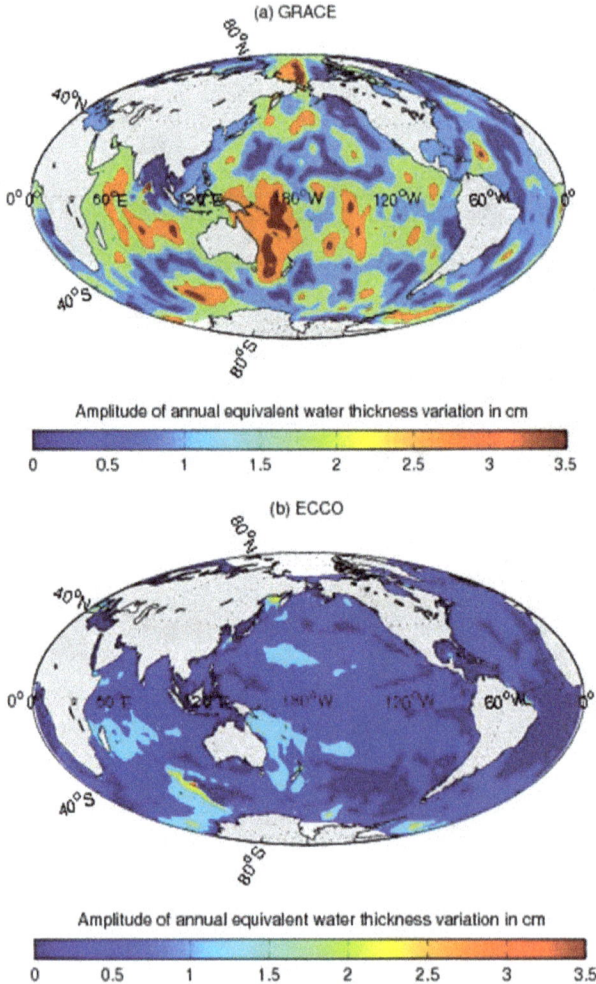

Figure 3. Amplitude of annual OBP variations from GRACE and ECCO.

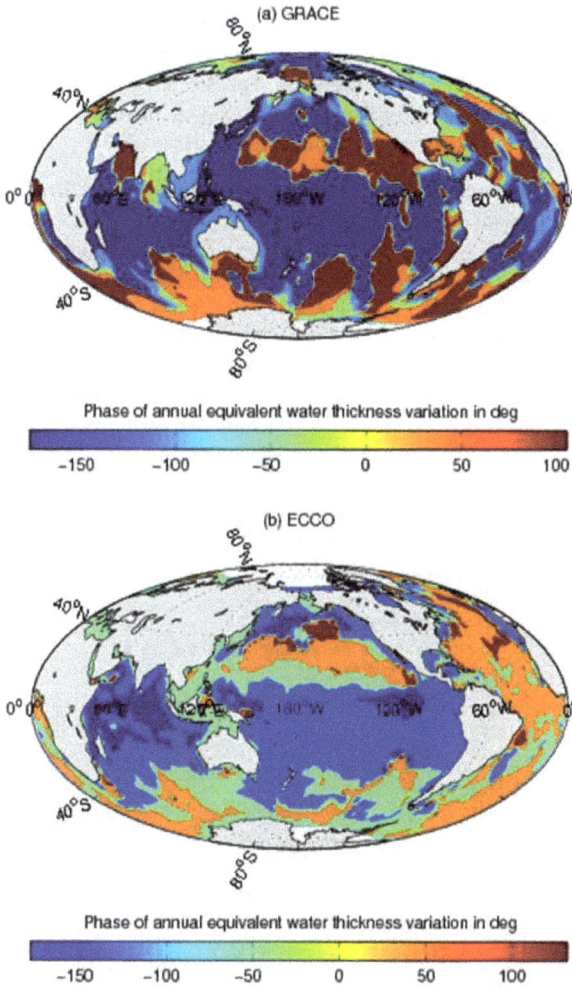

Figure 4. Phase of annual OBP variations from GRACE and ECCO.

4.2.2 Secular OBP variation

Current sea level rise is due mianly to human-induced global warming, which will increase sea level over the coming century and longer periods. One is the steric sea change (i.e. thermal expansion) by the thermal expansion of water due to increasing temperatures, which is well-quantified. The other is non-steric sea level change (i.e. eustatic sea level change) relat-

ed to mass changes through the addition of water to the oceans from the melting of continental ice sheets and fresh water in rivers and lakes. However, the eustatic sea level change
is more difficult to predict and quantify due to high uncertain estimates of the Antarctic and
Greenland mass and terrestrial water reservoirs. The Satellite-based GRACE observations
provide a unique opportunity to directly measure the global ocean mass change (equivalently ocean bottom pressure), which can qualify the OBP change.

The secular OBP variations are analyzed from the almost 10-year monthly GRACE OBP time
series (August 2002- December 2011) at 1 ×1 grid. After we check the OBP time series, some
anomaly of OBP time series are found between the end of 2004 and early of 2005 near Southeast Asia. Figure 5 shows the non-seasonal mass change time series as the equivalent water
thickness in centimeter (cm) at grid point (90.5°E, 2.5°S). It has clearly shown a sudden jump
of non-seasonal mass change between the end of 2004 and early of 2005. While two largest
earthquakes occurred during these time recorded in about 40 years. One is the Sumatra-
Andaman earthquake (Mw = 9.0) on December 26, 2004, and the other one is the Nias earthquake (Mw =8.7) on March 28, 2005. The Sumatra-Andaman earthquake raised islands by up
to 20 meters [16] and the ruptures extended over approximately 1800 km in the Andaman and
Sunda subduction zones [6]. A number of researchers found gravity anomalies from GRACE
before and after the Sumatra-Andaman and Nias earthquakes associated with the subduction and uplift, which agreed with model predictions [e.g., 14]. Therefore, the co-seismic gravity
effects should be removed for further analyzing the secular OBP variations.

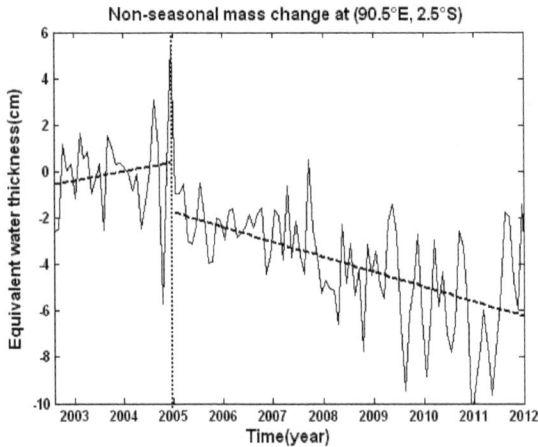

Figure 5. Non-seasonal mass change time series at point (90.5°E, 2.5°S).

Figure 6 shows the trend distribution of secular OBP variations (equivalent water thickness)
in cm/yr, ranging from -1.0 to 0.9±0.2 cm/yr, where the upper panel a) is from GRACE and
the bottom panel (b) is from ocean model ECCO. Both show significant subsidence of OBP
in Atlantic and uplift in northwest Pacific, but the amplitude from GRACE is significantly

larger. The mean OBP time series in Pacific and Atlantic from GRACE and model ECCO also show similar opposite secular OBP variations (Figure 7), reflecting secular exchange of
Pacific and Atlantic water. However, the secular change of OBP from GRACE in the Indian
sea is subsiding at larger amplitude, while that from ECCO is a little uplift. It needs to be
further investigated using long-term satellite observations and other data in the future.

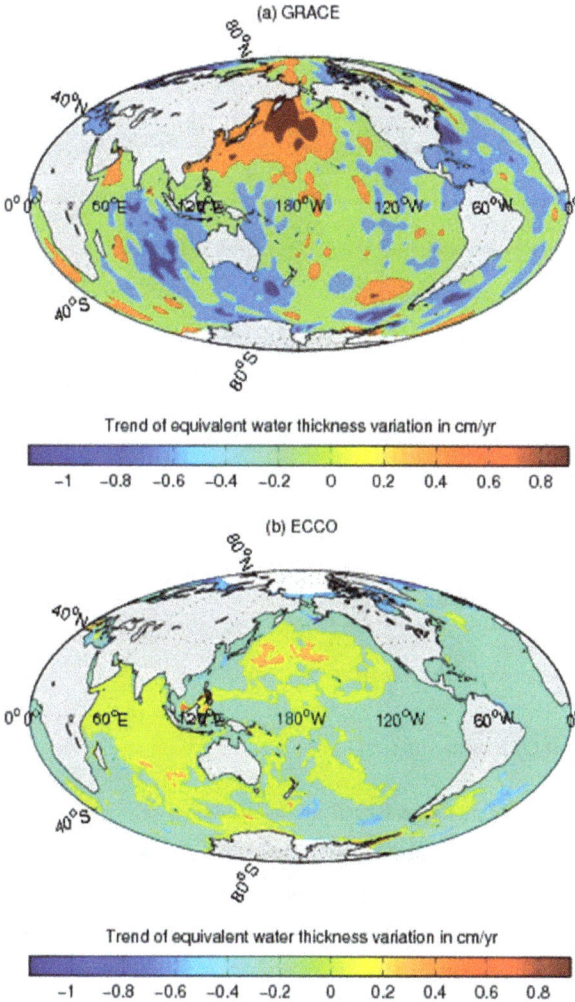

Figure 6. OBP Trend as equivalent water thickness variation in cm/yr.

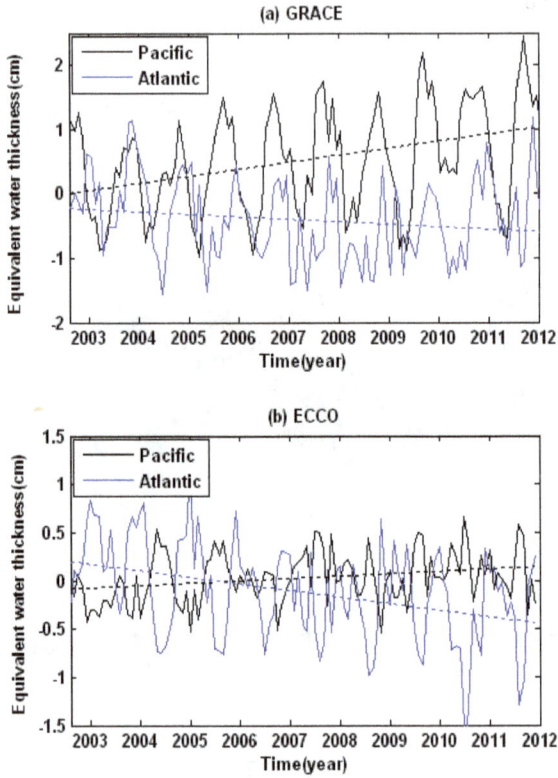

Figure 7. Mean OBP time series from GRACE and ECCO in Pacific and Atlantic.

4.2.3 High frequent OBP variations

The unmodelled OBP residuals (observed minus modelled seasonal terms) reflect the high frequency variation, mainly the high frequent and noise components. We estimate the higher frequency variability by taking the root-mean-square (RMS) of the OBP time series after removing the constant, trend, annual and semi-annual variations as the best-fit sinusoid:

$$RMS = \sqrt{\frac{1}{N}\sum_{t=1}^{N}(OBP_o^t - OBP_M^t)^2} \qquad (5)$$

where OBP_o^t is the OBP from GRACE or ECCO at time t, OBP_M^t is the best fitted value at time t from A*sin($2\pi(t-t_0)/p +\varphi$)+B+C(t-t$_0$), and N is the total observation number. The RMS of high-frequency OBP variations at globally distributed grid sites are shown in Figure 8. The high frequency variability of OBP from GRACE ranges from 0 to 3.4 cm with mean amplitude of about 2.0 cm, primarily due to in high frequent OBP variations and noise components of GRACE data processing, while the high frequency variability of OBP from ECCO is ranging from 0 to 2.3 cm with mean amplitude of about 0.7 cm, particularly smaller and smoother in tropical regions. Both have shown the similar higher frequency variability in high-latitude, especially in southern high latitude areas.

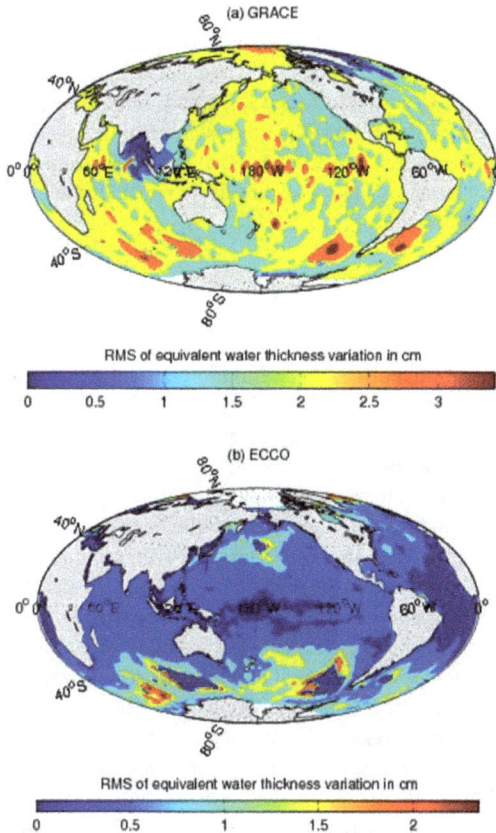

Figure 8. The root-mean-square (RMS) of OBP after removing the constant, trend, annual and semi-annual variations terms.

4.3 Discussions

Although GRACE can well estimate global larger-scale mass transport and redistribution in the Earth system, but it is still subject a number of effects, such as orbital inclination of GRACE, hardware noise and data processing methods. Therefore, the terrestrial water storage and ocean bottom pressure need to be further improved. In addition, the accuracy of geophysical models, post-glacial rebound and tide model also affect the GRACE results. For ocean bottom pressure variation, although 2) found that the seasonal mode of OBP variation in the North Pacific from GRACE data agreed qualitatively with the ocean model, while the secular trend and mean high frequency variability of global OBP from GRACE are higher than that from ECCO by 2-3 times. On one hand, the leakage of land hydrology signals will involve in GRACE-derived OBP estimates (30). Other reasons are the aliasing errors of OBP fluctuations, including atmospheric model and glacial isostatic adjustment (GIA) model. These can affect the tendency, seasonal and high-frequent variations with larger amplitude in the GRACE data than the ECCO estimates, while leakage effect at semi-annual period is less (26), but the GIA will largely affect the OBP trend. In addition, the tides can dealiase errors of 1 cm over most of the oceans [28], and such errors may affect GRACE OBP estimates during non-tidal models corrections. Finally, the instrument noises may affect GRACE solutions [28]. Therefore, one needs to further consider the instrument noise effects and tide aliasing errors in the future.

5. Conclusion

In this Chapter, the mass transport and redistribution in the Earth system are studied using monthly GRACE data. Seasonal and secular changes of global terrestrial water storage in the past 10 years are investigated from GRACE data as well as compared with GLDAS model. The results have shown that the global terrestrial water storages have obvious seasonal changes and long-term trend. The annual amplitude can reach up to 20cm in South America's Amazon River Basin and almost about 10cm in the Niger, Lake Chad and Zambezi River Basins in Africa, the Ganges and the Yangtze River region in Southeast Asia. The maximum terrestrial water storage normally appears in Sep-Oct, and the minimum terrestrial water storage normally appears around in Mar-Apr. The long-term variations of terrestrial water storage are also clear in some areas. For example, the terrestrial water storage is decreasing at about -15.5mm/y in Northwest India due to groundwater depletion, increasing at about 20.5mm/yr in Amazon River Basin of South America due to the flood, and reducing at about -9.8mm/yr in La Plata region due to recent drought. In addition, the secular TWS changes are also significant due to glacier melting, such as in Antarctica, Greenland, Canadian Islands, Alaska, Himalayan and Patagonia glaciers. These results indicate that the satellite gravity could well monitor terrestrial water storage changes and their responses to extreme climate events.

For ocean areas, strong seasonal variability in GRACE OBP at both annual and semi-annual periods are found, coinciding well with model ECCO results but the model amplitudes are much weaker. Phase patterns tend to match well at annual and semi-annul period. The secular global OBP variations are ranging from -1.0 to 0.9±0.2 cm/yr. The mean OBP time series in Pacific and Atlantic from GRACE and model ECCO both show similar opposite secular OBP variations, reflecting secular exchange of Pacific and Atlantic water. However, the secular change of OBP from GRACE in the Indian sea is down at larger amplitude, while that from ECCO is a little uplift. It needs to be further investigated using long-term satellite observations and other data in the future. In addition, on a global scale, the monthly OBP time series from GRACE have a stronger high-frequent variability than the ocean general circulation model (ECCO), particularly in tropical regions, but both have shown the similar higher frequency variability in high-latitude, especially in southern high latitude areas.

Some uncertainties at the secular, annual, semi-annual and high frequency periods might be from GRACE instruments noises and data processing strategies. It needs to further improve OBP estimates from GRACE by removing data noise from aliasing or combining other data in the future. With the launch of the next generation of gravity satellite with improving the measurement accuracy, data processing methods and geophysical model, and extending the observation time, it will get more high-precision global terrestrial water storage and global ocean bottom pressure to get more detailed information of global mass transport and distribution.

Acknowledgements

The author thanks the GRACE and ECCO team for providing the data. The GRACE data are available at http://grace.jpl.nasa.gov. This work was supported by the National Keystone Basic Research Program (MOST 973) Sub-Project (Grant No. 2012CB72000), National Natural Science Foundation of China (NSFC) (Grant No.11043008), Main Direction Project of Chinese Academy of Sciences (Grant No.KJCX2-EW-T03), and National Natural Science Foundation of China (NSFC) Project (Grant No. 11173050).

Author details

Shuanggen Jin[1*]

Address all correspondence to: sgjin@shao.ac.cn

1 Shanghai Astronomical Observatory, Chinese Academy of Sciences, China

References

[1] Baker, Robert. M. L. (1960). Orbit determination from range and range-rate data. *The Semi-Annual Meeting of the American Rocket Society, Los Angeles.*

[2] Bingham, R. J., & Hughes, C. W. (2006). Observing seasonal bottom pressure variability in the North Pacific with GRACE. *Geophys. Res. Lett.*, 33, L08607, doi: 10.1029/2005GL025489.

[3] Chao, B., Au, A., Boy, J., & Cox, C. (2003). Time-variable gravity signal of an anomalous redistribution of water mass in the extratropic Pacific during 1998-2002. *Geochemistry Geophysics Geosystems*, 4(11), 1096, doi:10.1029/2003GC000589.

[4] Chambers, D. P., Wahr, J., & Nerem, R. S. (2004). Preliminary observations of global ocean mass variations with GRACE, Geophys. *Res. Lett.*, 31, L13310, doi: 10.1029/2004GL020461.

[5] Cheng, M, & Tapley, B. D. (2004). Variations in the Earth's oblateness during the past 28 years. *J. Geophys. Res.*, 109, B09402, doi:10.1029/2004JB003028.

[6] Chlieh, M., Avouac, J. P., Hjorleifsdottir, V., et al. (2007). Coseismic slip and afterslip of the great (Mw 9.15) Sumatra-Andaman earthquake of 2004. *Bull. Seismol. Soc. Am.*, 97(1A), S 152-S173.

[7] Dickey, J. O., Bentley, C. R., Bilham, R., et al. (1999). Gravity and the hydrosphere: new frontier. *Hydrological Sciences Journal*, 44(3), 407-415.

[8] ESA, Reports for Mission Selection. (1999). Gravity Field and Steady-State Ocean Circulation Mission. SP-1233(1), *ESA Publication Division, ESTEC*, Noordwijk, The Netherlands (available from web site, http://www.esa.int/livingplanet/goce.

[9] Flechtner, F. (2007). AOD1B Product Description Document for Product Releases 01 to 04, GRACE 327-750. *CSR publ. GR-GFZ-AOD-0001 Rev. 3.1, University of Texas at Austin*, 43.

[10] Frappart, F., Calmanta, S., Cauhopéa, M., et al. (2006). Preliminary results of ENVISAT RA-2- derived water levels validation over the Amazon basin. *Remote Sensing of Environment*, 100, 252-264.

[11] Fukumori, I., Lee, T., Menemenlis, D., et al. (2000). A dual assimilation system for satellite altimetry. *paper presented at Joint TOPEX/Poseidon and Jason-1 Science Working Team Meeting, NASA, Miami Beach, Fla.*, 15-17, Nov.

[12] Gill, A., & Niiler, P. (1973). The theory of seasonal variability in the ocean. *Deep-Sea Research*, 20, 141-177.

[13] Han, D., & Wahr, J. (1995). The viscoelastic relaxation of a realistically stratified earth, and a further analysis of post-glacial rebound. *Geophysical J. Int.*, 120, 287-311.

[14] Han, S. C., Shum, C. K., Bevis, M., et al. (2006). Crustal dilatation observed by GRACE after the 2004 Sumatra-Andaman earthquake. *Science*, 313(5787), 658-666, doi:10.1126/science.1128661.

[15] Heiskanen, W. A., & Moritz, H. (1967). Physical Geodesy. *Freeman, San Francisco*.

[16] Hopkin, M. (2005). Triple slip of tectonic plates caused seafloor surge,. *Nature*, 433(3), doi:10.1038/433003b.

[17] Jin, S.G., Chambers, D.P., & Tapley, B.D. (2010). Hydrological and oceanic effects on polar motion from GRACE and models. *J.Geophys. Res.*, 115, B02403, doi: 10.1029/2009JB006635.

[18] Jin, S.G., Zhang, L, & Tapley, B. (2011). The understanding of length-of-day variations from satellite gravity and laser ranging measurements. *Geophys. J. Int.*, 184(2), 651-660, doi:10.1111/j.1365-246X.2010.04869.x, 1365-246.

[19] Leuliette, E, Nerem, R, & Russell, G. (2002). Detecting time variations in gravity associated with climate change. *J. Geophys. Res.*, 107, B62112, doi:10.1029/2001JB000404.

[20] Losch, M., Adcroft, A. J., & Campin, M. (2004). How sensitive are coarse general circulation models to fundamental approximations in the equations of motion? *J. Phys. Oceanogr.*, 34, 306-319.

[21] Luthcke, S. B., Zwally, H. J., Abdalati, W., et al. (2006). Recent Greenland ice mass loss by drainage system from satellite gravity observations,. *Science*, 314, 1286-1289.

[22] Marshall, J., Adcroft, A., Hill, C., et al. (1997). A finite-volume, incompressible, Navier Stokes model for studies of the ocean on parallel computers. *J. Geophys. Res.*, 102, 5753-5766.

[23] Paulson, A, Zhong, S, & Wahr, J. (2007). Inference of mantle viscosity from GRACE and relative sea level data. *Geophys. J. Int.*, doi:10.1111/j.1365-246X.2007.03556.x, 171, 497-508.

[24] Park, J. H., Watts, D. R., Donohue, K. A., et al. (2008). A comparison of in situ bottom pressure array measurements with GRACE estimates in the Kuroshio Extension. *Geophys. Res. Lett.*, 35, L17601, doi:10.1029/2008GL034778.

[25] Pollitz, F. F. (2006). A new class of earthquake observations. *Science*, 313(619), doi: 10.1126/science.1131208.

[26] Ponte, R. M., Quinn, K. J., Wunsch, C., et al. (2007). A comparison of model and GRACE estimates of the large-scale seasonal cycle in ocean bottom pressure. *Geophys. Res. Lett.*, doi:10.1029/2007GL029599, 34, L09603.

[27] Ramillien, G., Cazenave, A., & Brunau, O. (2004). Global time variations of hydrological signals from GRACE satellite gravimetry. *Geophys. J. Int.*, 158(3), 813-826.

[28] Ray, R. D., & Luthcke, S. B. (2006). Tide model errors and GRACE gravimetry: Towards a more realistic assessment. *Geophys. J. Int.*, 167, 1055-1059.

[29] Reigber, Ch., Schwintzer, P., & Luhr, H. (1999). The CHAMP geopotential mission. *Boll. Geof. Teor. Appl.*, 40, 285-289.

[30] Rietbroek, R., Le Grand, P., Wouters, B., et al. (2006). Comparison of in situ bottom pressure data with GRACE gravimetry in the Crozet-Kerguelen region. *Geophys Res Lett.*, 33, L21601.

[31] Rodell, M., Houser, P. R., Jambor, U., et al. (2004). The Global Land Data Assimilation System. *Bull. Amer. Meteor. Soc.*, 85(3), 381-394, doi: BAMS-85-3-381.

[32] Rodell, M., Velicogna, I., & Famiglietti, J. S. (2009). Satellite-based estimates of groundwater depletion in India. *Nature*, 460, 999-1002, doi: 10.1038/nature08238.

[33] Simons, M., & Hager, B. H. (1997). Localization of the gravity field and the signature of glacial rebound. *Nature*, 390(6659), 500-504.

[34] Swenson, S, Chambers, D, & Wahr, J. (2008). Estimating geocenter variations from a combination of GRACE and ocean model output. *J. Geophys. Res.*, 113, B08410, doi: 10.1029/2007JB005338.

[35] Swenson, S, & Wahr, J. (2002). Methods for inferring regional surface-mass anomalies from Gravity Recovery and Climate Experiment (GRACE) measurements of time-variable gravity. *J. Geophys. Res.*, 107, B92193, doi:10.1029/2001JB000576.

[36] Swenson, S C., & Wahr, J. (2006). Post-processing removal of correlated errors in GRACE data. *Geophys. Res. Lett.*, 33(L08402), doi:10.1029/2005GL025285.

[37] Tapley, B. D., Bettadpur, S., Ries, J. C., et al. (2004). GRACE measurements of mass variability in the Earth system. *Science*, 305(568), 503-505.

[38] Tapley, B. D., & Reigber, Ch. (2001). The GRACE Mission: Status and future plans. *Eos Trans, AGU*, 82(47), Fall Meet. Suppl., G41 C-02.

[39] Velicogna, I., & Wahr, J. (2006). Measurements of time-variable gravity show mass loss in Antarctica,. *Science*, 311, 1745-1756, doi:10.1126/science.1123785.

[40] Wahr, J., Molenaar, M., & Bryan, F. (1998). Time-variability of the Earth's gravity field: Hydrological and oceanic effects and their possible detection using GRACE. *J. Geophys. Res.*, 103(32), 205-30.

[41] Washington, W. M., Weatherly, J. W., Meehl, G. A., et al. (2000). Parallel climate model (PCM) control and transient simulations. *Climate Dynamics*, 16, 755-774.

[42] Watkins, M., & Bettadpur, S. (2000). The GRACE mission: challenges of using micron-level satellite-to-satellite ranging to measure the Earth's gravity field. *Proc. of the International Symposium on Space, Dynamics, Biarritz, France, Center National d'Etudes Spatiales (CNES), Delegation a la Communication (pub1.)*.

[43] Wolff, M. (1969). Direct Measurement of the Earth's Gravitational Potential Using a Satellie Pair. *J. Geophys. Res.*, 74(22), 5295-5300, doi:10.1029/JB074i022p05295.

[44] Yoder, C. F., Williams, J. G., Dickey, J. O., et al. (1983). Secular variation of Earth's gravitational harmonic J coefficient from Lageos and non-tidal acceleration of Earth rotation. *Nature*, 303, 757-762.

Geodetic Atmosphere

Chapter 5

Ionospheric Sounding and Tomography by GNSS

Vyacheslav Kunitsyn, Elena Andreeva,
Ivan Nesterov and Artem Padokhin

Additional information is available at the end of the chapter

1. Introduction

Studies of the ionosphere and the physics of the ionospheric processes rely on the knowledge of spatial distribution of the ionospheric plasma. Being the propagation medium for radio waves, the ionosphere significantly affects the performance of various navigation, location, and communication systems. Therefore, investigation into the structure of the ionosphere is of interest for many practical applications. Existing satellite navigation systems with corresponding ground receiving networks are suitable for sounding the ionosphere along different directions, and processing the data by tomographic methods, i.e. reconstructing the spatial distribution of the ionospheric electron density.

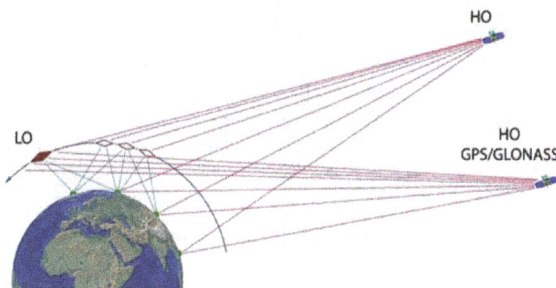

Figure 1. The geometry of the satellite radio sounding the near-Earth environment

In this Chapter, radio tomography (RT) methods based on low-Earth-orbiting (LO) and high-Earth-orbiting (HO) navigational satellites as well as radio occultation (RO) methods that use the data of quasi tangential sounding are considered. The "old" LO navigational systems (American Transit and Russian Tsikada) allow the receivers to determine their locations everywhere on the Earth's surface, but not continuously. The time gap between the neighboring positioning determinations depends on the number of operational satellites in orbit. The "new" HO systems (GPS/GLONASS) are suitable for continuous worldwide positioning measurements. As far as spatial coverage is concerned, all satellite navigation systems are global. They are further referred to here as the Global Navigation Satellite Systems (GNSS). Note that at present, the term GNSS is mainly applied to HO navigation systems (the American GPS and Russian GLONASS systems which are currently operational as well as European Galileo, Chinese BeiDou systems and Japanese QZSS).

Figure 1 depicts satellite radio probing of the near-Earth's environment that includes the atmosphere, the ionosphere, and the protonosphere. Transmitters onboard the LO and HO satellites and the ground receivers provide the sets of rays intersecting the earth's near-space and allow determining the group and phase paths of the radio signals (in the case of LO systems, only the phase paths) along the corresponding rays. The receivers onboard the LO satellites that receive the radio transmissions from the HO satellites are also suitable for determining the group and phase paths of the signal along the set of the rays that are quasi-tangential to the Earth's surface. These measurements are suitable for sounding the near-space environment along various directions and calculating the integrals (or the differences of the integrals) of the refraction index in the medium. This set of integrals can be inverted by the RT procedure for the parameters of the medium. In the case of ionospheric sounding, the integrals of the refraction index are reduced to the integrals of the ionospheric electron density.

2. The methods of GNSS-based radio tomography

Methods of the satellite ionospheric radio tomography are being successfully developed at present [1-8]. Since the early 1990s, RT methods based on the LO navigation systems have been operational. In recent years, RT studies based on the measurements using HO navigation systems have been extensively conducted [6-8]. Further in the text, various types of radio tomography are referred to as low-orbital RT and high-orbital RT (LORT and HORT).

2.1. Low orbital ionospheric radio tomography

Present-day navigational systems are based on the low orbiting satellites flying in near-circular orbits at an altitude of about 1000-1150 km. These utilize chains of ground-based receivers, which capture RT data along different rays. In RT experiments, the phase difference between two coherent signals transmitted from the satellite at the frequencies of 150 and 400 MHz is recorded at a set of receiving stations on the ground. The receivers are arranged in a chain parallel to the ground projection of the satellite track, the distance between the neighboring receivers being typically a few hundred kilometers. The reduced phases ϕ recorded at the receiving sites are the input data for the RT imaging. The integrals of the electron density N

along the rays linking the ground receivers with the onboard satellite transmitter are proportional to the absolute (total) phase Φ [1, 2], which includes the unknown initial phase ϕ_0:

$$\alpha \lambda r_e \int N d\sigma = \Phi = \phi_0 + \phi \tag{1}$$

Here, λ is the wavelength of the satellite radio signal, $d\sigma$ is the element of the ray, and r_e is the
classical electron radius. The scaling coefficient α (of the order of unity) depends on the
sounding frequencies used. Equation (1) can be recast in the operator form [4] that includes
the typical uncorrelated measurement noise ξ:

$$PN = \Phi + \xi \tag{2}$$

where P is the projection operator mapping the two-dimensional (2D) distribution N to the set
of one-dimensional (1D) projections Φ. Thus, the problem of tomographic inversion is reduced
to the solution of the linear integral equations (2) for the electron concentration N. One of the
probable ways to solve (2) is to discretize (approximate) the projection operator P. This yields
the corresponding system of linear equations (SLE) with the discrete operator L:

$$LN = \Phi + \xi + E, \ E = LN - PN \tag{3}$$

where E is the approximation error that depends on the solution N itself. Note that equations
(2) and (3) are equivalent if the approximation error E is known. However, in the case of
reconstructing the data of a real RT experiment, E is not known, and, in fact, quite a different
SLE is actually solved:

$$LN = \Phi + \xi \tag{4}$$

The system (4) is not equivalent to SLE (3). In other words, the difference between the solutions
of (3) and (4) ensues from the difference in both the quasi-noise component and the correlated
(in time and rays) approximation error E. For SLE (4) to be solved, the absolute phase Φ
together with ϕ_0 should be known. The errors in ϕ_0 estimated by the different receivers can
result in the contradictory and inconsistent data, which leads to low-quality RT reconstructions. In order to avoid this difficulty, a method of phase-difference radio tomography (RT
based on the difference of the linear integrals along the neighboring rays) was developed [9],
which does not require the initial phase ϕ_0 to be determined. The SLE of the phase-difference
RT is determined by the corresponding difference:

$$AN = LN - L'N = \Phi - \Phi' = D + \xi \tag{5}$$

where $LN = \Phi$ is the initial SLE and $L'N = \Phi'$ is the system of linear equations along the set of neighboring rays.

There are numerous algorithms, both direct and iterative, that solve SLEs (4) and (5). At present, in the problems of ray radio tomography of the ionosphere, iterative algorithms are most popular, although non-iterative algorithms are also used. These algorithms utilize a singular value decomposition with its modifications, regularization of the root mean square (RMS) deviation, orthogonal decomposition, maximum entropy, quadratic programming, Bayesian approach, etc. [3-7]. Extensive numerical modeling and LORT imaging of numerous experimental data revealed the efficient combinations of various methods and the algorithms that yield the best reconstructions.

"Phase-difference" LORT provide much better results and higher sensitivity compared to "phase only" methods. This is confirmed by reconstructions of the experimental data as well [4, 7]. The horizontal and vertical resolution of LORT in its linear formulation is 20-30 km and 30-40 km, respectively. If the refraction of the rays is taken into account, the spatial resolution of LORT can be improved to 10-20 km [7].

2.2. High orbital ionospheric radio tomography

Deployment of the global navigational systems (GPS and GLONASS) in USA and Russia offers the possibility to continuously measure trans-ionospheric radio signals and solve the inverse problem of radio sounding [6-8]. In the near future, there are plans to launch the European Galileo and Chinese BeiDou satellite systems. Signals of the present-day GPS/GLONASS are continuously recorded at the regional and global receiving networks (e.g., the network operated by the International GNSS Service, IGS, which comprises about two thousand receivers). These data are suitable for reconstruction of the ionospheric electron density, the total electron content (TEC).

Inverse problems of radio sounding based on the GPS/GLONASS data, which pertain to the tomographic problems with incomplete data, are inherently high-dimensional. Due to the relatively low angular velocity of the high-orbiting satellites, allowance for the temporal variations of the ionosphere becomes essential. This makes the RT problem four-dimensional (three spatial coordinates and time) and exacerbates incompleteness of the data: every point in space is not necessarily traversed by the rays that link the satellites and the receivers, therefore the data gaps arise in the regions where only few receivers are available. The solution of this problem requires special approaches [10].

The methods of ionospheric sounding typically analyze the phases of the radio signals that propagate from the satellite to the ground receiver at two coherent multiple frequencies. For example, in the GPS-based soundings, these frequencies are $f_1 = 1575.42$ MHz and $f_2 = 1227.60$ MHz. The corresponding data (L_1 and L_2) are the phase paths of the radio signals measured in the units of the wavelengths of the sounding signal. Another parameter that can be used in the analysis is the pseudo-ranges (the group paths of the signals), which is the time taken by the wave-trains at the frequencies f_1 and f_2 to propagate from the source to the receiver. The phase delays L_1 and L_2 are proportional to the total electron content, TEC, the integral of electron density along the ray between the satellite and the receiver:

$$TEC = \left(\frac{L_1}{f_1} - \frac{L_2}{f_2} \right) \frac{f_1^2 f_2^2}{f_1^2 - f_2^2} \frac{c}{K} + const,$$
(6)

where $K = 40.308$ m^{3}s^{-2} and $c = 3 \cdot 10^8$ m/s is the speed of light in vacuum. Note that, by using the phase delay data, it is only possible to calculate the TEC value up to a certain constant indicated as the additive term in formula (6). The relationship (6) is similar to formula (1) with the unknown constant in the right-hand side of the system.

TEC values can also be derived from the pseudo-ranges P_1 and P_2 [11]:

$$TEC = \frac{P_2 - P_1}{K \left(\frac{1}{f_2^2} - \frac{1}{f_1^2} \right)}$$
(7)

However, compared to phase data, the pseudo-range data are strongly distorted and contaminated by noise. The noise level in P_1 and P_2 is typically 20-30% and even higher, while in the phase data it is below 1% and rarely reaches a few percent. Therefore, for HORT, the phase data are preferable.

Most authors [6] solve the HORT problem using a set of linear integrals. In that approach, it is assumed that the TEC data are sufficiently accurately determined from the phase and group delay data (6, 7). However, the absolute TEC (7) is determined with a large uncertainty in contrast to the TEC differences that are calculated highly accurately. Therefore, the phase-difference approach was applied in this case, too [10, 12]. In other words, instead of the absolute TEC, its corresponding differences or the time derivatives dTEC/dt were used as input data for the RT problem.

The problem of the 4-D GNSS-based radio tomography can be solved by the approach developed in 2-D LORT. In this approach, the electron density distribution is represented in terms of a series expansion of the certain local basis functions; in this case, the set of the linear integrals or their differences is transformed into SLE. However, in contrast to 2-D LORT, here it is necessary to introduce an additional procedure interpolating the solutions in the area with missing data. The implementation of this approach in the regions covered by dense receiving networks (e.g., North America and Europe) with a rather coarse calculation grid and suitable splines of varying smoothness [10,12] has proved highly efficient.

Another approach seeks sufficiently smooth solutions of the problem so that the algorithms provide a good interpolation in the area with missing data. For example, let us consider a Sobolev norm and seek a solution that minimizes this norm over the infinite set of solutions of the initial (underdetermined) tomographic problem (5):

$$AN = D, \quad \min_{AN=D} \|f - f_0\|_{W_n^2} \tag{8}$$

Here, function f is the solution with a given weight .

Practical implementation of this approach faces difficulties associated with solution of the constrained minimization problem. The direct approach utilizing the method of Lagrange's undetermined multipliers gives SLE with high-dimensional (due to the great number of rays) matrices, which do not possess any special structure that would simplify the solution. Therefore, we solve this minimization problem by an iterative method [10] that is a version of the SIRT technique, with additional smoothing (by filtering) of iterative increments over the spatial variables. This method allows for use of a-priori information that can be introduced both through the initial approximation for the iterations and through weighting coefficients that determine the relative intensity of electron density variations at different heights.

Computer-aided modeling shows that quasi-stationary ionospheric structures can be reconstructed with reasonable accuracy, although HORT has a significantly lower resolution than LORT. As a rule, the vertical and horizontal resolution of HORT is 100 km at best, and the time step (the interval between two consecutive reconstructions) is typically 20 - 60 minutes. In regions covered by dense receiving networks (Europe, USA, and Alaska), the resolution can be improved to 30-50 km with a 10 - 30 minute interval between consecutive reconstructions. Resolution of 10-30 km with a time step of 2 minutes can only be achieved in the regions with very dense receiving networks (California and Japan).

3. Testing and validation of ionospheric radio tomography

In numerous experiments, RT images of the ionosphere have been compared with corresponding parameters (vertical profiles of electron density and critical frequences) measured by ionosondes [13-18, 4, 19, 7]. In most cases, the RT results closely agree with ionosonde data within the limitations of the accuracy of both methods. An example of such a comparison with the world's first RT reconstruction of an ionospheric trough is presented in Figure 2. Here, the dots show the vertical profile of electron density according to measurements by an ionosonde in Moscow, and the solid line displays the corresponding profile calculated from the RT reconstruction for April 7, 1990 (22:05 LT).

A comparison of a few hundred ionospheric RT cross-sections in the region of the equatorial anomaly with observations by two ionosondes in October-November 1994 [19] is illustrated in Figure 3. The distributions of electron density were reconstructed from RT measurements at the low-latitude Manila-Shanghai chain, which included six receivers arranged along the meridian 121.1±1ºE within the latitude band between 14.6ºN and 31.3ºN. One ionosonde was installed at 25.0ºN, 121.2ºE 7.5 km of Chungli almost in the middle of the chain. Another ionospheric station was located in the southernmost part of the chain in Manila (14.7ºN, 121.1ºE).

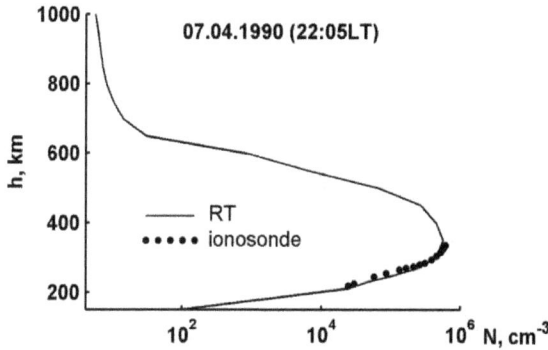

Figure 2. Vertical profiles of electron density in Moscow at 22:05 LT on April 7, 1990, depicting both radio tomography and ionosonde data.

Using RT reconstructions, maximal electron densities and plasma frequencies f_0F2 were calculated in the vicinities of the ionospheric stations. These parameters were then compared to the corresponding values determined from ionosonde measurements. These two data sets are compared in Figure 3. The points that lie on the bisectix of the right angle correspond to the case where the RT-based and the ionosonde-based critical frequencies exactly coincide. We also calculated the normalized root mean square (rms) deviations of the RT-based critical frequencies f_0F2 from the corresponding values inferred from the ionosonde measurements.

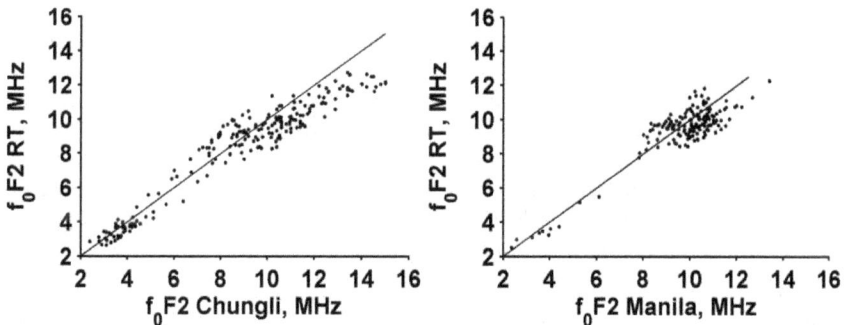

Figure 3. The comparison between the plasma frequencies f_0F2 calculated from the RT reconstructions and the corresponding values derived from the measurements by the ionosondes in (left) Chungli and (right) Manila for October-November 1994.

A detailed analysis reveals the following points.

1. The scatter in f_oF2 in Chungli is larger than in Manila. The normalized rms deviations for the ionosondes in Chungli and Manila are 11.2% and 8.9%, respectively.

2. In the case of high electron concentrations, especially for f_oF2 above 13 MHz, the experimental points tend to saturation: the critical frequencies f_oF2 derived from RT fall short of the corresponding values calculated from the ionosonde measurements.

These features indicate that strong spatial gradients in electron density typical in the region of the equatorial anomaly can cause the discrepancies in the plasma frequencies calculated from RT and ionosonde measurements.

In experiments on vertical pulsed sounding of the ionosphere, the signal is not reflected from directly overhead. Even in the case of vertical sounding of a horizontally stratified ionosphere, the ordinary wave tends to deviate toward the pole, and in the point of reflection, it becomes perpendicular to the local geomagnetic field [20]. Therefore, in the general case, reflection does not occur vertically above the sounding point but somewhat away from overhead.

Zero offsets are only observed at the equator, while in the region of the Chungli ionosonde, the offset can be ~10 km. In addition, Chungli is located close to the maximum of the equatorial anomaly, which is marked by very strong gradients. Therefore, the sounding ray of the Chungli ionosonde will significantly deflect before having been reflected backwards. Therefore, the values of f_oF2 recorded by the Chungli ionosonde will by no means be the actual f_oF2 values exactly overhead. These considerations will help us to interpret the results of comparison of plasma frequencies.

First, at high plasma frequencies (f_oF2 higher than 13 MHz), the experimental points in Figure 3 fall below the bisectrix. This area relates to the stage of a completely mature anomaly with a well-developed crest and, therefore, with strong gradients in electron density. It is due to these gradients that the values of f_oF2 determined from the ionosondes are, on average, overestimated compared to the actual critical frequencies f_oF2 overhead.

Second, values of f_oF2 at Chungli demonstrate a larger scatter than at Manila. Since Chungli is located in the central part of the RT chain, the most reliable RT reconstructions are expected at the latitudes near the middle segment of the chain (close to Chungli) rather than on its margins (Manila). This contradicts the actual results shown in Figure 3. On the other hand, Chungli is located in the vicinity of the peak electron density within the crest of the anomaly, in the area of strong gradients, where errors of vertical sounding (associated with deflection of the reflected ray) are most probable. Therefore, large discrepancies in the Chungli region are quite probable, which is confirmed by Figure 3.

It is worth noting that the ionosonde measurements during geomagnetically disturbed periods are often unstable because the ionosphere experiences significant transformations that alter the radio propagation conditions. In particular, the electron density N in the D-region ionosphere sharply increases, and, due to strong radio absorption, most ionograms do not show any reflections. The examples in Figures 4 and 5 illustrate a comparison of the critical frequencies calculated from the phase-difference RT reconstructions above the Cordova-Gakona-Delta chain in Alaska with those derived from ionosonde measurements in Gakona during the geomagnetic storms in October 2003 (Figure 4) and June 2004 (Figure 5). The interval from October 29 to 31, 2003 was marked by the strongest recorded geomagnetic storm. During the main phase of the storm, the 3-h Kp index attained its maximum possible value (9). Figure 4 shows that the ionosonde data are fragmentary and, starting from October 28, 2003 (13:00UT) through October 31, 2003 (19:30 UT), i.e., exactly during the peak of the geomagnetic storm, they are missing altogether. The ionosonde observations at Gakona are discontinuous during the geomagnetically disturbed period in July 2004 (Kp = 8 and Kp = 7 on July 25 and July 27, respectively). On some days, the reflections are almost absent (e.g., from July 23 through July 27). In contrast to the ionosondes, which are essentially HF radars, RT methods continue to be suitable for imaging the ionosphere even during strongly disturbed solar and geophysical conditions, because the high sounding frequency used in RT applications (150 MHz) allows one to neglect the absorption.

Comparison of the RT results with ionosonde data can only be implemented in terms of critical frequencies or vertical profiles of electron density up to the ionospheric peak height and only for points located close to the ionosondes. In other words, the single-point ionosonde measurements cannot be inverted into the two-dimensional (2D) ionospheric cross sections. Therefore, it is of particular interest to compare RT images with measurements by incoherent scatter (IS) radars, which are also suitable for reconstructing 2-D cross-sections of the ionosphere over an interval of a few hundred kilometers.

Pryse and Kersley [21] made a preliminary attempt to compare the ionospheric cross-section derived from measurements by two receivers with EISCAT IS radar data in Tromso (Norway). The RT reconstructions coarsely reproduced the horizontal gradient in electron density determined from the EISCAT. Since that work, the results of these two independent methods of the ionospheric research, namely, RT and IS radar measurements, have been intercompared for numerous RT experiments [22-25, 15, 26-28].

Below, we present the results of the Russian-American Tomography Experiment (RATE'93) carried out in 1993. RATE'93 was one of the first experiments in which RT images of the ionosphere were intercompared with IS data measured by the Millstone Hill radar (USA). The idea of this experiment was suggested by the American geophysicists from the Haystack Observatory and Philips Laboratory. The Russian team included scientists from the Polar Geophysical Institute and Moscow State University [24]. Four mid-latitude RT receiving sites were arranged along the 288ºE meridian in the northeastern USA and in Canada (Block Island (41.17ºN), Nashua, NH (42.47ºN), Jay, VT (44.93ºN), and Roberval, Canada (48.42ºN)).

Each receiving site was equipped with the Russian receivers, which recorded the signals of the Russian navigational satellites like Tsikada, and the receivers designed at the Philips Laboratory, which measured radio transmissions from the Transit satellites of the Navy Navigation Satellite System (NNSS). The IS radar at the Haystack observatory in Millstone Hill (42.6ºN, 288.5ºE) scanned the ionosphere in the latitude-height plane in a coordinated mode with RT observations. The Russian team reconstructed the distributions of the ionospheric electron density using the "phase-difference" approach while the American geophysicists used the "phase" RT method.

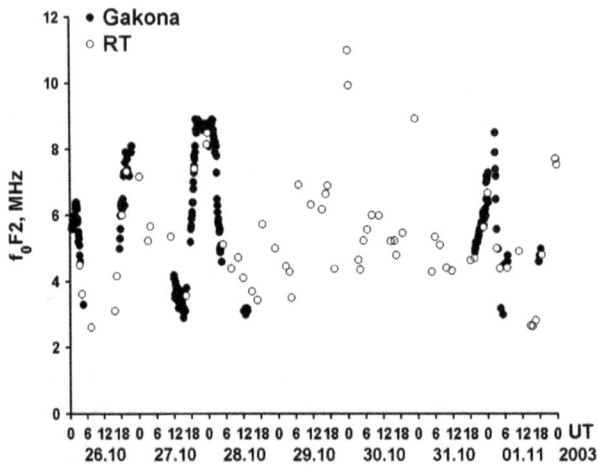

Figure 4. The comparison of the critical frequencies f_0F2 derived from the RT reconstructions and from the ionosonde measurements at Gakona during severe solar-geomagnetic disturbances from October 26 to November 1, 2003.

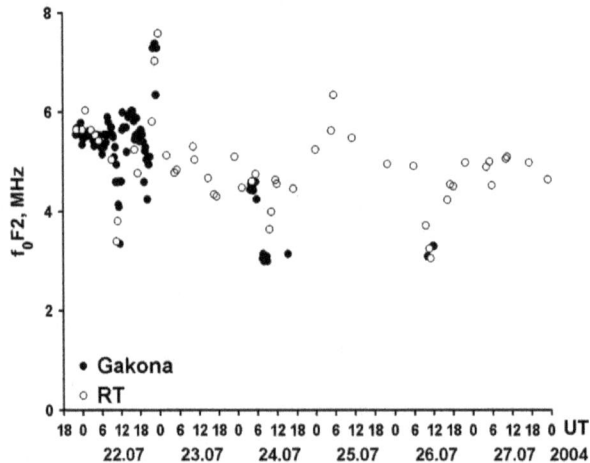

Figure 5. The comparison of the critical frequencies f_0F2 derived from the RT reconstructions and from the ionosonde measurements at Gakona during the geomagnetic and ionospheric disturbances from July 22 to July 27, 2004.

In this experiment, not only were the RT results compared with the IS radar data, but also the two different RT approaches (phase-difference and phase techniques) were assessed. The time of the experiment was chosen to coincide with expected solar activity [29]. A strong geomagnetic storm occurred between November 3 and 4, 1993. During this storm, the Ap index of planetary geomagnetic activity reached 111 nT, and during the main phase of the storm, the 3-h Kp index was 6.7.

The results of RATE'93 demonstrated the high quality of the RT ionospheric cross-sections reconstructed by the phase difference method [24]. It should be noted that RT cross-sections and radar ionospheric images are clearly similar in the case of a smooth, almost regular ionosphere with insignificant local extrema. Further in the text, the results are visualized in the coordinates "height h above the Earth's surface (in km) - latitude." Figure 6 presents the cross-sections of the ionosphere about 1 hour 45 minutes after the sudden storm commencement (at 23:00 UT on November 3, 1993). The ionospheric cross-section based on the IS radar data is displayed in Figure 6 (upper panel), and the phase-difference RT reconstruction is shown in Figure 6 (lower panel).

A characteristic trough appeared about 44°N, and the ionozation sharply increased in the height interval from 200 km to 300 km near 47°N, due to precipitation of low-energy particles between 46°N and 51°N [29]. Figure 6 shows that the RT cross-section closely agrees with the radar ionospheric image. However, it should be noted that radar measurements are limited to the height interval from 180 km (below which the concentrations are insignificant) to 600 km (above which the distortions and noise level are very high) [29].

The similarity of the RT cross-sections and radar ionospheric images (Figure 6) confirms that the increase in the radar signal in the bottom F-region ionosphere reflects the actual enhancement of electron density substantially below 300 km, but not the noise or coherent backscatter from the irregularities associated with the E-region electric fields [24, 29]. Moreover, both ionospheric cross-sections in Figure 6 clearly demonstrate the elevated F-region south of 45°N. The elevation increases with increasing distance to the trough. The elevation in the RT reconstruction attains 400-450 km and is larger than in the radar image.

Figure 6 (upper panel) shows the preliminary radar-based ionospheric cross-section as published in [24]. The difference between the reconstructions is 32.6%, which falls within the accuracy of both methods. However, it should be noted that the ionospheric cross-sections based on radar data and those reconstructed from the RT measurements do not correspond to the same time interval but are somewhat spaced in time. The time shift between them is 5 minutes, which is quite a significant value considering that the measurements were conducted during the period of active structural rearrangement of the ionosphere on November 4, 1993.

In their later paper, Foster and Rich [29] quote the final radar cross section that was reconstructed after scrupulous analysis of the measurements. On that image, the bottom of the ionospheric F-layer is obseved at the same height as in the RT image shown in Figure 6 (lower panel), i.e., at 400-450 km. According to the radar data, the F-layer remained elevated for a short lapse of time (~20 minutes).

In this experiment, a narrow (<2°) tilted trough was detected by both RT and radar observations at a latitude of 41°-42° at 04:56UT on November 4, 1993. Phase-difference tomography also revealed a border about 50 km in size on the northern wall of the trough, which has not been distinguished by the radar observations [24]. This means that the phase-difference RT method has higher horizontal resolution. The border clearly manifests itself in the phase of the signal recorded at the Nashua receiving site (Figure 7), where it produces a local maximum at about 41.5°-42°N. In contrast to the phase-difference RT, all the ionospheric images reconstructed by the phase RT are similar to each other and close to the PIM model [30]. Due to errors in the determination of the initial phases, the phase RT method does not even reveal the troughs themselves. The comparison shows that the cross-sections of electron density reconstructed by the phase-difference RT and the ionospheric images derived from the IS radar data agree within the accuracy of both methods. Moreover, compared to the IS method, phase-difference RT has a higher horizontal resolution. However, the IS radar revealed thin (<50 km) extended horizontal layers which are not resolved by radio tomography due to the insufficient base of the RT measurements (the distance between the outermost receivers in the experiment was only 800 km) and a quasi-tangential (to the Earth's surface) rays were absent.

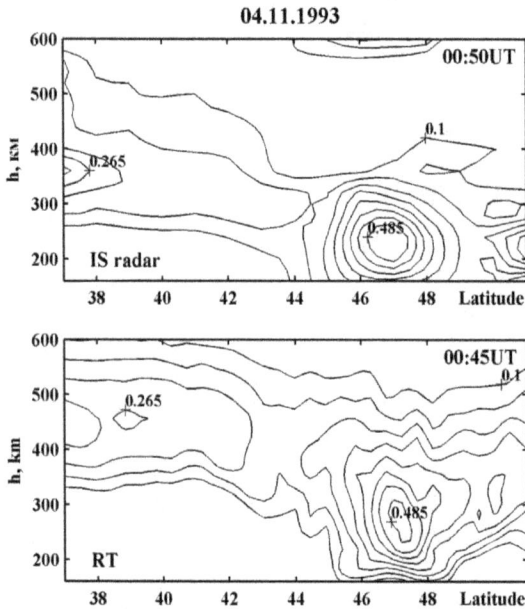

Figure 6. Cross sections of the ionosphere after sudden commencement of the geomagnetic storm (00:45 UT on November 4, 1993): (a) according to the radar data; (b) according to phase-difference RT.

This experiment also demonstrated that the method of phase-difference RT has noticeable advantages (in sensitivity and quality of the reconstructions) over the phase radio tomography which uses linear integrals. A comprehensive intercomparison between the results of phase tomography, phase-difference tomography, and IS radar studies is presented in the final summarizing paper [24].

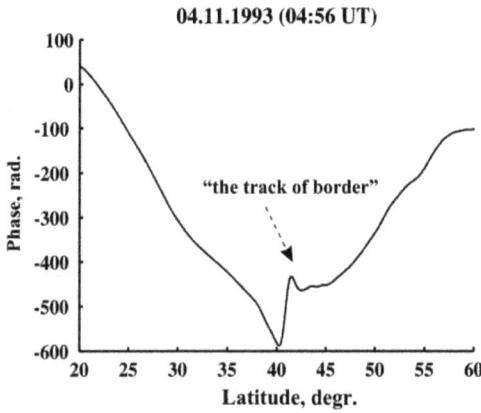

Figure 7. Phase of the satellite radio signal recorded at Nashua at 04:56 UT on November 4, 1993

Numerous comparisons have been carried out between the HORT results and the ionospheric measurements in Europe, Russia, America, and Southeast Asia. The diurnal behavior of critical frequencies derived from the HORT data quite closely agrees with the ionosonde data. Figures 8-9 present examples of a comparison of HORT-based critical frequencies with ionosonde measurements at Rome, Tromso, and Boulder during the disturbed period on March 9, 2012, when the Kp index reached 7. A similar comparison for Yamagawa is presented for the mega-earthquake in Japan. During geomagnetically quiet periods, discrepancies between the critical frequencies are typically far below 1 MHz.

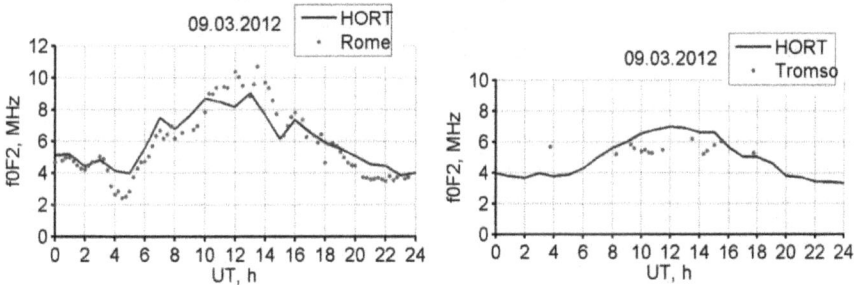

Figure 8. The comparison the HORT-based critical frequencies with the ionosonde measurements at Rome (left) and Tromso (right)

During severe geomagnetic disturbances, the RMS deviation of critical frequencies is about 1 MHz. These comparisons were carried out using data containing a few thousand samples. For example, for the interval of the strong Halloween storm from October 29 to November 1,

2003, with measurements from 13 North American ionosondes, the RMS deviation of the critical frequencies was 1.7 MHz [31].

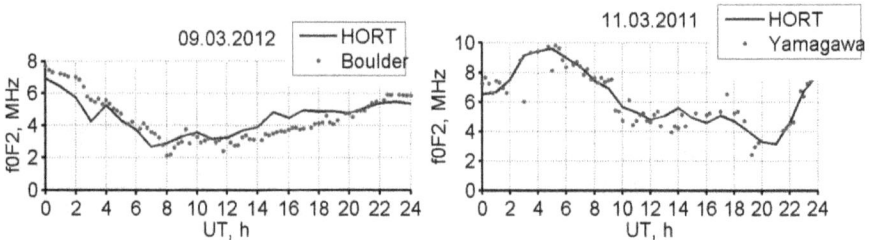

Figure 9. Comparison of HORT-based critical frequencies with ionosonde measurements at Boulder (left) and Yamagawa (right)

4. Examples of the GNSS-based ionospheric radio tomography

The world's first LORT images were reconstructed in March-April 1990 by geophysicists of the Moscow State University and Polar Geophysical Institute of the Russian Academy of Sciences [32]. One of the first RT cross sections of the ionosphere between Moscow and Murmansk is presented in Figure 10. The horizontal axis in this plot shows the latitudes, and the vertical axis, the heights. The ionospheric electron density is given in units of 10^{12} m^3. This image clearly shows a proto-typical ionospheric trough at about 63°--65°N and a local extremum within it. Further experiments revealed the complex and diverse structure and dynamics of the ionospheric trough [1-8]. In 1992, preliminary results in RT imaging of the ionosphere were obtained by colleagues from the UK [21]. The LORT-based studies and applications drew significant interest from geophysicists around the globe. At the present time, more than ten research teams in different countries are engaged in these investigations [3-7]. Series of LORT experiments carried out in Europe, America, and Southeast Asia during the last twenty years [3-8] have demonstrated the high efficiency of radio tomographic methods for study of diverse ionospheric structures.

Quite often, RT images of the ionosphere in different regions reveal the well-known wave structures - traveling ionospheric disturbances (TID). The example in Figure 11 depicts an RT ionospheric image with distinct TIDs having a typical slope of 45°, as recorded on the Moscow-Arkhangelsk RT link [33]. Here, the ionosphere is quite moderately modulated by the TIDs (the modulation depth is 25-30%). Such TIDs are often observed in RT reconstructions, as mentioned in [5, 34].

RT experiments in Southeast Asia at the low-latitude Manila-Shanghai chain revealed a series of characteristic features in the structure of the equatorial anomaly (EA) including the post-noon alignment of the mature core of the EA (the area close to the peak electron concentration) along the magnetic field axis; the significant asymmetry of the EA flanks, and the specific alteration of the thicker and thinner segments of the ionospheric F-layer [35, 19, 36]. The stable structural features of the EA observed in the RT experiments can be interpreted by analyzing ionospheric plasma flows and velocities in the region of the EA, which are caused by the fountain effect [7, 20, 35].

Imaging the E-region ionosphere by LORT methods is a far more difficult problem, because the contribution of the E-layer in the RT input is noticeably smaller than the contribution of the F-layer. However, if the size and geometry of the RT observation system are favorable for constructing a set of the rays that intersect the ionospheric F- and E-layers, it is possible to image the distribution of electron density in the E-layer [37]. An example of the LORT reconstruction of the F- and E-layers in the EA region is presented in Figure 12. The geomagnetic field lines are shown by the dashes. The mature core of the anomaly is oriented along the geomagnetic field; the edges of the EA crest are clearly asymmetric, and the thickness of the F-layer experiences distinct variations. The bottom side F-layer sags down due to the field-aligned plasma stream, i.e. the ionospheric plasma from the F-region penetrates down into the lower part of the E-layer in the interval of 24º-26ºN. A constriction ("isthmus") is formed in the area beyond the core of the anomaly (~ 28º-31º).

Today, more than a dozen LORT receiving networks (chains) are currently operational in different regions of the world (in Russia, contiguous United States and Alaska, Great Britain, Scandinavia and Finland, Greenland, Japan, and the Caribbean region [2-8]), which are extensively used for research and scientific studies. A new LORT system has been built in India, and the LORT system in Southeast Asia is being upgraded. The Russian transcontinental LORT chain includes nine receiving sites arranged along the Sochi-Moscow-Svalbard line. It is the world's longest LORT chain (about 4000 km in length) [38]. This chain is unique in the fact that its measurements cover a huge ionospheric sector stretching from the polar cap and auroral regions to the low latitudes. Due to this, measurements on this chain are suitable for studying the transfer of the disturbances between the auroral, subauroral, and low-latitude ionosphere, and for investigating the structure of the ionospheric plasma in different latitudinal regions as a function of solar, geophysical, and seasonal conditions. The Russian RT chain and the receiving RT chain in Alaska are located on the opposite sides of the Earth with a 12-hour time shift between them.

Various waves and wavelike structures are another specific feature often observed in LORT reconstructions. Figure 13 presents an example of a complex wave-like perturbation with a distinctive wave-front which was observed during the Halloween storm of 2003 in the Alaska region. A similar structure was also detected above the Russian LORT system (Moscow-Svalbard) during the geomagnetic storm in July, 2004 [7]. It is worth noting that the ionospheric plasma can be highly complex in its structure even in undisturbed conditions. This is illustrated by Figure 14 which shows the LORT cross section of the ionosphere between Sochi and Svalbard during geomagnetically quiet conditions (Kp < 1). Wavelike disturbances with a characteristic size of 50 km are seen above Svalbard (78º-79ºN). In the central segment of the image (59º-65ºN) the electron density decreases. In the southern part of the cross-section (42º-55ºN), wavelike structures with a spatial period of 100-150 km are apparent. A wide ionization trough in the interval of 62º-64ºN is observed on the LORT reconstruction in Figure 15. The local maximum at 65º-66ºN is almost merged with the polar wall of the trough. A spot of enhanced ionization is identified within the trough about 63º-64ºN latitude. And, wavelike disturbances are revealed throughout 66º-78ºN.

Besides being suitable for reconstructing the large-scale ionospheric phenomena of natural origin, LORT is also efficient for tracking artificial ionospheric disturbances. The LORT cross-section in Figure 16 shows wavelike structures that formed in the ionosphere within 30 minutes after launch of a rocket from the Plesetsk Cosmodrome. The cosmodrome is located approximately 63ºN (200 km) distant from the satellite ground track. These anthropomorphic disturbances have a very complex structure wherein large irregularities (200-400 km) coexist with smaller ionospheric features (50-70 km), and the slope of the "wavefront" is also varying. Wave disturbances generated by launching high-power rocket vehicles are described in [39] where it is shown that ignition of the rocket generates acoustic-gravity waves (AGW) which, in turn, induce corresponding perturbations in electron density. During RT experiments wtih the Moscow-Murmansk chain, long-lived local disturbances in the ionospheric plasma were also identified above sites where ground industrial explosions were carried out [40].

RT methods revealed generation of ionospheric disturbances by the Sura ionospheric heating facility, which radiated high-power HF waves, modulated with a 10-minute heat/off cycles [41]. Figure 17 shows an ionospheric cross-section through a typical heated area. A narrow trough in the ionization, aligned with the propagation direction of the heating HF wave, is identified. Traveling ionospheric disturbances associated with the acoustic gravity waves (AGWs) generated by the Sura heater are observed diverging from the heated area. Unfortunately, insufficient density of HO receivers in central Russia prevented us from obtaining high-quality HORT images of the ionosphere during this heating experiment; however, the data recorded by the few available receivers support presence of the AGWs [41].

Thus, LORT is capable of reconstructing nearly instantaneous 2-D snapshots of the electron density distribution in the ionosphere (which actually cover a time span of 5-15 minutes). The time interval between the successive RT reconstructions depends on the number of the operational satellites and, as of now is 30-120 minutes. The LORT method is also suitable for determining plasma flows by analyzing successive RT cross sections of the ionosphere [42]. An optimal LORT receiving system, consisting of several parallel chains located within a few hundred km of each other, would allow 3-D imaging of the ionosphere. The requirement for multiple receiving chains is the major limitation of LORT.

The reconstructions presented below illustrate the possibilities of newly developed HORT techniques. Figure 18 displays the evolution of the ionospheric trough above Europe in the evening on April 17, 2003. The TEC maps and the meridional cross sections along 21ºE show the trough widening against the background, with an overall nighttime decrease in electron density.

Figure 19 shows anomalous increases in electron density (up to 3 10^{12} m^{-3}) above the Arctic during the severe Halloween geomagnetic storm on October 29-31, 2003. The spots of increased electron density in the night sector are associated with the ionospheric plasma entrained by the anti-sunward convection stream from the dayside ionosphere. These areas with increased electron density are shaped as tongues with a patchy structure (upper panels), which can also be seen on the vertical cross sections (lower panels). The cross-sections cut the ionosphere along the lines indicated on the TEC maps. This spatial distribution of ionospheric plasma is a result of ionospheric plasma instabilities and the formation of wavelike disturbances. An example of imaging the Arctic ionosphere on December 16, 2006 is presented in Figure 20. Here, a characteristic ring-shaped structure encircles the pole, which is associated with convection and entrainment of ionospheric plasma from the dayside ionosphere into the night-side sector. Similar ring structures were observed in modeling of the Arctic ionosphere [43, 44].

Data recorded at the European GPS receiving network were used for imaging the ionosphere above Western Europe during the strong geomagnetic storm on October 28-31, 2003. The vertical TEC calculated from HORT reconstructions from 23:00 UT on October 30 to 02:00 UT on October 31 during the main phase of the storm are shown in Figures 21 and 22 as TECU isolines. During that time, the region was dominated by enhanced ionization. The electron density in the central part of the spot of increased ionization $((1.5-2) \bullet 10^{12} \text{m}^{-3})$ significantly exceeds typical daytime values. The size of the spot (at half-maximum VTEC) is about 1500-2000 km in the north-to-south direction and 2500-3000 km in the west-to-east direction. The spot is seen moving from west to east with a southward component.

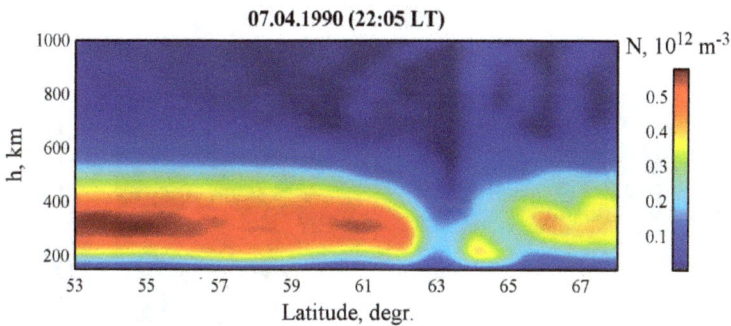

Figure 10. LORT image of the ionosphere (Moscow-Murmansk) on April 7, 1990 at 22:05 LT

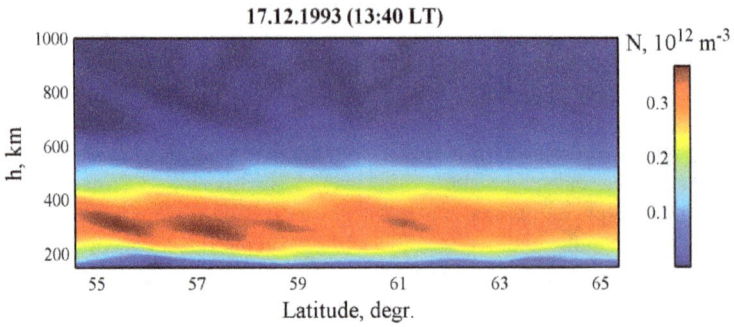

Figure 11. LORT image of the ionosphere (Moscow-Arkhangelsk) on December 17, 1993 at 13:40 LT

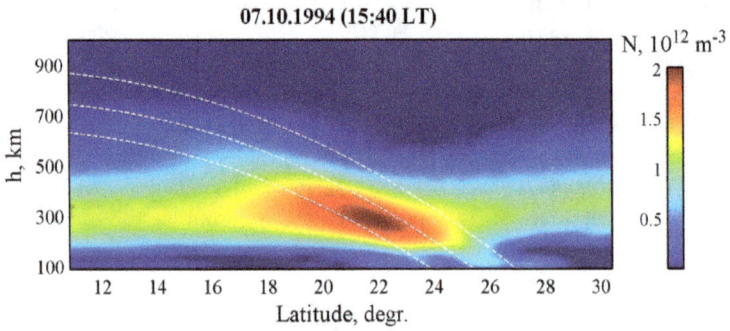

Figure 12. LORT image above Manila-Shanghai on October 7, 1994 at 15:40 LT

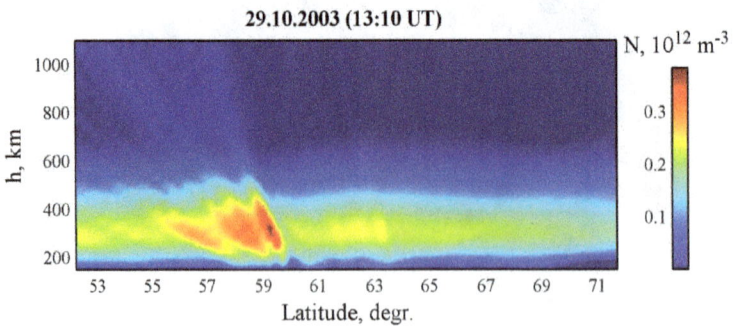

Figure 13. LORT image of the ionosphere above Alaska region on October 29, 2003 at 13:10 UT

16.04.2009 (06:20 LT)

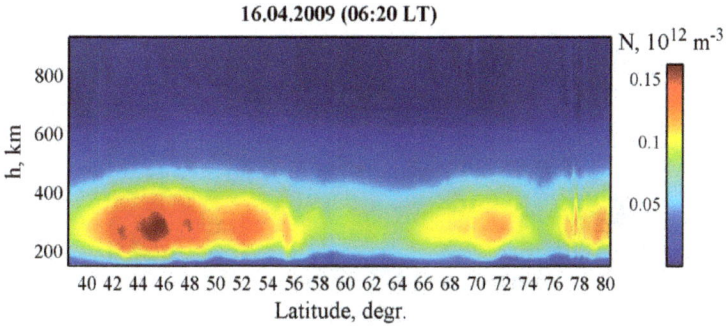

Figure 14. LORT image of the ionosphere (Sochi- Svalbard) on April 16, 2009 at 06:20 LT

11.04.2012 (04:08 LT)

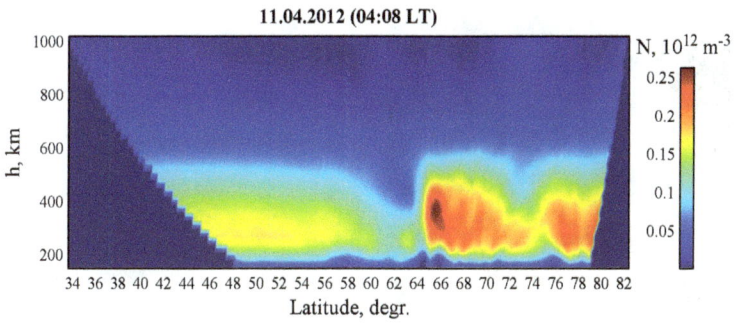

Figure 15. LORT image of the ionosphere (Moscow-Svalbard) on April 11, 2012 at 04:08 LT

18.02.1991 (06:06 LT)

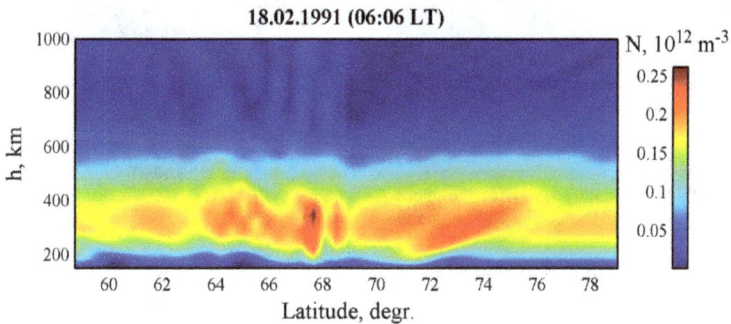

Figure 16. LORT image of the ionosphere (Moscow-Murmansk) on December 18, 1991 at 06:06 LT

18.08.2011 (22:48 LT)

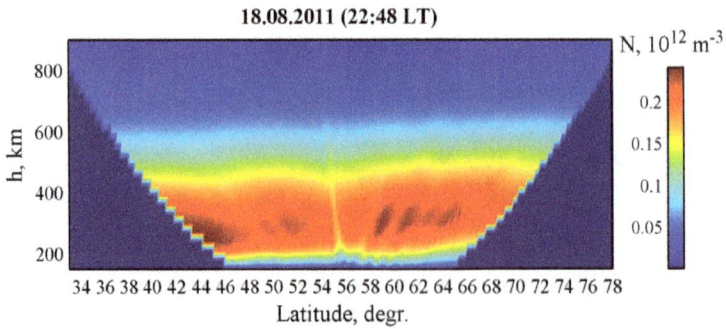

Figure 17. LORT-image of the ionosphere above Sura heating facility on August 18, 2011 at 22:48 LT

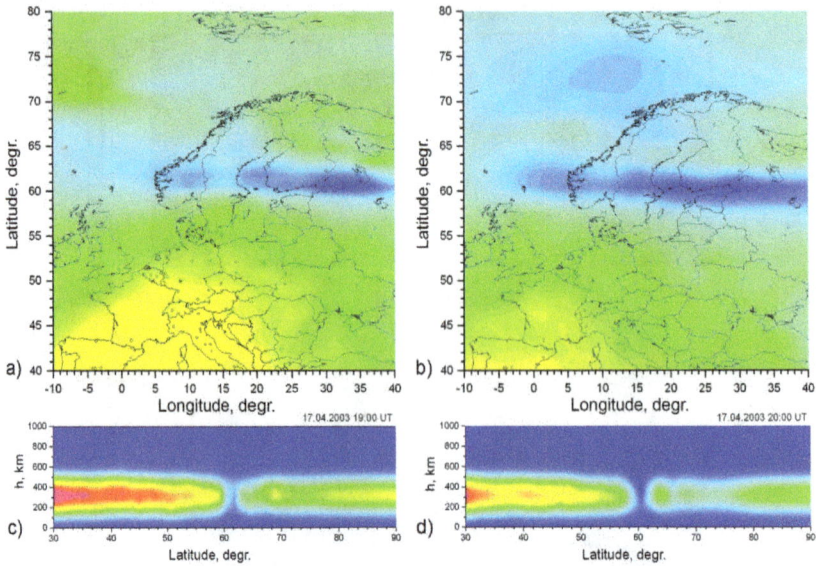

Figure 18. Ionospheric HORT reconstructions over Europe on April 17, 2003: (a, c) 19:00 UT, (b, d) 20:00 UT. (a, b): TEC maps in the latitude-longitude coordinates; the color scale is from 0 to 35 TECU (1 TECU=10^{16} m^{-2}). (c, d): Meridional cross sections along 21°E in the latitude-altitude coordinates; the color scale is from 0 to 0.6•10^{12} m^{-3}

The LORT reconstruction above at the Russian chain (Moscow-Murmansk) reveals a variegated multi-extremal distribution of electron density during the night of 30/31 October (Kp = 9). A spot of enhanced ionization attaining 1.5•10^{12}m^{-3} (which is a typical value for the equatorial anomaly), is observed at 70°-72°N (Figure 23). This spot is probably a result of the antisunward ionospheric convection combined with low-energy particle precipitation. Global two-cell ionospheric convection is responsible for formation of the tongues of ionization (TOI), which are fragments of the low- and mid-latitude dayside ionospheric plasma entrained through the cusp and polar cap into the night-side ionosphere and distributed along the night-side of the auroral oval. The TOI is associated with the spots of the increased ionization drifting from the north southwards and from the east westwards [45].

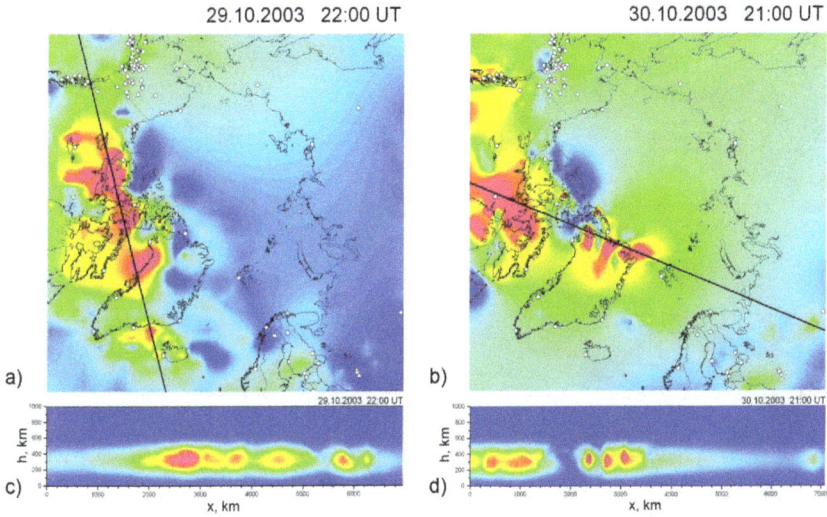

Figure 19. HORT reconstructions over the Arctic on October 29 and 30, 2003. (a, b) TEC maps, the color scale is from 0 to 60 TECU. (c, d) The cross-sections cut the ionosphere along the lines indicated on the TEC maps, x is is the distance on Earth's surface along highlighted lines and h is the height, the color scale is from 0 до $2.5 \bullet 10^{12}$ m^{-3}.

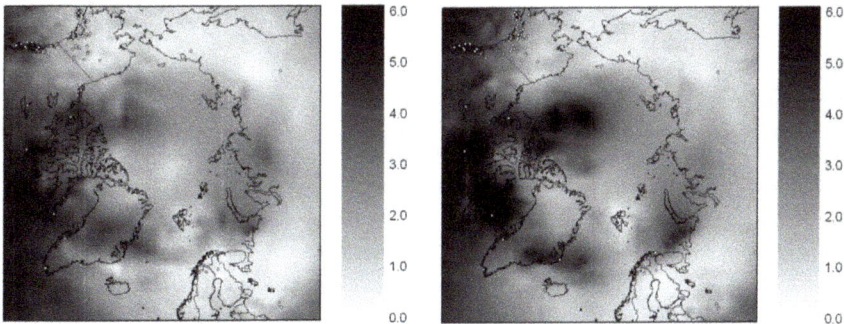

Figure 20. Vertical TEC in Arctic according to HORT on December 16, 2006 at 18:00 UT (left) and 19:00 UT (right)

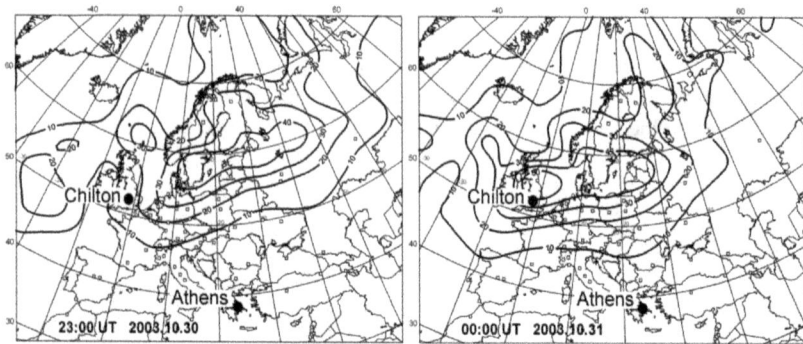

Figure 21. Contours of vertical TEC over European region on 30 October, 2003 at 23:00 UT (left) and at 24:00 UT (right)

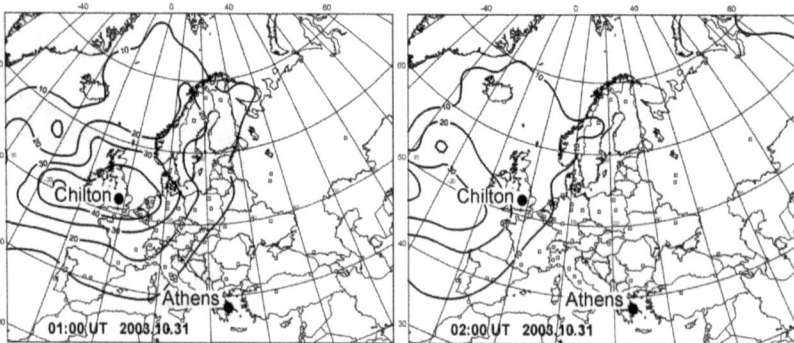

Figure 22. Contours of vertical TEC over European region on 31 October, 2003 at 01:00 UT (left) and at 02:00 UT (right)

The TOI observed on October 30, 2003, whose further evolution as the plasma moved above the northern Europe was traced in [46] from the LORT and HORT data and the European ionosonde measurements, is also analyzed in a number of other works. Mitchell et al. [47] compare the signatures of TOI that were observed on October 30, 2003 in the TEC distributions calculated by MIDAS GPS RT tool with the amplitude and phase fluctuations of the GPS signal recorded by the specialized receiver on Svalbard. Increased TEC (up to 70 TECU) is observed from 21:00 to 21:30 in a band north of and across Scandinavia towards Greenland at above 70ºN.

Ionospheric disturbances above Japan after the strongest Tohoku earthquake were analyzed by HORT methods with very high time resolution (2-3 minutes, a result of the 1200 ground stations in Japan) in [48]. The disturbances observed in the vertical TEC an hour after the main shock are shown in Figure 24. The TEC waves induced by the earthquake-generated AGWs are seen propagating outwards from the epicentral area. The spatial limits of the diagram correspond to the limited area within which the receiving network is sufficiently dense.

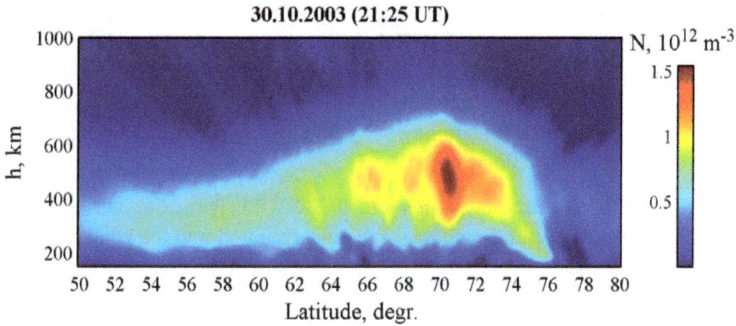

Figure 23. LORT image of the ionosphere (Moscow-Svalbard) on October 30, 2003 at 21:25 UT

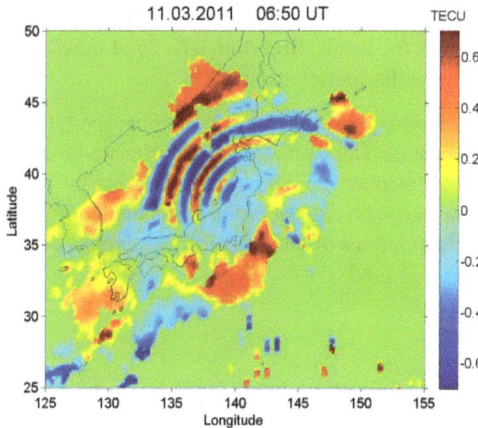

Figure 24. The diverging disturbance caused by the acoustic gravity waves generated by the Tohoku earthquake

We note that GPS/GLONASS-based remote sounding and HORT are being extensively developed and used in new practical applications [49-53].

5. Combination with other sounding techniques

Existing systems (FormoSat-3/COSMIC and a few others are capable of recording GNSS signals onboard low-Earth orbiting satellite platforms). These satellites effectively demonstrate the radio occultation technique (OT) which acquires quasi-tangential projections of electron density N [54-56]. The OT method, coupled with subsequent ground reception of the OT data, opens up the possibility of sounding the ionosphere in a wide range of different geometries of the transmitting and receiving systems. The OT method provides integrals of N over a set of quasi-tangential rays (the satellite-satellite links), and is a particular case of the RT method. It is, however, necessary to construct a procedure for synthesizing the occultation data into the general RT process [7, 8, 57]. The combination of RT and OT, when the RT data are supplemented by the satellite-to-satellite sounding (OT) data, would noticeably improve the vertical resolution of the RT reconstructions.

Existing ultra-violet (UV)-sounding systems (GUVI, SSULI, FormoSat-3/COSMIC) provide integrals of N squared. UV sounding data can be incorporated into the general tomographic iterative scheme. For example, at the first step of a reconstruction, the main iteration is accomplished with linear integrals and RT data alone. Then, based on the distribution of electron density N obtained in the first iteration, we run the iterative scheme for N squared with the UV input. Third, we transform the distribution of N squared derived at this step of the reconstruction into the distribution of N (the result of the second iteration); since this distribution can be used further. Thus, the odd iterations will work with the radio sounding data, while even iterations will use UV input. Overall, we obtain a tomographic methodology which uses both radio sounding and UV sounding data. However, in order to ensure convergence and to obtain high-quality final results, the experimental data of different kinds should be consistent and have commensurate accuracy, otherwise the additional iterations based on the "bad" data will degrade the result. Unfortunately, as of now, the UV data are far less accurate that the navigation radio sounding data.

Note that the RT methods described here refer to "ray" tomography [1] that neglects diffraction effects. In previous work, we developed methods for diffraction tomography and statistical tomography [1, 2, 4, 7]. Diffraction tomography is applicable for imaging the structure of isolated localized irregularities with allowance for diffraction effects. Statistical tomography reconstructs the spatial distributions of the statistical parameters of the randomly irregular ionosphere [7, 58].

6. Conclusions

This Chapter briefly outlines the results of tomographic studies of the ionosphere conducted with the participation of the authors. The methods applied in satellite radio tomography of the near-Earth plasma, including LORT and HORT, are described. During the last two decades, numerous RT experiments, studying the equatorial, mid-latitude, sub-auroral, and auroral ionosphere were carried out in different regions of the world (in Europe, USA, and Southeast

Asia). Examples of RT images of the ionospheric electron density based on data recorded in a series of RT experiments have been presented.

An RT system is a distributed sounding system: the moving satellites with onboard transmitters and receivers together with the ground receiving networks enable continuous sounding of the medium along different directions and support imaging of the spatial structure of the ionosphere. Satellite RT, utilizing a system of ground and satellite receivers, combined with traditional means of ionospheric sounding provides the basis for regional and global monitoring of the near-Earth plasma.

Acknowledgements

We are grateful to professors L.-C. Tsai and C.H. Liu, our colleagues from the Center for Space and Remote Sensing Research (National Central University, Chungli, Taiwan), and the University of Illinois at Urbana Champaign for providing experimental data. We acknowledge use of the worldwide ionosonde database accessed from the National Geophysical Data Center (NGDC) and Space Physics Interactive Data Resource (SPIDR). The authors are grateful to North-West Research Associates (NWRA) for providing experimental relative TEC data in the Alaska region. This work was supported by the Russian Foundation for Basic Research (grants 13-05-01122, 11-05-01157), the Ministry for Education and Science of the Russian Federation (project 14.740.11.0203), a grant of the President of Russian Federation (project MK-2544.2012.5), and M.V. Lomonosov Moscow State University Development Programme.

Author details

Vyacheslav Kunitsyn*, Elena Andreeva, Ivan Nesterov and Artem Padokhin

*Address all correspondence to: kunitsyn@phys.msu.ru

M. Lomonosov Moscow State University, Faculty of Physics, Moscow, Russia

References

[1] Kunitsyn, V. E, & Tereshchenko, E. D. Tomography of the Ionosphere. Moscow: Nauka; (1991). (In Russian).

[2] Kunitsyn, V. E, & Tereschenko, E. D. Radiotomography of the Ionosphere. IEEE Antennas and Propagation Magazine (1992). 34, 22-32.

[3] Leitinger, R. Ionospheric tomography. In: Stone R. (ed.) Review of Radio Science 1996-1999. Oxford: Science Publications; (1999). p.581-623.

[4] Kunitsyn, V. E, & Tereshchenko, E. D. Ionospheric Tomography. Berlin, NY: Springer; (2003).

[5] Pryse, S. E. Radio Tomography: A new experimental technique. Surveys in Geophysics (2003). 24, 1-38.

[6] Bust, G. S, & Mitchell, C. N. History, current state, and future directions of ionospheric imaging. Reviews of Geophysics (2008). 46, RG1003, doi: 10.1029/2006RG000212

[7] Kunitsyn, V. E, Tereshchenko, E. D, & Andreeva, E. S. Radio Tomography of the Ionosphere. Moscow: Nauka; (2007). (In Russian).

[8] Kunitsyn, V. E, Tereshchenko, E. D, Andreeva, E. S, & Nesterov, I. A. Satellite radio probing and radio tomography of the ionosphere. Uspekhi Fizicheskikh Nauk (2010). 180(5), 548-553.

[9] Andreeva, E. S, Kunitsyn, V. E, & Tereshchenko, E. D. Phase difference radiotomography of the ionosphere. Annales Geophysicae (1992). 10, 849-855.

[10] Kunitsyn, V. E, Andreeva, E. S, Kozharin, M. A, & Nesterov, I. A. Ionosphere Radio Tomography using high-orbit navigation system. Moscow University Physics Bulletin (2005). 60(1), 94-108.

[11] Hofmann-Wellenhof, B, Lichtenegger, H, & Collins, J. Global Positioning System: theory and practice. Berlin, NY: Springer; (1992).

[12] Kunitsyn, V. E, Kozharin, M. A, Nesterov, I. A, & Kozlova, M. O. Manifestations of helio-geophysical disturbances in October, 2003 in the ionosphere over West Europe from GNSS data and ionosonde measurements. Moscow University Physics Bulletin (2004). 59(6), 68-71.

[13] Kersley, L, Heaton, J, Pryse, S, & Raymund, T. Experimental ionospheric tomography with ionosonde input and EISCAT verification. Annales Geophysicae (1993). 11, 1064–1074.

[14] Kunitsyn, V. E, Andreeva, E. S, Razinkov, O. G, & Tereschhenko, E. D. Phase and phase-difference ionospheric radiotomography. International Journal of Imaging Systems and Technology (1994). 5(2), 128-140.

[15] Heaton, J, Pryse, S, & Kersley, L. Improved background representation, ionosonde input and independent verification in experimental ionospheric tomography. Annales Geophysicae (1995). 13, 1297-1302.

[16] Heaton, J, Jones, G, & Kersley, L. Toward ionospheric tomography in Antarctica: First steps and comparison with dynasonde observations. Antarctic Science (1996). 8, 297-302.

[17] Heaton, J, Cannon, P, Rogers, N, Mitchell, C, & Kersley, L. Validation of electron density profiles derived from oblique ionograms over the United Kingdom. Radio Science (2001). 36, 1149-1156.

[18] Dabas, R, & Kersley, L. Radio tomographic imaging as an aid to modeling of ionospheric electron density. Radio Science (2003). 38(3), doi:10.1029/2001RS002514.

[19] Franke, S. J, Yeh, K. C, Andreeva, E. S, & Kunitsyn, V. E. A study of the equatorial anomaly ionosphere using tomographic images. Radio Science (2003). 38(1), doi: 10.1029/2002RS002657.

[20] Yeh, K. C, & Liu, C. H. Theory of Ionospheric Waves. New York: Academic Press; (1972).

[21] Pryse, S, & Kersley, L. A preliminary experimental test of ionospheric tomography. Journal of Atmospheric and Terrestrial Physics (1992). 54, 1007-1012.

[22] Raymund, T, Pryse, S, Kersley, L, & Heaton, J. Tomographic reconstruction of ionospheric electron density with European incoherent scatter radar verification. Radio Science (1993). 28, 811-817.

[23] Kersley, L, Heaton, J, Pryse, S, & Raymund, T. Experimental ionospheric tomography with ionosonde input and EISCAT verification. Annales Geophysicae (1993). 11, 1064-1074.

[24] Foster, J, Buonsanto, M, Holt, J, Klobuchar, J, Fougere, P, Pakula, W, Raymund, T, Kunitsyn, V. E, Andreeva, E. S, Tereshchenko, E. D, & Khudukon, B. Z. Russian-American Tomography Experiment. International Journal of Imaging Systems and Technology (1994). 5, 148-159.

[25] Pakula, W, Fougere, P, Klobuchar, L, Kuenzler, H, Buonsanto, M, Roth, J, Foster, J, & Sheehan, R. Tomographic reconstruction of the ionosphere over North America with comparisons to ground-based radar. Radio Science (1995). 30(1), 89-103.

[26] Walker, I, Heaton, J, Kersley, L, Mitchell, C, Pryse, S, & Williams, M. EISCAT verification in the development of ionospheric tomography. Annales Geophysicae (1996). 14, 1413-1421.

[27] Nygren, T, Markkanen, M, Lehtinen, M, Tereshchenko, E, Khudukon, B, Evstafiev, O, & Pollari, P. Comparison of F-region electron density observations by satellite radio tomography and incoherent scatter methods. Annales Geophysicae (1996). 14, 1422-1428.

[28] Spenser, P, Kersley, L, & Pryse, S. A new solution to the problem of ionospheric tomography using quadratic programming. Radio Science (1998). 33(3), 607-616.

[29] Foster, J, & Rich, F. Prompt mid-latitude electric field effects during severe geomagnetic storm. Journal of Geophysical Research (1998). 103(11), 26367-26372.

[30] Raymund, T, Bresler, Y, Anderson, D, & Daniell, R. Model-assisted ionospheric tomography: A new algorithm. Radio Science (1994). 29, 1493-1512.

[31] Kunitsyn, V, Nesterov, I, Padokhin, A, & Tumanova, Y. Ionospheric Radio Tomography Based on the GPS/GLONASS Navigation Systems. Journal of Communications Technology and Electronics (2011). 56(11), 1269-1281.

[32] Andreeva, E. S, Galinov, A. V, Kunitsyn, V. E, Mel'nichenko, Yu. A, Tereshchenko E. D, Filimonov M. A, & Chernyakov, S. M. Radio tomographic reconstruction of ionisation dip in the plasma near the Earth. Journal of Experimental and Theoretical Physics Letters (1990). 52, 145-148.

[33] Oraevsky, V. N, Rushin, Yu. Ya, Kunitsyn, V. E, Razinkov, O. G, Andreeva, E. S, Depueva, A. Kh, Kozlov, E. F, & Shagimuratov, I. I. Radiotomographic cross-sections of the subauroral ionosphere along trace Moscow-Arkhangelsk. Geomagnetism and Aeronomy (1995). 35(1), 117-122.

[34] Cook, J, & Close, S. An investigation of TID evolution observed in MACE'93 data. Annales Geophysicae (1995). 13, 1320-1324.

[35] Andreeva, E. S, Franke, S. J, Yeh, K. C, & Kunitsyn, V. E. Some features of the equatorial anomaly revealed by ionospheric tomography. Geophysical Research Letters (2000). 27, 2465-2458.

[36] Yeh, K. C, Franke, S. J, Andreeva, E. S, & Kunitsyn, V. E. An investigation of motions of the equatorial anomaly crest. Geophysical Research Letters (2001). 28, 4517-4520.

[37] Andreeva, E. S. Possibility to reconstruct the ionosphere E and D regions using ray radio tomography. Moscow University Physics Bulletin (2004). 59(2), 67-75.

[38] Kunitsyn, V. E, Tereshchenko, E. D, Andreeva, E. S, Grigor'ev, V. F, Romanova, N. Yu, Nazarenko, M. O, Vapirov, Yu. M, & Ivanov, I. I. Transcontinental Radio Tomographic Chain: First Results of Ionospheric Imaging. Moscow University Physics Bulletin (2009). 64(6), 661-663.

[39] Ahmadov, R, & Kunitsyn, V. Simulation of generation and propagation of acoustic gravity waves in the atmosphere during a rocket flight. International Journal of Geomagnetism and aeronomy (2004). 5(2), 1-12, doi:10.1029/2004GI000064.

[40] Andreeva, E. S, Gokhberg, M. B, Kunitsyn, V. E, Tereshchenko, E. D, Khudukon, B. Z, & Shalimov, S. L. Radiotomographical detection of ionosphere disturbances caused by ground explosions. Cosmic Research (2001). 39(1), 13-17.

[41] Kunitsyn, V. E, Andreeva, E. S, Frolov, V. L, Komrakov, G. P, Nazarenko, M. O, & Padokhin, A. M. Sounding of HF heating-induced artificial ionospheric disturbances by navigational satellite radio transmissions. Radio Science (2012). RS0L15, doi: 10.1029/2011RS004957.

[42] Kunitsyn, V. E, Andreeva, E. S, Franke, S. J, & Yeh, K. C. Tomographic investigations of temporal variations of the ionospheric electron density and the implied fluxes. Geophysical Research Letters (2003). 30(16), doi:10.1029/2003GL016908

[43] Kunitsyn, V. E, & Nesterov, I. A. GNSS radio tomography of the ionosphere: The problem with essentially incomplete data. Advances in Space Research (2011). 47, 1789-1803.

[44] Kulchitsky, A, Maurits, S, Watkins, B, et al. Drift simulation in an Eulerian ionospheric model using the total variation diminishing numerical scheme. Journal of Geophysical Research (2005). 110, 1-14. A09310, doi:10.1029/2005JA011033

[45] Foster, J. C, Coster, A. J, Erickson, P. J, Holt, J. M, Lind, F. D, Rideout, W, Mccready, M, Van Eyken, A, Barnes, R. J, Greenwald, A, & Rich, F. J. Multiradar observations of the polar tongue of ionization. Journal of Geophysical Research (2005). 110, A09S31, doi:10.1029/2004JA010928.

[46] Kunitsyn, V. E, Kozharin, M. A, Nesterov, I. A, & Kozlova, M. O. Manifestations of heliospheric disturbances of October 2003 in the ionosphere over Western Europe according to the data of GNSS tomography and ionosonde measurements. Moscow University Physics Bulletin (2004). 6, 67–69.

[47] Mitchell, C, Alfonsi, L, De Franceschi, G, Lester, M, Romano, V, & Wernik, A. GPS TEC and scintillation measurements from the polar ionosphere during the October 2003 storm. Geophysical Research Letters (2005). 32, L12S03, doi: 10.1029/2004GL021644.

[48] Kunitsyn, V, Nesterov, I, & Shalimov, S. Japan Earthquake on March 11, 2011: GPS-TEC Evidence for Ionospheric Disturbances. Journal of Experimental and Theoretical Physics Letters (2011). 94(8), 616-620.

[49] Ma, X. F, Maruyama, T, Ma, G, & Takeda, T. Three-dimensional ionospheric tomogra12 phy using observation data of GPS ground receivers and ionosonde by neural network. Journal of Geophysical Research (2005). 110, A05308, doi: 10.1029/2004JA010797.

[50] Jin, S. G, & Park, J. U. GPS ionospheric tomography: acomparison with the IRI-2001 model over South Korea. Earth Planets Space (2007). 59(4), 287-292.

[51] Zhao, H. S, Xu, Z. W, Wu, J, & Wang, Z. G. Ionospheric tomography by combining vertical and oblique sounding data with TEC retrieved from a tri-band beacon. Journal of Geophysical Research (2010). 115, A10303, doi:10.1029/2010JA015285.

[52] Jin, S. G, Feng, G. P, & Gleason, S. Remote sensing using GNSS signals: Current status and future directions. Advances in Space Research (2011). 47, 1645–1653.

[53] Jin, S. G. GNSS Atmospheric and Ionospheric Sounding. In: Jin SG (ed.) Global Navigation Satellite Systems- Signal, Theory and Application. Rijeka: InTech; (2012). p. 359-381.

[54] Hajj, G, Ibanez-meier, R, Kursinski, E, & Romans, L. Imaging the ionosphere with the global positioning system. International Journal of Imaging Systems and Technology (1994). 5(2), 174-187.

[55] Kursinski, E, Hajj, G, Beritger, W, et al. Initial results of radio occultation of Earth atmosphere using GPS. Science (1996). 271(5252), 1107-1110.

[56] Liou, Y. A, Pavelyev, A. G, Matyugov, S. S, et al. Radio Occultation Method for Remote Sensing of the Atmosphere and Ionosphere. Ed. Y.A. Liou. Rijeka: InTech; (2010)

[57] Andreeva, E. S, Berbeneva, N. A, & Kunitsyn, V. E. Radio tomography using quasi tangentional radiosounding on traces satellite-satellite. Geomagnetism and Aeronomy (1999). 39(6), 109-114.

[58] Tereschenko, E. D, Kozlova, M. O, Kunitsyn, V. E, & Andreeva, E. S. Statistical tomography of subkilometer irregularities in the high-latitude ionosphere. Radio Science (2004). 39, RS1S35, doi:10.1029/2002RS002829.

Chapter 6

Effects of Solar Radio Emission and Ionospheric Irregularities on GPS/GLONASS Performance

Vladislav V. Demyanov, Yury V. Yasyukevich and
Shuanggen Jin

Additional information is available at the end of the chapter

1. Introduction

During the period of 2001-2010, several strong geomagnetic storms and direct solar radio emission interference deteriorated serious GPS performance. The L-band solar radio emission has recently been regarded as a potential threat to stable GPS and GLONASS performance. However, the threat has not been completely investigated or assessed so far. Furthermore, ionization anomaly at low latitudes along with the effect of the equatorial plasma "bubbles" increase the possibility of fading for transionospheric signals, especially during geomagnetic storms. Instabilities of ionospheric plasma on the "walls" of a bubble with electron density lower than the background value are also characterized by sharp gradients of electron density. For example, the walls of bubbles can be a source of ionospheric scintillations as well. There-fore, the spatial orientation of plasma bubble plays a decisive role in strengthening and weakening the amplitude and phase scintillations of satellite vehicle (SV) signals.

Although GPS ranging and positioning failures were investigated under such unfavorable geophysical conditions, but their exact nature remains unclear. This chapter is devoted to some features of GPS/GLONASS performance under specific geophysical events such as solar radio emission bursts and strong satellite signal scattering from the equatorial plasma "bubbles". As irregularities are elongated in the magnetic field direction, a ray path that is parallel to the magnetic field incorporates a greater path length through depleted regions containing plasma density irregularities. We propose a GPS method to detect mid-latitude field-aligned irregu-larities (FAIs) by line-of-sight angular scanning regarding the local magnetic field vector. Using GPS data of the Japanese GPS network (GEONET), we analyze occurrence of GPS-phase slips and positioning errors during the geomagnetic storm of February 12, 2000.

In addition, it is very important to provide the stable performance of GPS (GLONASS) receiver under strong geomagnetic storms. The scintillation index S4 is suggested as an adequate indicator of the current media state along the each satellite vehicle (SV) line-of-sight when the significant small scale ionospheric irregularities are present. To prove this idea we further conduct positioning measurements on a basis of combined GPS/GLONASS receiver (MRC-19L) as well as the measurements of the current ionosphere state during the strong magnetic storm on November 9, 2004.

2. Solar radio emission power threat on GPS and GLONASS

Until recently the L-band solar radio emission was not considered as a potential threat to the stable performance of satellite radio navigation systems such as GPS and GLONASS (Chen et al., 2005; Jin et al., 2008). Power threshold of the solar radio emission at the level of 40.000 sfu (solar flux units), which still provided steady performance of GPS, was found by Klobuchar et al. (1999). However only several solar radio bursts with the power level higher than 40.000 sfu have been observed over the last 40 years (Cerruti et al., 2006; Chen et al., 2005). According to Carrano et al. (2007), the signal-to-noise ratio decreases by 10-30 dB depending on the angular position of the sun relative to the directional pattern of receiver antenna under the direct influence of the solar radio emission. It causes failure in signal tracking of many visible navigation satellites on the earth's dayside for up to 1 hour.

Detailed investigation of the direct of solar radio emission interference on the GPS equipment performance indicated that the unsafe threshold of the solar radio emission power should be reduced to 4.000-12.000 sfu. A specific value of this threshold should be determined according to the type of signal tracking algorithms which are utilized in GPS/GLONASS user equipment. As it was proven by Afraimovich et al. (2008) and Cerruti et al. (2006), many short-term failures in measurements of radio navigation parameters were observed in GPS and GPS/GLONASS receivers all over the world during strong solar radio bursts on December 6 and 13, 2006. Some failures in measurements of radio navigation parameters were recorded even when the solar radio flux power was as much as $3 \bullet 10^3$ sfu (Afraimovich et al., 2008).

Nevertheless, a detailed analysis of separate solar radio bursts impact on the GPS/GLONASS user navigation equipment is required in order to estimate an extent of deterioration of the positioning systems on a global scale. Especially, maximum allowable power of the solar radio emission, which provides satisfactory signal-to-noise ratio at the navigation receiver input, is also very important. Such research would allow us to reestimate GPS and GLONASS noise immunity and make necessary improvements into navigation receivers development accord-ing to the known impact of the solar radio emission. The aim of section 2 is to evaluate the unsafe threshold of solar radio emission power for GPS/GLONASS receivers on a basis of theoretical and experimental analysis of the potential noise immunity of GLONASS standard precision code (SP) and high precision code (HP) with comparison to GPS CA (P(Y)) dual - frequency receivers.

2.1. The solar radio emission power at GPS/GLONASS receiver antenna output

The steady operation of a GPS/GLONASS receiver under the influence of intensive radio interference depends on the characteristics of radiofrequency (RF) path of navigation receiver within which the main filtering and amplification of satellite signals take place. The exact measurement of solar radio emission power, which affects the radiofrequency path input, is therefore very important. This measurement is necessary to design optimal algorithms for primary processing and filtering of radio-navigation parameters in the processor of navigation receiver.

Computing of solar radio emission power at the radiofrequency path input should begin with consideration of the receiver antenna directional pattern. The antenna directive gain (AD) is defined as ratio between the power specified for a real antenna $P(\theta, \alpha)$ and the power specified for a reference isotropic antenna (P_0), provided that signal powers at the observation point are equal.

$$D(\theta,\alpha) = P(\theta,\alpha) / P_0 \qquad (1)$$

where θ and α are azimuth and elevation of line-of-sight (LOS).

First we should set the characteristics of directional properties of a real receiving antenna. The characteristics are the receiving antenna power gain $G(\theta)$ relative to the ideal isotropic antenna (Table 1, column 2) and the antenna directive gain $D(\theta, \alpha)$ (Table 1, column 3)

$$G(\theta) = 10 \cdot \lg(D(\theta,\alpha)) \qquad (2)$$

Elevation of signal reception θ (deg.)	G(θ) (dB)	D(θ, a) =P(θ, a)/P$_0$
0< θ <5	$-7.5 \le G(\theta) \le -5$	$0.1775 \le D(\theta, a) \le 0.316$
5< θ <15	$-4.5 \le G(\theta)$	$0.354 \le D(\theta, a)$
θ >15	$-2 \le G(\theta)$	$0.63 \le D(\theta, a)$

Table 1. Directive characteristics of a navigation receiver antenna (Kaplan, 1996)

Another property of the receiving antenna to be considered for performing calculations is the antenna effective area. AD values is related to the antenna effective area by

$$D(\theta,\alpha) = 4\pi \cdot A_e(\theta) / \lambda^2 \qquad (3)$$

where λ is the signal wavelength and A_e is the antenna effective area. Equation (3) implies that the real receiving antenna has AD in the azimuthal plane equal to that of the standard isotropic antenna, i.e. the value of $D(\theta, \alpha)$ does not depend on the azimuth α of the received signal. Its elevation dependence has already taken into account in Table 1. Thus, we can calculate the magnitudes of effective area of the real receiver antenna at two operating frequencies of GPS/ GLONASS and different elevations. Table 2 contains calculation results for GPS operating frequencies. In our further consideration we assume that the A_e values for GLONASS receivers are close to GPS ones.

$\theta,°$	$A_e(\theta)$, m^2	
	$\lambda_{L1}=0.1903$ m	$\lambda_{L2}= 0.2442$ m
$0°< \theta <5°$	$5.099 \cdot 10^{-4} \le A_e(\theta) \le 9.077 \cdot 10^{-4}$	$8.409 \cdot 10^{-4} \le A_e(\theta) \le 1.497 \cdot 10^{-3}$
$5°< \theta <15°$	$1.01 \cdot 10^{-3} \le A_e(\theta)$	$1.677 \cdot 10^{-3} \le A_e(\theta)$
$\theta >15°$	$1.809 \cdot 10^{-3} \le A_e(\theta)$	$2.984 \cdot 10^{-3} \le A_e(\theta)$

Table 2. Antenna effective area

After determining the main characteristics of the receiving antenna, we can compute the power of the solar radio emission at the receiving antenna output. In order to do this the following specifications and assumption were taken into account:

Solar radio emission flux with the power of 1 sfu is equal to the power spectral density corresponding to the interval of 1 Hz of this flux power spectrum passing through the area of 1 m^2, i.e. 1 sfu=10^{-22} W m^{-2} Hz^{-1} (Chen et al., 2005).

Radio emission of the solar flare relative to an actual satellite signal is considered as the white Gaussian noise. Generally the noise power P_n within the given frequency band from F_1 to F_2 is obtained from the noise spectrum $S(f)$ as follows

$$P_n = \int_{F_1}^{F_2} S(f)df \tag{4}$$

The solar radio emission intensity N_0 in the frequency band ΔF_n of the satellite signal is constant throughout the band. So the radio emission power of a solar flare within the frequency bands of GPS and GLONASS signals can be computed in the similar way as follows

$$P_n = \Delta F_n \cdot N_0 \tag{5}$$

The solar radio noise has the right hand elliptical polarization and undergoes attenuation proportional to the polarization mismatch factor of 3.4 dB (at the frequency L1) and 4.4 dB (at the frequency L2) when passing through the antenna (ICD-GPS-200c, 1993);

When passing through the atmosphere, the solar radio emission in the GPS (GLONASS) frequency band undergoes the maximum attenuation of -2 dB (ICD-GLONASS, 2002; ICD-GPS-200c, 1993);

Thus, without considering polarization loss and attenuation in the atmosphere the power of the solar radio noise P_n at the receiving antenna output can be defined as:

$$P_{n,GPS(GLN)} = \Delta F_{n,GPS(GLN)} \cdot k \cdot N_0 \cdot A_e(\theta) \tag{6}$$

where k is the rate of the solar radio emission flux and $N_0 = 10^{-22}$ W m^{-2} Hz^{-1}.

The value of the solar radio emission power also depends on the sun zenith angle. In Eq. (6), this dependence is expressed in implicit form in terms of $A_e(\theta)$ (Table 2). Since it is more convenient to use power units (dBW) for further analysis, the solar radio noise power at the receiving antenna output may be converted to these units as

$$L_{n,GPS(GLN)} = 10 \cdot \lg(P_{n,GPS(GLN)}) \tag{7}$$

If we take into account the polarization loss and attenuation in the atmosphere at the frequencies L1 and L2 (ICD-GPS-200c, 1993) we get $L_{n1,GPS(GLN)} = 10 \cdot \lg(P_{n1,GPS(GLN)}) - 2 - 3.4$ and $L_{n2,GPS(GLN)} = 10 \cdot \lg(P_{n2,GPS(GLN)}) - 2 - 4.4$, at the frequencies L1 and L2, respectively.

Finally, we can compute the solar radio noise power at the receiving antenna output for sun elevation >15º at the central solar radio emission frequency f=1415 MHz (we assumed λ=0.212 m, $A_e = 2.25310^{-3}$ m^2, polarization loss = -3.4dB) in Table 3.

The front-end passband of the GPS receiver radio path (ΔF_{GPS}) is 3 MHz (Kaplan, 1996), while the passband of the GLONASS receiver for the channel of each separate satellite is only 0.5 MHz (ΔF_{GLN}) (Perov and Kharisov, 2005). Hence power of solar radio emission should be considered only in these narrow frequency bands for GPS and GLONASS correspondingly (Table 3).

The solar radio emission flux k, sfu L_n, dBW	1	10^2	10^3	10^4	10^5	10^6
$L_{n,GPS}$, dBW (ΔF_{GPS}=3 MHz)	-187.1	-167.1	-157.1	-147.1	-137.1	-127.1
$L_{n,GLN}$, dBW (ΔF_{GLN} =0.5 MHz)	-194.8	-174.8	-164.8	-154.8	-144.8	-134.8

Table 3. Rate of the solar radio emission flux

Here we should provide some special explanation about the front-end passband of the GPS and GLONASS radiofrequency chain. Generally the front-end band width should be twice the chipping rate - 1.023 MHz for GPS and 0.511 MHz for GLONASS (Perov and Kharisov, 2005; Tsui, 2005). However, depending on the navigation receiver specification the bandwidth can be set significantly larger or lower. Moreover, we should take into account that a navigation receiver utilizes the same RF path in order to process CA and P(Y) code signals concurrently at the same current frequency (1.5 GHz for GPS and 1.6 GHz for GLONASS). Hence we can not set RF front-end band width too narrow because it can cause severe phase distortion of the P(Y) or HP signal. On the other hand, we should keep the bandwidth narrow enough for effective suppressing of external radio noise including the solar radio emission. The specific choice depends on the developer of the GPS/GLONASS receiver specification, while it is unknown for us exactly. In order to evaluate the effect of solar radio emission on navigation receivers we used some averaged values of the RF front-end bandwidth (Table 3).

Generally, it is known that GLONASS utilizes frequency division multiple access (FDMA) technology to separate the signals of particular GLONASS satellites. It requires to set narrower RF front-end bandwidth in comparing to GPS one. The main expected consequence is lower integral solar radio noise power at the AD converter input of the navigation receiver. A comparative plot between powers of GPS and GLONASS received signals and the solar radio emission P_{SFU} at the output of receiving antenna can be made on the basis of above mentioned reasons. Fig. 1a gives the ratio of powers of the solar radio emission flux and GPS signal components and Fig. 1b illustrates the same for GLONASS.

The horizontal lines in Fig. 1a indicate levels of minimum (red lines) and maximum (black lines) power of GPS signal components at the receiving antenna output (ICD-GPS-200c, 1993). The power levels for the components of coarse acquisition code – CA (P(CA)) and encrypted P(Y)- code (P(P(Y))) at the L1 frequency, and for the P(Y) component at the L2 frequency are presented by dashed lines, dots and solid lines, respectively. The slant line shows power values of the solar radio emission flux P_{SFU}. Fig. 1b has the same notations for the components of the standard precision code and high precision code of GLONASS signals at frequencies L1 and L2 (ICD-GLONASS, 2002). As power values of some components are close, these lines partially mix in the diagram. The power values of the solar radio emission flux in terms of sfu are plotted along the horizontal axis with the logarithmic scale.

It is obvious from Fig. 1 for a solar radio noise within 10^2 - 10^4, the power of the solar radio noise is compared with power of the satellite signal at the receiving antenna output and exceeds it. When the solar radio emission flux is 10^6 sfu, the level of the solar radio noise exceeds the signal by 26-39 dB at the GPS receiving antenna output and by 20-32 dB at the GLONASS receiving antenna output.

2.2. Unsafe threshold of the solar radio emission power for GPS and GLONASS

Based on the above mentioned estimates, we can determine the unsafe threshold of solar radio noises at which the signal-to-noise ratio at the receiving antenna output is insufficient for stable tracking of satellite signals. First, we should take into account the fact that a significant gain in the signal-to-noise ratio is observed due to the use of correlation processing of the received

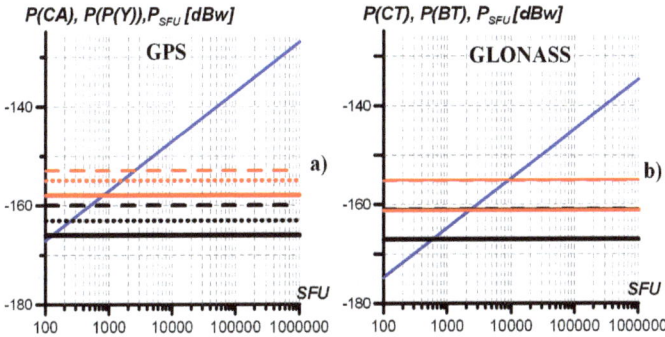

Figure 1. Power level of the solar radio emission at the receiving antenna output

signals if the structure of the pseudonoise ranging code is known at the receiving site. This gain (in terms of dBW) can be calculated as follows (Kaplan, 1996)

$$SN_{cor} = 10 \cdot \lg(\frac{F_{PRN}}{2 \cdot \Delta F_{PD}}) \tag{8}$$

where F_{PRN} is the frequency of elementary pulse in pseudorandom sequences of ranging code: F_{PRN} =1.023 MHz for the CA code and 10.23 MHz for the P(Y) code of GPS, or 0.511 MHz for the SP code and 5.11 MHz for the HP code of GLONASS, respectively; ΔF_{PD} is the predetector passband that is found from the lowest modulation frequency of the satellite signal by the service information data (50 Hz) (Kaplan, 1996; Perov and Kharisov, 2005).

Since the CA (SP) code structure is always known, it is obvious that GPS and GLONASS maximum noise immunity under the influence of a powerful solar radio emission takes place for the CA (SP) code at the main operating frequency L1. On the other hand, L2 GPS signal is only modulated by the encrypted code. "Semicodeless" or "codeless" processing algorithms are widely used in dual-frequency receivers in order to extract the P(Y) or HP-code signal components at the L2 GPS (GLONASS) frequency. Usage of these algorithms with no sufficient data on the P(Y) or HP -code structure reduces the stability of signal tracking at the L2 frequency under the influence of external radio emission (Skone and Jong, 2001). Significant fading of the signal-to-noise ratio while the encrypted signal is extracting and tracking at the L2 frequency may take place depending on the type of correlation processing algorithm. The level of these losses in the signal-to-noise ratio ΔSN_{cor} is within 14 - 17 dB (if "semicodeless" algorithms are utilized) and 27 - 30 dB (with the use of "codeless" algorithms) (Chen et al., 2005).

Access to the HP signal component of GLONASS and to the P(Y) code in the GPS system is not for common use, and in our examination we suppose that the correlation losses of HP signal tracking at L2 GLONASS frequency are equal to ones of GPS. It is known, that integration time periods in phase and code tracking loops of GPS (GLONASS) receivers inside the "integrate and dump" module are approximately the same: 1 ms – in the phase tracking loop and 20 ms – in the code tracking loop. The origin of the distortion of the correlation integral under "semicodeless" or "codeless" processing technique for both GPS and GLONSASS is the same. Hence we suppose that correlation losses ΔSN_{cor} should also be approximately the same.

When the satellite signal is locked the coherent tracking of the carrier frequency phase and code delay of the signal starts. The noise immunity of a navigation receiver is defined by the noise immunity level of the phase lock loop (PLL) of the receiver (Kaplan, 1996). That is why the unsafe threshold of solar radio emission, causing the satellite signal tracking loss, should be determined from the level of minimum acceptable signal-to-noise ratio that provides stable performance of the PLL.

The discrimination characteristics of the phase discriminator has a limited linear section, therefore severe requirements are imposed on the maximal acceptable carrier phase filtering error $\sigma_{\varphi,max}$. If the error level is exceeded, phase filtration is divergent and the signal tracking loss is observed. Magnitude of filtering error σ_φ of the carrier phase depends on many factors such as receiver thermal noises, short-term instability of reference generator frequency and phase fluctuation, caused by dynamic impact on the navigation satellite. With all the above consideration, we can determine filtering error magnitude of the carrier phase as follows (Kaplan, 1996),

$$\sigma_\phi = \sqrt{\sigma_T^2 + \sigma_F^2} + \frac{\sigma_S}{3} \leq \sigma_{\phi,max} \tag{9}$$

where σ_T, σ_F, and σ_S are root-mean-square (RMS) of the carrier phase filtration (in degrees) error caused by thermal noises, short-term instability of the reference generator frequency, and dynamic stress of the receiver, respectively. The magnitudes of separate components of (9) can be calculated from the following formulae (Kaplan, 1996),

$$\sigma_T = \frac{360}{2 \cdot \pi} \sqrt{\frac{\Delta F_{PLL}}{cn_0} \cdot \left[1 + \frac{1}{2T_{COR} \cdot cn_0}\right]} \tag{10}$$

where ΔF_{PLL} is the noise bandwidth of PLL, cn_0 is the signal-to-noise ratio at the receiver input, expressed by the power ratio determined on the basis of the receiver sensitivity and, T_{COR} is the pre-detection integration time. The rms of error for the short-term instability of reference generator frequency is

$$\sigma_F = 160 \cdot \frac{\sigma_F(\tau) \cdot F_c}{\Delta F_{PLL}} \tag{11}$$

where F_c is the satellite signal carrier frequency. The rms of error for the dynamic stress of the receiver is

$$\sigma_S = 0.4828 \cdot \frac{dR^3/dt^3}{(\Delta F_{PLL})^3} \tag{12}$$

where dR^3/dt^3 is the maximum dynamic stress of the receiver along the "satellite-receiver" line of sight. Expressions (11) and (12) are written for the third-order loop filter which is typically used for signal phase tracking in navigation receivers.

From (9) - (12) we can find an expression which defines the minimal allowable signal-to-noise ratio at the receiver input (CN_{thr} - in terms of dBW) for the maximum allowable value of the phase filtering error ($\sigma_{\varphi,max}$ - in terms of degrees) as follows,

$$CN_{thr} = -10 \cdot \lg \left[T_{COR} \cdot \sqrt{1 + \frac{2B}{T_{COR}\Delta F_{PLL}}} - T_{COR} \right]$$
$$B = (\tfrac{2\pi}{360})^2 \cdot \left[(\sigma_{\phi,max} - \tfrac{\sigma_S}{3})^2 - \sigma_F^2 \right] \tag{13}$$

Next we can determine an equivalent signal-to-noise ratio at the receiver input under the influence of solar radio emission using the following equation (Kaplan, 1996),

$$CN_{eq} = -10 \cdot \lg \left[10^{-0.1CN_0} + \frac{10^{0.1JS}}{r \cdot Q \cdot F_{PRN}} \right] \tag{14}$$

where $CN_0 = 10 \cdot \lg(P_{min})$ is the signal-to-noise ratio in terms of dBW at the receiver input, determined for the minimum power of the received signal, Q is the parameter of the spectral distribution of the external radio emission relative to the desired signal spectrum ($Q = 1$ for a narrow-band interference and $Q = 2$ for a wide-band Gaussian interference), JS is the relationship between the jamming solar radio emission power and the satellite signal power (dBW) and r is the coefficient considering distortion of the correlation integral when using "semi-codeless" or "codeless" technique for encrypted P(Y) or HP-code signal extracting.

The coefficient r can be calculated as follow

$$r = \frac{2\Delta F_{PD}}{F_{PRN}} 10^{0.1(SN_{cor} - \Delta SN_{cor})} \tag{15}$$

Here the gain in the signal-to-noise ratio due to the correlation processing SN_{cor} can be determined from (8), while the value of the correlation processing losses ΔSN_{cor} is defined as difference between signal-to-noise ratio of ideal and real correlation receivers. The approximate ΔSN_{cor} values were given above for "semi-codeless" and "codeless" correlation techniques (Chen et al., 2005).

It is interesting to know how solar radiation flux is related to the "ambient noise floor" of the GPS/GLONASS receiver. It would allow us to assess how many "extra sfus" are required to get a certain signal-to-noise ratio decrease with noise figures of the given radio frequency chain. In order to achieve this goal we need to determine the value of the unjammed signal-to-noise ratio CN_0 in terms of dBW taking into account all the total ambient noise figure sources as follows (Kaplan, 1996),

$$CN_0 = P_{\min} + G(\beta) - 10\lg(m \cdot T_0) - N_f - L \tag{16}$$

where P_{min} - is the minimal received signal power in dBW, determined according to the Interface Control Documents of GPS or GLONASS, $G(\theta)$ is the receiving antenna power gain (Table 1), the term $10 \cdot \lg(m \cdot T_0)$ is the ambient thermal noise density at the temperature of T_0 (K), m=1.38·10^{-23} (W·s/K) is the Bolzman's constant, N_f is the noise figure of receiver that includes antenna and cable losses in units of dB and L denotes the implementation losses, including AD converter loss in dB.

Finally, we can determine the unsafe threshold of solar radio emission power, which could cause GPS and GLONASS navigation satellite signals tracking loss under the given characteristics of the navigation receiver performance. Table 4 presents standard conditions of the GPS and GLONASS receiver performance (Kaplan, 1996; Perov and Kharisov, 2005) with estimated threshold signal-to-noise ratio.

Characteristics	Parameter value
Noise bandwidth of the third-order phase lock loop (Kaplan, 1996)	ΔF_{PLL} =18 Hz
Allan deviation oscillator phase noise (Kaplan, 1996)	$\sigma_f(\tau) = 10^{-10}$
Maximal line-of-sight jerk dynamics	$dR^3/dt^3 = 0$ m^3/s^3
Pre-detection integration time (Kaplan, 1996)	$T_{COR} = 20$ ms
Front-end passband of GPS receiver (Kaplan, 1996)	$\Delta F_{GPS} = 3$ MHz
Front-end passband of GLONASS receiver (for each satellite) (Perov and Kharisov, 2005)	$\Delta F_{GLN} = 0.5$ MHz
Correlation losses for "semicodeless" correlation techniques (Chen et al., 2005)	$\Delta SN_{cor} = 17$ dB
Correlation losses for "codeless" correlation techniques (Chen et al., 2005)	$\Delta SN_{cor} = 27$ dB
Maximum value of the phase filtering error (Kaplan, 1996)	$\sigma_{\varphi,max} = 15°$
Receiver thermal noise power under T=290 K° (Kaplan, 1996)	$N = -204$ dBW

Table 4. The characteristics of the navigation receiver performance

The maximum allowable phase filtering error value $\sigma_{\varphi,max} = 15°$ is determined with Monte Carlo simulation of GPS receiver phase lock loop performance under the combined dynamic and signal-to-noise ratio conditions (Kaplan, 1996). Considering the characteristics in Table 4 and using Eq. (13), we have determined the minimum allowable signal-to-noise ratio at the receiver input, CN_{thr} for frequencies L1 and L2 of GPS and GLONASS. In the case under consideration, the CN_{thr} values of GPS and GLONASS turned out to be very close to each other: 24.59 dB and 24.56 dB for L1 and L2, respectively.

The values of unjammed CN_0 ratio were computed with Eq. (.16), taking in account that T_0 =290K, N_f = 4 dB, L= 2dB (Kaplan, 1996) and $G(\theta)$=-2dB (θ>15°, Table 1). Corresponding values of equivalent signal-to-noise ratio at the receiver input CN_{eq} under the direct influence of solar radio emission (14) are calculated using data of solar radio emission power (Table 3), which are transformed into a jam-to-noise ratio in terms of dBW (JS in the equation 14).

Fig. 2 presents results of these calculations for L1 and L2 signals of GPS and GLONASS. The horizontal solid lines in all figures show the minimum allowable signal-to-noise ratio at the receiver input CN_{thr} and the horizontal dotted and dash-dotted lines stand for unjammed CN_0 values for CA (SP) code (pink lines), P(Y) or the HP code at frequency L1 (blue lines) and for P(Y) or the HP code at frequency L2 (black lines), respectively. The other curves show the CN_{eq} values for the CA (SP) code (pink curves), P(Y) or the HP code at frequency L1 (blue dotted curves) and P(Y) or the HP code at frequency L2 (black dashed curves).

Three cases were considered:

1. the ranging code on the receiving side is well known, and there are no correlation losses (Fig. 2a,b);

2. "semicodeless" (Fig. 2c, d) and "codeless" processing (Fig. 2e, f) for signal extraction with using unknown code.

From Fig. 2 we can conclude that the signal of C/A (SP) code at L1 frequency turns out to be the most resistant to the influence of the solar radio emission. The equivalent signal-to-noise ratio for the SP signal (GLONASS) in the case under consideration is higher than the CN_{thr} critical level, even under the influence of the solar radio emission flux of 10^6 sfu (Fig. 2a and b). At the same time, the CN_{eq} value for the C/A signal (GPS) reduces more noticeably and can drop below CN_{thr} level when solar radio noise power is just about 10^6 sfu (Fig. 2a). Thus we can expect the failure in the C/A signal tracking when the power level of solar radio noises is ≈ 10^6 sfu. In our opinion this conclusion can be explained with the idea that the front-end pass-band of the GPS receiver radio path is wider than that one of GLONASS receiver particular satellite radio channels. Hence, integral power of the solar radio noise which penetrates into a GPS receiver is higher in comparison to the GLONASS one.

When high-precision ranging codes (P(Y) and HP) on the receiving side are known, stable tracking of these signal components in GPS and GLONASS receivers is provided with a high quality even under the influence of solar radio noises of more than 10^6 sfu (Fig. 2a and b). However signal tracking of these components can fail when the power level of solar radio emission is more than 38.000 sfu (the P(Y) component at frequency L2, Fig. 2c) when the "semi-codeless" correlation processing is used. Since the power of P(Y) and HP signal components at frequency L1 is considerably higher, the tracking of these components may fail when the rate of the solar radio emission flux is more than 100.000 sfu (Fig. 2c and d).

The situation is the worst when the "codeless" processing of encrypted signals is utilized (Fig. 2e and f). One can see that the P(Y) signal tracking fails when the solar radio noise powers are about 4000 sfu and 10.000 sfu at frequencies L2 and L1, respectively. The corresponding power levels of the solar radio emission flux that can cause the failure of the GLONASS high precision signal tracking at frequencies L1 and L2 are 10.000 and 13.500 SFU, respectively.

Note that these estimations have been obtained for relatively favorable initial conditions, assuming relatively good Allan deviation factor and no dynamic stress or vibrations. We have also ignored effects of amplitude and phase ionospheric scintillations, which may cause significant fading of signal amplitude at the receiver antenna output. The multipath-propagation effect of signals in the reception point has also been ignored. Nevertheless, we have obviously proven the negative effect of powerful solar radio emission on the GPS and GLONASS performance.

2.3. Experimental statistics of GPS phase slips and counts omission during powerful solar flares

We use GLOBDET software developed at the ISTP SB RAS to process GPS data from the global network of dual-frequency receivers (Afraimovich, 2000). For analysis we used data in RINEX (Gurtner and Estey, 2009) format. We collected data over 1500 GPS sites from IGS network (http://sopac.ucsd.edu/other/services.html). For December 6, 2006 we used RINEX files from the CORS network (262 sites at ftp://www.ngs.noaa.gov/cors/rinex/). We also employ data from the Japanese GPS network GEONET (about 1225 stations) for December 13, 2006.

Fig. 3 shows the experimental geometry of GPS measurements during the solar flare on December 6 and 13, 2006. the GPS sites are marked by dots. Stars indicate the location of sunlit points for the solar flare on December 6 and 13, 2006.We calculate the 30 s series of the L1-L2 phase difference on two GPS frequencies f1 and f2 along lines of sight (LOS) of "receiver–satellite" to confirm a slip in measurements of the L1-L2 phase difference (Afraimovich et al., 2002). These data for each GPS satellite are then averaged over a period of dT=5 min at all chosen sites. It allows us to calculate the average number of observations $S(t)$ and slips $N(t)$ for all n LOS. Further we calculate the average relative density of slips $P(t) =100\% \bullet N(t)/S(t)$, and determine the maximal value P_{max}. If some count in a RINEX file is absent, the number of slips is equated to that of expected observations, so the density of slips becomes equal to 100 %.

Failures in L1–L2 make precision positioning in the dual-frequency mode impossible. The positioning is generally impossible if the signal at both GPS operating frequencies is not received at all. In order to estimate a probability of such failures for all LOS we define a number of counts omission $M(t)$ for each 30 s observation epoch. Considering $S(t)$ as an expected number of observation within the current epoch, we define counts omission density as $W(t) = M(t)/S(t)$. we also determine the corresponding maximum value W_{max}. The 30 s time resolution of the $W(t)$ rows allows us to conduct a detailed analysis of time behavior of $W(t)$ values under the solar radio emission flux variations.

In order to compare GPS and GLONASS noise immunity, we compute the relative losses-of-lock density $Q(t)$ in percent of the main signal parameters: L1, L2 (signal carrier phase), and C1, P1, P2 (C/A (SP) and P(Y)(HP) code delay) at $f1$ and $f2$ GPS and GLONASS frequencies. a

FULL P(Y), BT-CODE ACCESS

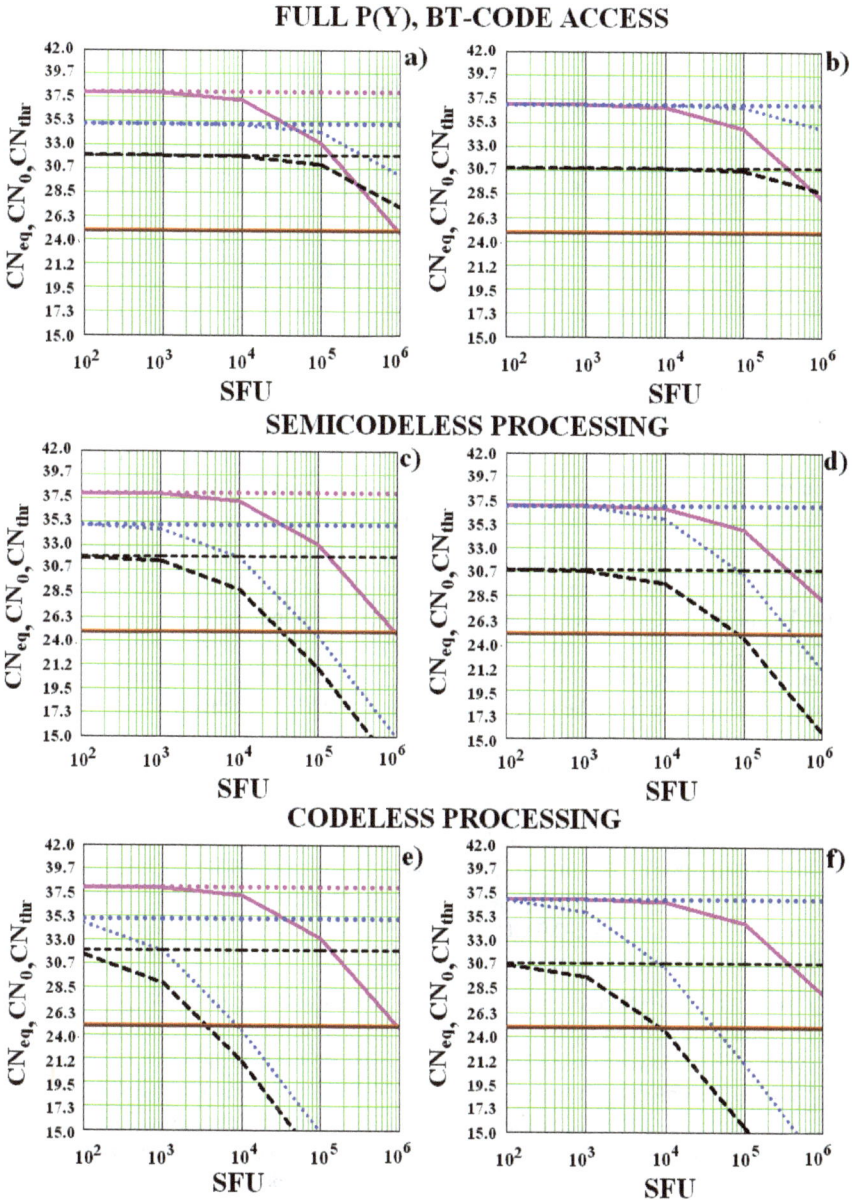

SEMICODELESS PROCESSING

CODELESS PROCESSING

Figure 2. An equivalent signal-to-noise ratio for GPS (a, c, e) and GLONASS (b, d, f) navigation receiver input under direct influence of the solar radio emission in units of sfu

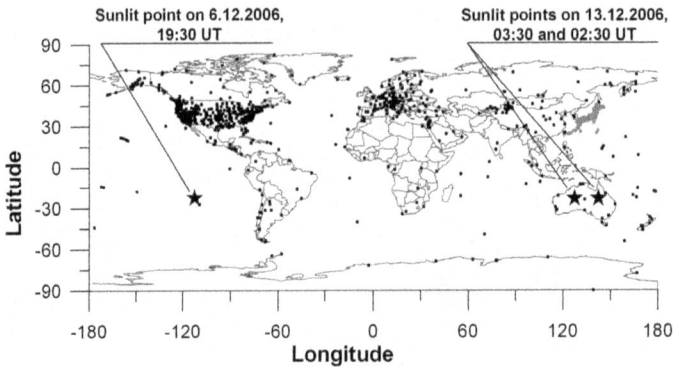

Figure 3. GPS measurement geometry during the solar flare on December 6 and 13, 2006. The GPS sites are marked by dots. Asterisks show the location of sunlit points for the solar flare

measurement slip is an event when the current epoch count of corresponding GPS parameters equals to zero or this count was absent in the RINEX file.

Unfortunately, there on December 6 and 13, 2006 there were only 44 combined GPS/GLONASS sites available. Using these data we were able to investigate GPS measurement slips in more details. We utilized the combined GPS/GLONASS data set to conduct the comparative analysis of GPS and GLONASS noise immunity under the direct solar radio emission interference. Due to uneven distribution of GPS/GLONASS sites on the earth surface there were only 4 and 7 sites within the sunlit side of the earth on December 6 and 13, respectively.

2.4. GPS phase slips and counts omission as a result of solar radio bursts on December 6, 2006

According to the data from the Owens Valley Solar Array (OVSA), the power of solar radio emission on December 6, 2006 in the GPS frequency band exceeded 10^6 sfu. The background emission level is about $\sim 10\text{--}10^2$ sfu. Fig. 4e shows the right handed circular polarization (RHCP) radio emission spectrum at 1.2–2.0 GHz, registered at the Solar Radio Spectrograph OVSA. The planetary index of geomagnetic activity was Kp~4.

Fig. 4a presents the $P(t)$ time dependences on the earth sunlit side (200–300° E; –80+80° N). these $P(t)$ data were derived from n=12793 LOS observations for all observable GPS satellites which were recognized with their pseudo random noise (PRN) code numbers at the elevation $\theta > 10°$ during the observation time from 18:00 to 20:00 UT (heavy black line). Fig..4a shows significant increasing of the $P(t)$ within 19:30–19:40 UT above the background level of slips, which usually does not exceed $P_{max} \sim 0.2\text{--}0.3$ % for such weakly disturbed ionosphere (Afraimovich et al., 2009a). Sharp increases of $P(t)$ values corresponded to an abrupt increasing of the solar radio emission flux at the very period of time.

The maximum relative density value of slips P_{max}=18.5% exceeded the background one in about ~50 times. At the same time, the average density $P(t)$ on the earth night side for $\theta > 10°$ and

n=3521 LOS did not exceed the background one. Unfortunately, the time resolution of $P(t)$ dependence, dT=5 min appeared to be insufficient to display the fine time structures of the radio emission flux (Fig.4e), obtained with the resolution lower than 1 s.

Nevertheless, the concurrence in the form of envelopes of the phase slip distribution and solar radio flux is obvious. The W_{max} values, observed from 19:15 to 19:45 UT, can reach 82% and 69% (PRN12, n=50 GPS sites; and PRN24, n=299 GPS sites). It can be seen from Fig. 4b that the sharp increase in phase slips and number of counts omission is totally consistent with the moments of the most powerful solar radio bursts (moments T1, T2, T3). Deep fading of the signal-to noise ratio at the L1 GPS frequency was observed during the same periods of time. It proves the idea that such sharp fading of GPS signal-to noise ratio is caused by direct interference of solar radio emission in the 1-2 GHz frequency band.

2.5. GPS phase slips and counts omission as a result of solar radio bursts on December 13, 2006

According to the data from the Learmonth Solar Radio Spectrographs, the total flux $F(t)$ of radio emission on December 13, 2006 exceeded 10^5 sfu at 1415 MHz (Fig. 5e). Sharp increasing of the solar radio flux power can be noted within the time periods 02:20 to 02:28 UT (time interval A) and 03:30 to 03:37 UT (time interval B). The horizontal line marks the spectrograph amplitude saturation level of about ~ 110000 sfu.

According to the data from the Nobeyama Radio Polarimeters (http://solar.nro.nao.ac.jp/norp/html/event/20061213_0247/norp20061213_0247.html), the RHCP solar radio emission power exceeded $1.47 \cdot 10^5$ sfu at 1 GHz at 02:28:09 UT and $2.57 \cdot 10^5$ sfu at 2 GHz at 03:35:51 UT on December 13, 2006. Since there were too few GPS sites on the earth sunlit side (40-200 E; -80+80 N) on December 13, 2006 (http://sopac.ucsd.edu/other/services.html/) we used the data set from the Japanese network GEONET which comprises 1225 GPS permanent sites. Fig. 5a, b shows the dependences P(t) for the December 13, 2006 flare over Japan for some satellites which were observed from 02:15 to 03:45 UT. Maximum values W_{max} can run to 50% (PRN13) and 27% (PRN16). The sharp increase of count omissions coincide with the impulses of solar radio emission during the time intervals A and B.

For the December 13 flare, on the earth sunlit side (40°-200° E; -80°+80° N) Fig. 5c and d present $W(t)$ values of counts omission which were registered for all satellites observed from 02:15 to 03:45 UT. Obviously, the maximum values W_{max} can reach 50% and 39% (PRN28, n=16 GPS sites; and PRN08, n=23 GPS sites). It has shown that the sharp increase of slips and count omissions completely coincide with the impulse solar radio bursts during A and B periods, including the fine time structure of solar radio burst.

2.6. GPS phase slips and counts omission as a result of solar radio bursts on October 28, 2003

it is especially interesting to assess GPS measurement slips density caused by the weaker solar radio burst on October 28, 2003. the power of this burst was by 3 orders of magnitude less than the solar radio bursts on December 6 and 13, 2006.

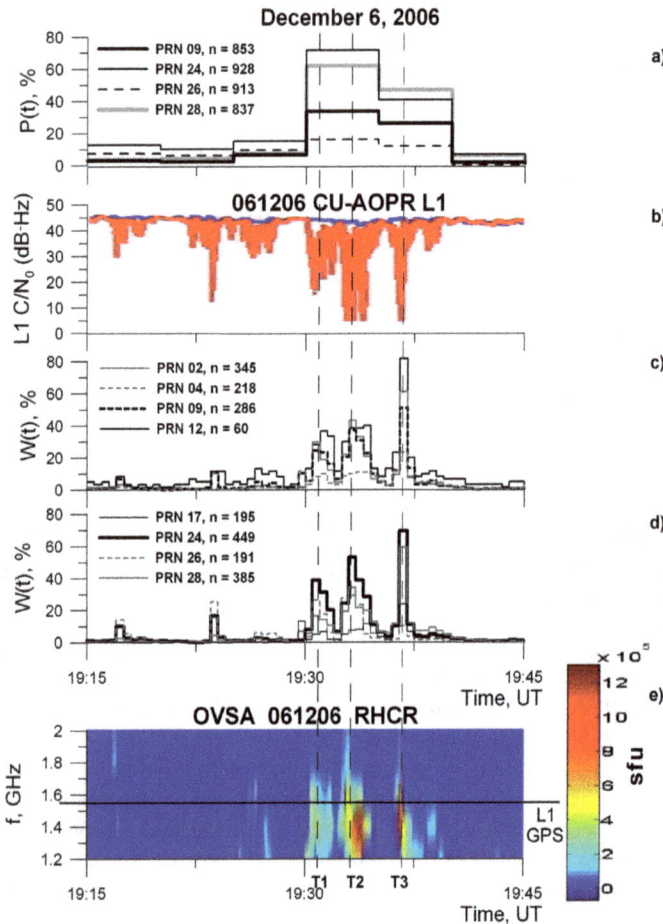

Figure 4. GPS L1-L2 phase slips on December 6, 2006

According to the data from the Trieste Solar Radio Spectrograph, Italy, the RHCP solar radio noise level exceeded 3×10^3 sfu at 1420 MHz on October 28, 2003 (Fig.6e). There are two solar radio bursts when the power of radio emission flux exceeded the level of 3×10^3 sfu: within the time periods from 11:05 to 11:08 UT (time interval A) and from 11:40 to 12:00 UT (time interval B).

Fig. 6d (black line) presents $P(t)$ dependences for the October 28, 2003 flare on the earth sunlit side (330–120° E; –80+80° N) derived from n=2452 LOS observations, for all observable satellites at the elevation angle $\Theta > 10°$ during the time period from 11:00 to 12:00 UT. A significant excess of the background level, $P_{max} \sim 0.2$–0.5 %, was observed from 11:02 to 11:10 UT. This

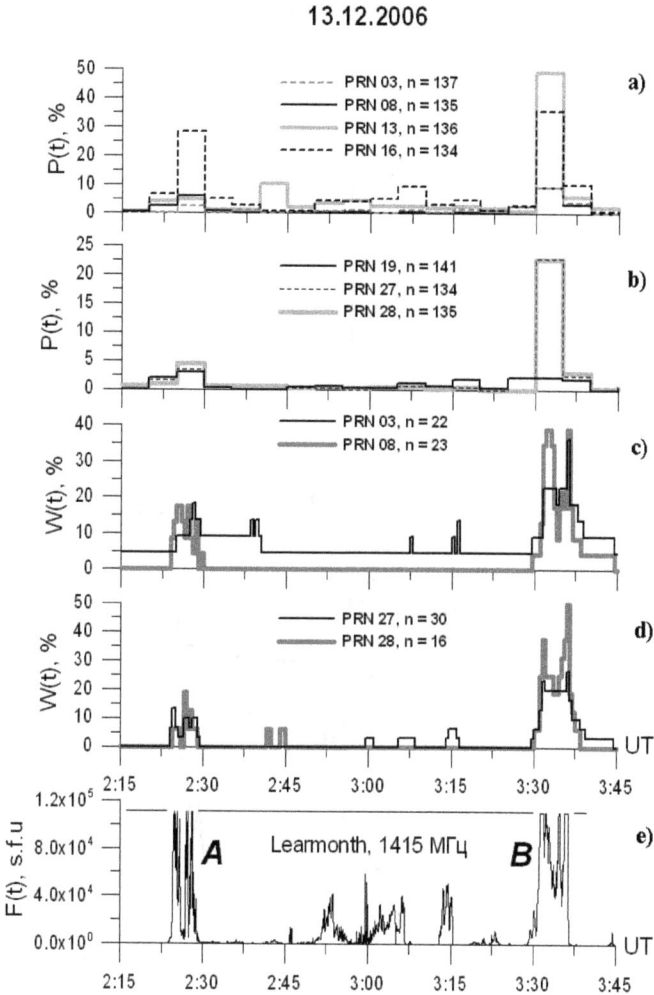

Figure 5. GPS L1-L2 phase slips on December 13, 2006

event corresponds to an abrupt increase of the solar radio emission flux during time interval A. The maximum value P_{max}=1.7% exceeds the background one in about ~3-4 times. At the same time, the average density of slips on the night side of the earth for $\Theta > 10°$ (n=12070 LOS) does not exceed the background one (Fig. 6d, thin gray line).

more substantial evidence on GPS-functioning quality deterioration can be found by estimating the average relative density of slips for separate GPS satellites. In Fig. 6a-c, the $P(t)$

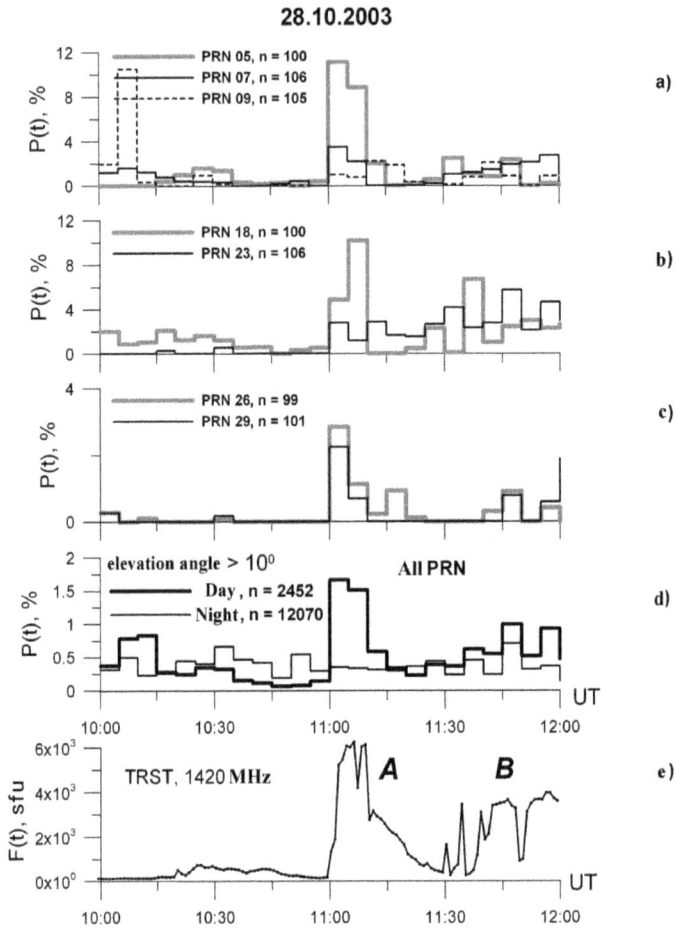

28.10.2003

Figure 6. GPS L1-L2 phase slips on October 28, 2003

dependences are given for some satellites observed during the time period 10:00 to 12:00 UT. The maximum values P_{max} can reach 11% and 10.2 % (for PRN05 and PRN18, n=100), whereas the value P_{max} =2.3 % for satellite PRN29 is close to P_{max}=1.7 %, which was determined for all satellites. the sharp increasing of slips density and count omissions happened simultaneously with the most powerful solar radio bursts for the time intervals A and B. Though power of the solar radio burst on October 28, 2003 is by 2–3 orders of magnitude less than that on December 6 and 13, 2006, the maximum values of phase slips are smaller by only in 5–10 times. Unfortunately the combined GPS/GLONASS data set was not enough in order to analyze the solar radio burst effect on October 28, 2003.

2.7. Comparative analysis of GPS and GLONASS performance

It is known that the basic principles of GLONASS and GPS functioning are almost identical from the viewpoint of estimation of the signal power. Hence, a comparative analysis of the GPS and GLONASS receiver noise immunity under direct interference of the solar radio emission is of obvious interest. For example, in the case of GLONASS, the normalized minimum power should not be less than –157 dBW at the main operating frequency of GLONASS (1600 MHz) and –163 dBW at the auxiliary frequency 1250 MHz (ICD-GLONASS, 2002). The corresponding standard for GPS determines these values as –163 dBW at the main frequency of GPS (1545.42 MHz) and –166 dBW at the second operating frequency (1227.6 MHz) (ICD-GPS-200c, 1993).

As we can conclude from the Fig. 1 signal-to-noise ratio at GLONASS receiver antenna output is 7 dB lower than the GPS one under the same solar radio emission power. It seems that we should expect lower noise immunity for GLONASS receivers under the same level of solar radio emission interference. Nevertheless, Fig. 2 convincing proves that GLONASS noise immunity should be higher. Our experimental results confirm that GPS receivers presented lower noise immunity under solar radio bursts interference on December 6 and 13, 2006.

Fig. 7 presents relative densities Q(t) of L1, C1, L2, P1, and P2 measurement failures respectively, which were computed for all observed GPS satellites (thick gray curves) and GLONASS satellites (thin black curves) registered within the sunlit zone on December 6 and 13, 2006. Symbols A (December 6) and B (December 13) mark the time intervals when the maximal level of solar radio emission power was >10^6 sfu (December 6) and > 10^5 sfu (December 13), respectively. As one can see, there is high reliability of L1 and C1 measurements on the main operating frequency of both GPS and GLONASS systems even in the periods "A" and "B" (Fig. 7a,e). No failures of L1 and C1 parameter were detected on December 6 (Fig. 7a). Only coincident short failures of L1 and C1 measurements were found simultaneously for both GPS and GLONASS at 03:34 UT (Fig. 7e). The results are in good agreement with the idea that if we know the ranging code structure exactly we do not have correlation losses and the unsafe level of solar radio emission is about 10^6 sfu for both GPS and GLONASS.

The situation is much worse when we utilize the codeless method in order to extract the P(Y) or HP code, in which case we have correlation losses is about ~27 dB. We can see sharply increasing Q(t) values by up to 35% on December 6, 2006 in P1, P2 and L2 parameters of GPS in the very period of sharply increasing solar radio emission power (Fig. 7b,c and d). The situation turned out to be more dramatic on December 13, 2006. One can see the Q(t) values of P1, P2 and L2 parameters exceed 50% within periods "A" and "B" (Fig 7f,g and h).

More important result consists of the fact that the maximum Q(t) value of all signal parameters of GLONASS, except for L1 and C1, is lower than the one for GPS by a factor of 2-4. In our opinion this advantage is due to the fact that a GLONASS receiver can perform its function more reliably under conditions of the powerful solar radio interference because of the narrower front-end passband of the GLONASS receiver for the separate GLONASS satellites compared to the GPS receivers. Unfortunately, the small statistics did not allow us to get more statistically significant assessments. We will further investigate it in the near future.

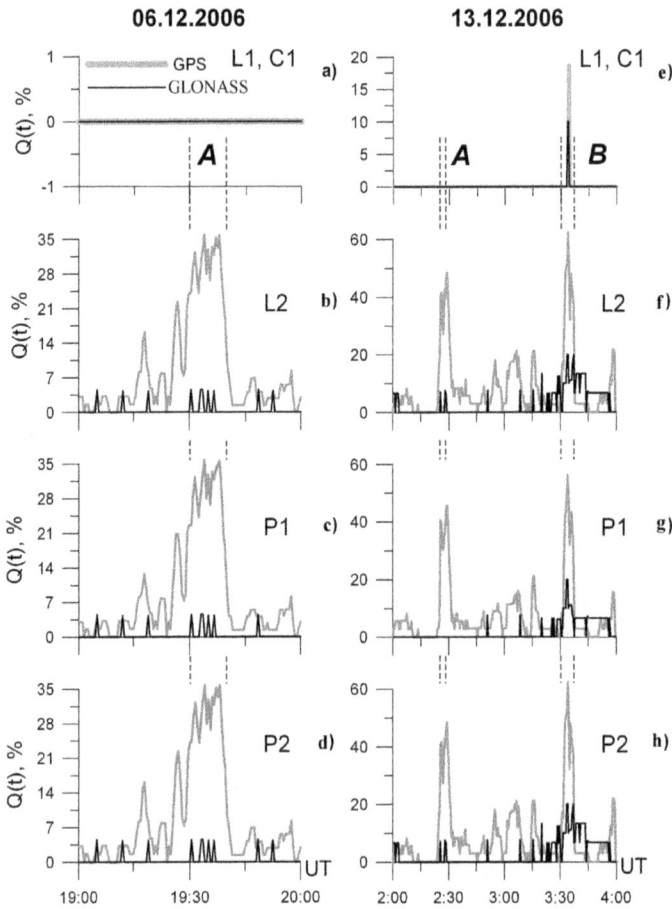

Figure 7. A comparative analysis of GPS and GLONASS noise immunity under powerful solar radio emission interference on December 6 and 13, 2006

As the whole results in Fig. 7 prove our theoretical assessments of the unsafe level of solar radio emission when GPS (GLONASS) signal tracking failures appear. The comparison between global failures of the GPS and GLONASS shows that the unsafe level of solar radio emission for GLONASS is higher than GPS. Thus, we have obtained the "unsafe" power level of the solar radio emission that could cause failures of GPS and GLONASS signal tracking. It was found that signal tracking started to fail when the solar radio emission power is about 4000 sfu at the GPS frequency L2 and 10.000 sfu at the frequency L1 when the "codeless" correlation processing technique is utilized. These assessments for GLONASS turned out to be 10.000 and 13.500 sfu at frequencies L1 and L2, respectively. Hence the GLONASS navigation receivers are more resistant to intensive solar radio emission under considered conditions. In our opinion this occurs because the GLONASS receiver radio path is characterized with a narrower front-end passband in order to provide effective extraction of signals of the particular GLONASS satellites.

Our theoretical assessments are proven by experimental statistics of GPS phase slips and counts omission that was found during powerful solar flares condition on December 6 and 13, 2006 and, especially, during the much weaker solar radio burst condition on October 28, 2003. It is very important to emphasize that although power of the solar radio burst on October 28, 2003 is by 2–3 orders of magnitude less than that on December 6 and 13, 2006, the maximum values of phase slips are only 5–10 times less. Particular slips of L1-L2 phase measurements started to appear under the interference of solar radio flux when power was just a bit higher than 10^3 sfu.

Experimental results have shown that for over 10–15 minutes the high-precision GPS positioning was partially disrupted on all sunlit sides of the earth on December 6 and 13, 2006. The high level of phase slips and count omissions resulted from the wideband solar radio noise emission. The statistics of phase slips obtained in this study for all sunlit sides of the earth confirms the suppression effect of GPS receiver performance during the December 6, 2006 flare with more reliability than the previously published, which were obtained and discussed at only several GPS sites by Carrano et al. (2007) and Cerruti et al. (2006).

In general, our results are in good agreement with earlier results by other authors, which indicate that direct impact of the solar radio emission can cause failure in GPS signal tracking of navigation receivers, even if the solar radio emission power is relatively low (about 10^3 sfu). It proves that solar radio noises of more than 10^3 sfu can have a negative influence on the GPS/GLONASSS performance.

3. Ionospheric super-bubble effects on GPS performance

Strong scintillations of amplitude and phase of transionospheric radio signals occur due to signal scattering on intensive small scale irregularities (Yeh and Liu, 1982). The size of such irregularities is the order of the first Fresnel zone radius, 150-300 m for 1.2-1.5 GHz frequency band. Scintillation can have an adverse effect on GPS signals and cause a GPS receiver to lose of signal lock in some extreme cases. Positioning quality deterioration can appear as a direct consequence of such physical phenomenon.

Over recent years, extensive studies of mid-latitude GPS phase fluctuations and phase slips under geomagnetic disturbances condition have been made (Skone, 2001; Conker et al., 2003; Ledvina et al., 2002, 2004; Ledvina and Makela, 2005; Afraimovich et al., 2002, 2003, 2009b; Astafyeva et al., 2008; Meggs et al., 2006). The expansion of the auroral oval equatorward is known to be accompanied by an increase of a number of slips in satellite signal tracking and GPS positioning deterioration in the mid-latitude region (Afraimovich et al., 2002, 2003, 2009b).

Ionospheric irregularities and steep gradients often occur at high-latitude ionosphere which is strongly disturbed due to auroral substorms. Existence of the ionization anomaly at low-latitudes, along with the well-known effect of the equatorial plasma bubble formation during the evening hours, increases the possibility for transionospheric signal fading in this region as well. Though the mid-latitudes are considered comparatively quiet, strong perturbations of plasma density are often observed at mid-latitudes as well. This is related to expanding either the auroral oval or the equatorial anomaly, especially during geomagnetic storms.

The plasma bubbles are associated with equatorial spread F (ESF) processes. A plasma bubble develops along the Earth's magnetic field line, elongated in the meridional direction, but it is much narrower in the zonal one. The plasma bubble has a finite height, and the poleward limit is determined by its equatorial height. The equatorial bubbles are not often observed at mid-latitudes. Observations have shown that in rare cases the ESF density depletions can reach high altitudes and extend to the equatorial anomaly latitudes. For example, satellite measurements showed that the ESF plasma depletions may reach apex altitude of 2000-6000 km (Burke et al., 1979; Obara and Oya, 1994; Ma and Marayama, 2006; Huang et al., 2007). Nevertheless, it is not fully understood yet how high ESF bubbles can emerge.

Although the ESF plasma bubble is a common phenomenon and it has been studied for years, precise observed data of ionospheric scintillations and loss of lock to GPS receivers at mid-latitude due to plasma bubbles affectation are still limited. A post sunset bubble manifested by loss of L2 signal lock was observed at mid-latitudes (~30°-34°N, ~130°-134°E) during the main phase of the 12 Feb 2000 storm (Ma and Marayama, 2006). The bubble had unusually large latitudinal extension reaching mid-latitude of 36.5°N (31.5°N magnetic latitude), indicating an apex height of ~2500 km.

However, in most papers there are no data regarding the space geometry of field-aligned irregularities (FAIs). For example, Ma and Marayama (2006) have identified their observations as bubbles using only the following criteria: the occurrence time (post sunset), the availability of phase slips, the total electron content (TEC) and DMSP local density depletion. But the above criteria are insufficient to identify the bubbles. It is necessary to obtain some direct evidence of the obliqueness of the scattering structure (i.e. FAI) along the magnetic field line.

What is the physical nature of the above phenomenon? It is necessary to obtain some direct evidence of the obliqueness of the scattering structure along the magnetic field line. We propose to use the additional information regarding angular characteristics of the scattering process. Our main idea was to test the relation between the LOS direction r and the direction of magnetic field line vector $\vec{B}$ at the altitude h_{max} of the ionospheric F2 layer maximum where the intensive irregularities causing the scattering of GPS signals and losses of L1 and L2 phase lock are located.

The goal of Section 3 is to describe our method of GPS detecting the mid-latitude FAIs, using loss of L2 phase lock, and to estimate their characteristics by example of the unusual February 12, 2000 events. Furthermore, it will further analyze the dependency of GPS positioning quality on the orientation of signal propagation relative to geomagnetic field lines during the registered super-bubble on February 12, 2000 on the basis of from the Japanese GPS network GEONET.

3.1. Mid-latitude field-aligned irregularities from GPS

Ma and Marayama (2006) have used only 300 GPS receivers selected homogeneously from GEONET to analyze loss of L2 signal lock for each visible satellite. We carried out our analysis for the total number of GPS GEONET receivers (~ 1000). We determine the azimuth α_S and the elevation θ_S of the LOS between a GPS site and a satellite and analyze L2 phase losses-of-lock density.

We use the data regarding L2 phase losses-of-lock only as initial. The lower signal to noise ratio at L2 is generally due to the fact that the L2 power at the GPS satellite transmitter output is 6 dB less than the fundamental frequency f1 with the C/A code (ICD-GPS-200c, 1993). Phase loss-of-lock at L2 may also be caused by a lower signal to noise ratio (SNR) at L2 frequency when commercial semi-codeless or codeless receivers are used. These receivers have no access to the military «Y» code at L2 GPS frequency. Hence they have to use a specific processing technique in order to extract L2 phase and ranging measurements. As a result we get significant correlation losses and the SNR at L2 is 13-17 dB lower in compare to the full code access mode at least. Thus the loss of signal lock at L2 may be considered as a sensitive indicator of the trancionospheric signal scintillations for all kinds of GPS receivers on the global scale under geomagnetic storm conditions (Afraimovich et al., 2002, 2003, 2009b). Such data may be an important addition to the data obtained by a few specialized monitors of ionospheric scintillations in the L range at mid-latitude as well (Kintner and Ledvina, 2005).

On the first step, we determine the coordinates of the ionospheric pierce point S_i at an altitude of F2 maximum h_{max} in the geodetic coordinate system ($\phi_p,\ \ell_p$)

$$\varphi_P = \arcsin\left(\sin\varphi_B \cos\psi_P + \cos\varphi_B \sin\psi_P \cos\alpha_S\right)$$
$$\ell_P = \ell_B + \arcsin\left(\sin\psi_P \sin\alpha_S \sec\varphi_P\right)$$
$$\psi_P = \frac{\pi}{2} - \theta_S - \arcsin\left(\frac{R_z}{R_z + h_{max}}\cos\theta_S\right) \tag{17}$$

Where R_z is the Earth radius;.($\phi_B,\ \ell_B$) are GPS site' geodetic coordinates (latitude and longitude), ($\phi_p,\ \ell_p$) are the geodetic coordinates of ionospheric pierce point. Ionospheric pierce point corresponds to altitude h_{max} =300 km.

Fig. 8 presents our scheme of the LOS angular scanning. The axes h, Y, and X are directed, respectively, zenithward, southward (Y) and eastward (X). The arrows LOS indicate the direction $\vec{r}$ along LOS and the magnetic field line direction $\vec{B}$; θ_s and θ_B - the elevations of vectors $\vec{r}$ and $\vec{B}$, respectively; γ - the angle between the vectors $\vec{B}$ and $\vec{r}$. L_{II} and L are the longitudinal and cross size of large scale bubble, elongate along magnetic field line and containing the set of small scale irregularities. The LOS1 direction (magnetic normal) corre-

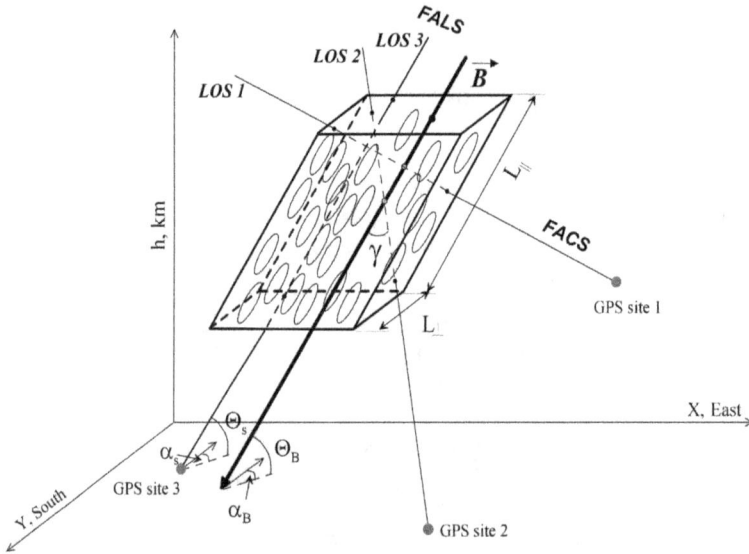

Figure 8. The scheme of the LOS angular scanning

sponds to the field of across scattering, F ACS ($\gamma{\sim}90°$) and LOS3 direction (magnetic zenith) is near the field of aligned scattering, FALS ($\gamma{\sim}0°$).

We calculate magnetic field line direction at altitude h_{max} in the ionospheric precise point S_i using the International Geomagnetic Reference Field model IGRF-10 (http://ngdc.noaa.gov/IAGA/vmod/igrf.html) and determine the angle γ between the vectors $\vec{r}$ and $\vec{B}$. Then we calculate the histograms $N(\gamma)$ of number of L2 phase losses-of-lock on the dependence of the angle γ value. The angular histogram bin $\Delta\gamma$ equals to 1 degree for all our results.

Fig. 9 presents the L2 phase losses-of-lock statistics on 12 Feb 2000: In the Fig 9a the map of sub-ionospheric points with the L2 losses-of-lock is presented (11:00-14:00 UT, PRN 7, 20773 30-sec counts, black dots; PRN 13, 25474 counts, gray dots; PRN16, 2758 counts, blue dots; and PRN24, 3175 counts, red dots). Fig. 9b shows the time dependence of the numbers of L2 losses-of-lock $N(t)$. The scale for PRN16 is shown on the right. Fig. 9e shows the angular trajectories of the sub-ionosphere points for GEONET GPS site 3054 (34.7°N, 137.7°E) during whole day February 12, 2000. The filled circle and filled band mark the LOS angular field near FALS and FACS band, respectively, with angular window of about 10°; violet curve correspond the γ = 90°. Fig. 9c is the same as Fig. 9e, but for all PRN during whole day 12 February 2000.

One can see the good likeness of Fig. 9a to the Fig. 4 from the paper by Ma and Marayama (2006). But here we can see the detailed distributions of losses of L2 phase lock for selected PRN. The number of L2 phase losses-of-lock $N(t)$ runs up to unexpectedly high value (up to

230 receivers for PRN 13) in ten minutes (Fig. 9b). The total number of L2 losses-of-lock is very significant; it corresponds to a very high level of scintillations, caused by GPS signal scattering on the intensive small scale irregularities. The effects of GPS signal multipath may be the alternative reason of such a deep radio-signal decay which causes the L2 signal lock.

Fig. 9f (blue line) shows the total histograms N(γ) for all PRN during the whole day February 12, 2000. First of all, the total number of L2 losses-of-lock is very significant (ΣN = 149948). Furthermore, the greatest number corresponds to the low value of the angle γ and the angle γ near 90°.

For the adequate angular distribution calculation it is necessary to take into consideration the real distribution of all LOS angular counts, the background function S(γ). Fig. 9f (black line) gives the total histograms S(γ) for all PRN during whole day February 12, 2000 (ΣN = 17961082). The most probable value S equals of about 64°; the least probable value S corresponds to the FALS. Fig. 9f presents the procedure of normalization of distribution N(γ): the blue, black and red lines are the initial series N(γ), the background function S(γ), which equal to the total amount of 30-s samples for chosen value of γ, and the normalized distribution Q(γ)= N(γ)/S(γ), respectively. One can see that the normalized distribution Q(γ) differs from initial series N(γ) significantly; the most probable value of γ for all L2 phase losses-of-lock during the whole day 12 Feb is close to the FALS.

The Fig. 9d shows the normalized distribution of L2 losses-of-lock Q(γ). The scale for PRN16 and PRN24 is shown on the right. Let us analyze the time interval 11:00-14:00 in details. Fig. 9d presents the normalized phase losses-of-lock distribution P(γ) of angle γ for chosen satellites (PRN07, PRN13, PRN16, PRN24); the total numbers of L2 phase losses-of-lock are indicated earlier. Corresponding LOS trajectories of the sub-ionosphere points for GPS site 3054 (34.7°N, 137.7°E) are shown in Fig. 9e. One can see that the L2 losses-of-lock are registered near the FALS (PRN 7 and PRN 13) and near the FACS (PRN 16 and PRN 24).

Similar angular dependencies are obtained for other time intervals (05:00-09:00 UT and 16:00-21:00 UT). Above mentioned results confirm well with data of investigation of magnetic field orientation control of GPS occultation observations of equatorial scintillation during detailed LEO CHAMP, SAC-C and PICOSat measurements, realized by Anderson and Strauss (2005). Inclination of LEO orbits allows authors to study magnetic field dependence in wide range of the angle between the occultation ray path and the magnetic field.

Fig. 10 presents the comparison between our the normalized distribution P(γ) of angle γ between the direction $\vec{r}$ along LOS and the magnetic field line direction $\vec{B}$ obtained from GEONET data, with the occurrence frequency for occultation events with maximum S4 greater than 0.09 versus the angle between the occultation ray path and the magnetic field from CHAMP and SAC-C measurements (Anderson and Straus, 2005). SAC-C naturally has the largest number of occultationevents and greatest probability of occurrence because it orbits half the time within the post-sunset MLT sector where equatorial scintillation is most likely to occur. CHAMP has the smallest because of its lower altitude and the smaller Fresnel scale as discussed earlier. Despite these differences, the data from all three satellites clearly indicate a strong dependence of the occurrence of equatorial scintillation on the angle between the

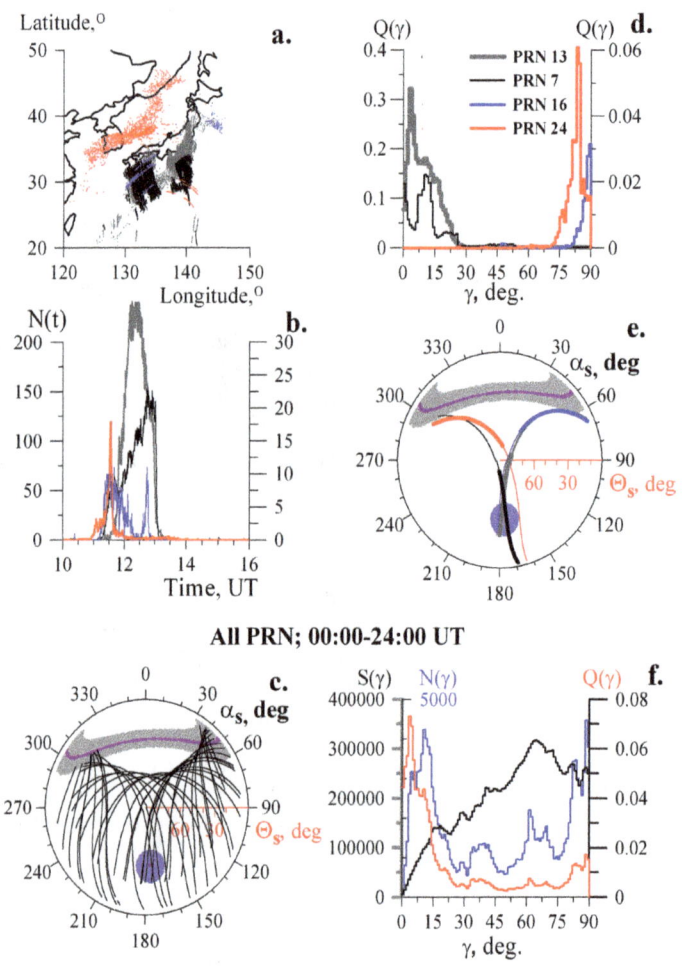

Figure 9. The statistics of the L2 phase losses-of-lock on 12 February 2000

occultation ray path and the magnetic field. One can see that the maximum values of the S4 occurrence probability and the normalized distribution $P(\gamma)$ correspond the small values of "B Field/Occultation Angle" and the angle γ. More narrow distribution $P(\gamma)$ caused by more hard condition for GPS LOS hit in the field of aligned scattering, corresponding to the exact direction of vector $\vec{B}$.

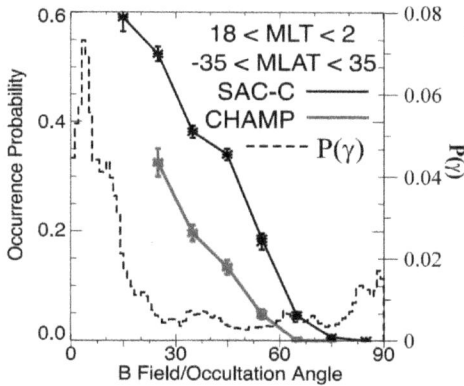

Figure 10. Comparison between our normalized distribution P(γ) and the occurrence of maximum S4 from CHAMP measurements (Anderson and Straus, 2005).

We agree with the clear explanation by similar dependence of the occurrence of GPS scintillation on the angle between the occultation ray path and the magnetic field (in our case the angle γ) presented by Anderson and Straus (2005): " ... we must examine the structure of ionospheric bubbles and the instabilities associated with them that lead to scintillation. Bubbles are formed in the bottomside of F region and drift upwards under the influence of horizontal electric fields. The plasma drifts in the post sunset hours are faster near the F peak than at other altitudes, leading to a "C shape" when viewed with imagers from the ground (Woodman and LaHoz, 1976), or from high altitude (Kelley et al., 2003). However, when viewed at a single altitude, the bubbles are nominally stretched out in latitude, or along magnetic field lines (McClure et al., 1977). So the walls of the bubbles are extended vertically upward and stretched in the north-south direction along the magnetic field lines (see Fig.8). Ionospheric interchange instabilities responding to the steep density gradient on the walls of the bubble in the F region produce the electron density irregularities responsible for radiowave scintillation. An occultation with a ray path which is perpendicular to the magnetic field cuts across the bubbles perpendicular to the two walls of the bubble and passes through the regions of density irregularities on the walls only briefly (see Fig. 8). However, an occultation with a ray path that is parallel to the magnetic field may pass along the edge of a bubble and the bubble wall, thus experiencing a path that may remain within the region of density irregularities for a significantly longer period than for occultation events with ray path perpendicular to the magnetic field. Put another way, the "slab thickness" of the irregularity region encountered by the GPS signals is larger when the ray path is aligned more closely to the direction of elongation of the bubble regions. Since scintillation strength increases with irregularity slab thickness (Yeh and Liu, 1982), the observed increase in scintillation occurrence for field-aligned conditions is reasonable".

The scattering in the field of aligned scattering, was studied well in numerous theoretical investigations and registered earlier many times in equatorial (Kintner et al., 2004) and high

(Wernik et al., 1990; Maurits et al., 2008) latitudes. Our results confirm the main conclusion of these investigations about the control over scintillation by the magnetic field direction. Of special importance are the investigations by Anderson and Straus (2005) where, during radio occultation observations, the scintillation increase at the angle decrease between the ray/LOS and the magnetic field was shown. However, the behavior of scintillations is determined not only by the angle between the field and the ray/LOS (off-B angle; Wernik et al., 1990), but also by some other factors: the elevation angle, the irregularity morphology, etc. This results in emergence of scintillations at large angles between the propagation and the magnetic field (see Fig. 10) which is in good agreement with the observational data, presented by Maurits et al. (2008).

The basic difference of our approach to investigating scintillations is in using of a rather substantial database on the information about malfunctions in the GPS receiving network. As it follows from the above comparison with the direct measurements of scintillations, such an indirect approach with using a close association of scintillations with GPS malfunctions, provides a possibility to monitor scintillations and scintillation-related FAI with an accuracy sufficient enough.

The effect of powerful HF radiation on the ionosphere has been investigated by using signals from high-orbiting GPS/GLONASS satellites (Tereshchenko et al., 2008). For the first time, they detected a quasi-stationary effect of magnetic zenith which leads to a decrease in the electron content and formation of electron density irregularities extended along the magnetic field lines. These authors exhibited the efficiency of GPS/GLONASS satellite signal application to investigate the ionosphere affected by HF radiation. But it is very important to compare the above results with the statistics of background TEC variation and scattering of GPS signals caused by natural field-aligned disturbances.

The scattering peculiarities of the transionospheric signal (when propagating along the magnetic field line) are necessary to take into account in all the radio occultation experiments where inhomogeneous media in the Ionosphere (Anderson and Straus, 2005), solar corona (Hewish and Symons, 1969; Pätzold et al., 1975), planets (Zhuk, 1980) and interstellar medium are investigated (Manchester and Taylor, 1977).

Our results are important for ionospheric irregularity physics development and the transionospheric radio wave propagation modeling. They are especially important in connection with the approaching of solar maximum which will produce strong ionospheric storms, increasing of the background ionization level and level of ionosphere plasma irregularities (Kintner et al., 2009). spatial-temporal features and modeling of mid-latitude FAIs are the goals of our future investigations.

3.2. GPS positioning errors during occurrence of a super-bubble

Figure 11 shows maps of sub-ionospheric points for those satellites for which L2 phase carrier measurements were experiencing slips during 5 minutes of observations. The points of appearance of GPS phase losses-of-lock indicate the values of angle γ by color. We found that before and after the bubble's observation at 10:00 and 13:45 UT respectively, there occurred

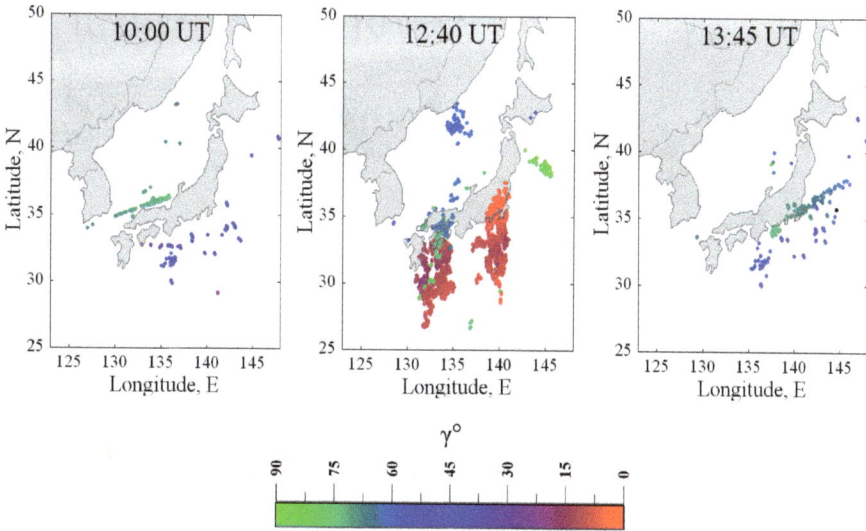

Figure 11. Sub-ionospheric points with L2 phase slips. The color scale identifies the size of the angle γ.

few phase losses of phase lock with weakly pronounced dependency on angle γ. The losses-of-lock occurred for the satellites with an angular orientation of LOS for a large range of γ ≈ 35…80°. The situation completely changed by 12:40 UT, when the bubble oriented along the geomagnetic field lines. During this time, we observed significant growth of up to 10 times in the number of L2 phase losses-of-lock. It is important to note that the majority of the observed lock losses corresponded to values γ ≈ 0…20°. In addition, we can see from the figure that these losses were concentrated within two regions, one in the area of evolution and drift of the bubble as was mentioned by Ma and Maruyama (2006), and the other is directed northeastward.

Positioning was carried out by what is called the Navigation Solution epoch by epoch using the dual-frequency ionospheric-free linear combination and the broadcast ephemeris as available from the RINEX files. The positioning errors are then calculated from by

$$\sigma(t_i) = \sqrt{\Delta x_i^2 + \Delta y_i^2 + \Delta z_i^2}, \tag{18}$$

where $\Delta x_i, \Delta y_i, \Delta z_i$ are the differences of the estimated geocentric position and the published GEONET coordinates. An event with $\sigma(t) > 500$ m is considered to represent an instant of a jump in the position estimate.

The distributions of positioning errors calculated for the GEONET stations for 10:00 UT, 12:40 UT and 13:45 UT are depicted in Fig. 12. At 10:00 UT, before the arrival of the bubble, the positioning errors of $\sigma(t) > 100$ m occurred at 60 GPS receivers, i.e. 6.3% of the GEONET stations.

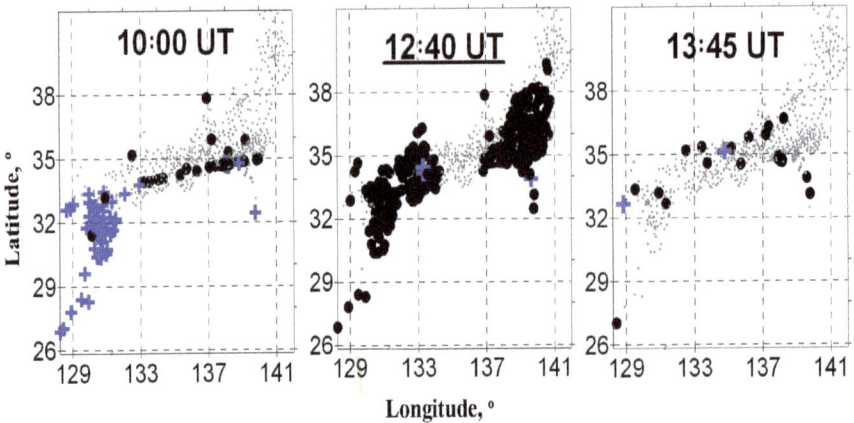

Figure 12. Positioning errors and jumps at GEONET GPS stations before the bubble appearance (10:00 UT), during the geomagnetic disturbance maximum (12:40 UT), and after the bubble disappearance (13:45 UT)

For the same time, positioning jumps were observed at 27 stations (2.8% of the stations). Note that the positioning errors mostly occurred on the southwest of the Japanese Islands, while the location of the positioning jumps was more sporadic (Fig. 12, left panel). By the time of the bubble arrived, at 12:40 UT, the number of the positioning errors $\sigma(t)>100$ m decreased to 0.4% (4 receivers), whereas the percentage of GPS positioning jumps increased to 33% (313 receivers). Comparison of Fig. 11 and 12 shows that at 12:40 UT the location of the positioning slips coincided with the location of L2 phase losses-of-lock, i.e. the losses-of-lock and jumps were concentrated in the region of the super-bubble. After the disappearance of the bubble by 13:45 UT only 2 GPS receivers (0.2%) registered positioning errors with $\sigma(t)>100$ m and 25 stations (2.6%) registered positioning jumps. Note that the jumps occurred mostly in the central part of Japan. Thus, before and after the bubble's appearance, the number of positioning jumps does not exceed 2.6%, whereas during the bubble occurrence that number reaches 33%.

Another important conclusion comes from Fig. 11 and 12. Although the bubbles are not known to propagate farther than 36.5° of latitude (Ma and Maruyama, 2006), their effects on the positioning quality may affect GPS users located more to the north. For instance, we observed occurrence of positioning jumps up to latitude 38.7° (Fig. 12).

In order to further study the relationship between occurrence of positioning jumps and latitude, we analyzed the temporal variations of $\sigma(t)$ at 12 GPS receivers located in the region of bubble observation. Fig. 13 shows the level of positioning errors between 6 and 16 UT at GPS stations within 31° and 41° latitude. For easy of interpretation of Fig. 13, we set $\sigma(t)=0$ whenever a positioning jump was occurred. The figure shows once again that the aspect conditions influence the positioning quality of GPS. From these stations, the northernmost ones (0027, 0024, 0026, bottom panels) did not reveal any indications of the effects of bubbles.

For GPS receivers located below about 38°N numerous positioning jumps occurred between 11-13 UT. This is due to the fact that the bubble did not reach the region of the northern stations.

Thus, using the Japanese GEONET, we observed numerous GPS phase losses-of-lock and positioning errors during the main phase of geomagnetic storm of February 12, 2000. The good coverage of the GEONET made it possible to perform for the first time such a detailed study on spatial-temporal dynamics of radio navigation failures associated with the evolution of electron density irregularities in the equatorial ionosphere.

Numerous GPS L2 phase losses-of-lock occurred during the time of the super-bubble obser-vations from 11 to 13 UT. Moreover, appearance of GPS losses-of-lock was found to depend on the angle γ between the station-satellite LOS and the geomagnetic field lines, with the maximum value of GPS phase losses-of-lock corresponding to $\gamma=0°$ and 90°. The maximum value of loses in the magnetic zenith region reached 32% and was observed at PRN13. The maximum density of phase losses-of-lock corresponded to the regions with maximum amplitude of TEC variations (and maximum TEC gradients), i.e. the regions of development and drift of the super-bubble as was mentioned by Ma and Maruyama (2006). Before and after the bubble's observations, i.e. before 11:00 and after 13:00 UT, few phase losses-of-lock occurred and their dependency on angle γ was weak.

Analysis of GPS positioning quality showed similar behavior. By the time of the bubble's appearance, 33% of GPS receivers experienced GPS positioning jumps with $\sigma(t) > 500$ m. Around 13:00 UT the positioning quality was worse than 100 m almost throughout all of Japan.

We also found latitudinal dependency of the number of positioning slips. There were practi-cally no slips registered at the northernmost GPS receivers of the GEONET. Thus, starting from latitude 38° N there were few positioning slips. In contrast to that, the site 0080 located at 34.3° N (Fig. 13) practically did not operate from 11 to 16 UT. It is important to note that, as the station latitude increases, the $\sigma(t)$ error generally decreases.

It was found that bubbles may propagate far on the north, although they are a typical feature of the equatorial ionosphere within ±20° relative to the geomagnetic equator. As we showed above, due to the emergence of aspect conditions for certain navigation satellites, the posi-tioning quality deterioration caused by the bubble is also possible at mid latitudes up to 38-39° N. The latter observation could serve as a probable explanation for the sharp worsening of positioning quality during the October 29-31, 2003 powerful geomagnetic storm at a number of GPS stations in California (30-40° N) reported by Afraimovich et al. (2009). It should be noted that the registered GPS phase losses-of-lock correspond to elevations much higher than the cut-off elevation value of 10°. Therefore, the registered losses-of-lock were not associated with low elevation effects.

Our results verify and supplement the results by Ma and Maruyama (2006) and Anderson and Straus (2005) and show that the aspect conditions for GPS signal propagation relative to electron density ionospheric irregularities aligned to the geomagnetic field may have a considerable effect on the performance of GPS.

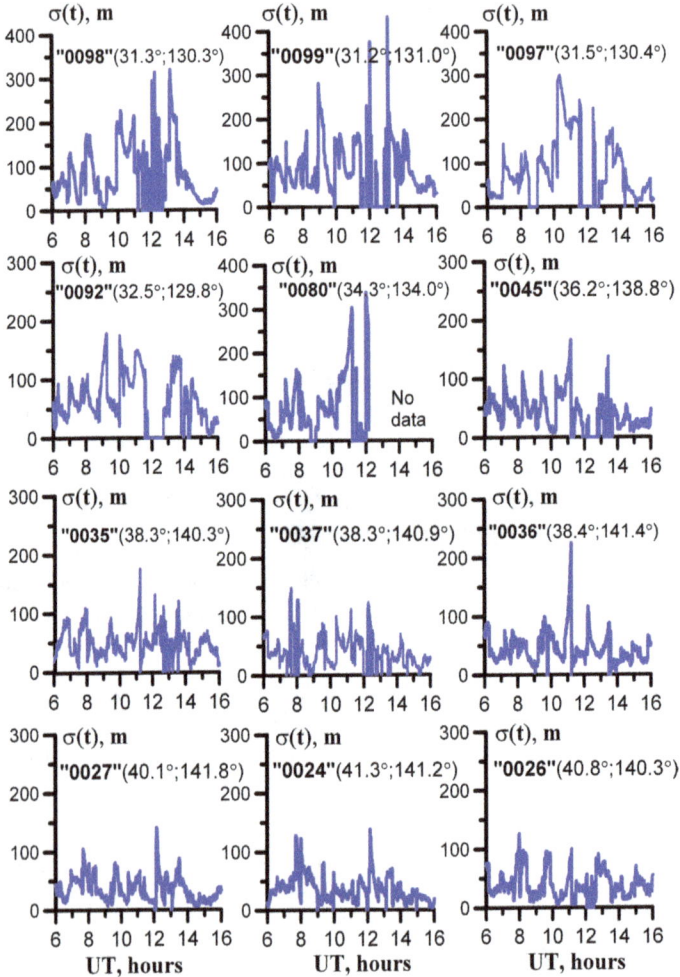

Figure 13. GPS positioning errors of 12 GPS GEONET stations from 6 to16 UT on February 12, 2000. The station names and the coordinates are indicated on the panels, respectively.

4. Real-time alert of sudden radio-propagation media disturbances for GNSS users

Upgrading the navigation support using global navigation satellite systems (GNSS) requires taking into account of more important effect factors, such as the large-scale disturbance of the near-Earth space environment. Influence of geomagnetic and sporadic disturbances of the near-Earth space environment on GNSS users includes effects of regular and irregular refraction in the near-Earth space environment, interference fading (scintillation) of transionospheric signals, and dispersion distortions.

Effects of regular and irregular refraction as well as their compensation methods by the GNSS equipment algorithms were studied quite well. Application of differential modes, along with two-frequency simultaneous measurements of radio navigation parameters, allows us to compensate sufficiently their influence on operation of GNSS consumer's equipment. However, according to recent research, it has precisely irregular effects of disturbances of the near-Earth space environment that can cause deterioration in the positioning accuracy or the positioning error due either to a sharp deterioration in constellation geometry. The effect of short-term fading and dispersion distortions is equivalent to the decrease in signal-to-noise ratio in the input of phase and code tracking loops. This may result in a sharp deterioration in measurement precision of radio navigation parameters or satellite vehicle (SV) signal tracking failure (Doherty et al., 2004).

As deduced from the findings (Kintner et al., 2001), influence of multiscale ionospheric disturbances on the GNSS operation is complicated and can not be predicted. This raises the question of the near-real-time control of signal propagation medium along the line of sight of a SV in the field of view at a given time. The aim of the near-real-time control is to give immediate notice to GNSS users, allowing the GNSS consumer's equipment to be adopted for sharply changing conditions of the SV signal propagation caused by geomagnetic and sporadic disturbances of the near-Earth space environment. This can be done through the adaptation of phase and code tracking loops. Removal of SV with abnormally strong scintillations of signal parameters from the solution of navigation problem can also be done to accomplish this. In the latter case, we can consider the possibility of giving notice of the propagation medium state as part of Receiver Autonomous Integrity Monitoring (RAIM) algorithms.

The aim of Section 4 is to analyze the possibility using the index of ionospheric scintillations of transionospheric signal power (S4) for online control of the integrity of measurements of radio navigation parameters with the use of signals from some SV under disturbed geomagnetic conditions of propagation medium of GNSS signals.

4.1. Connection between the scintillation index and TEC variations during multiscale ionospheric disturbances

According to modern opinion, electron density (ED) disturbances in the ionosphere are the superposition of compression - rarefaction waves with periods from tenths of minutes to 1-2 hours. Fig. 14 presents typical logarithmic power spectra of travelling ionospheric disturbances (TIDs) $\lg S^2$ obtained from GPS observations on July 29, 1999 (313 LOS) under quiet geomagnetic conditions and during the strong magnetic storm of July 15, 2000 (197 LOS) (Afraimovich and Perevalova, 2006). k is the spectrum slope, F is the TEC variation frequency, T is the period of variation and Λ is the wavelength (mean chosen velocity was 100 m/s).

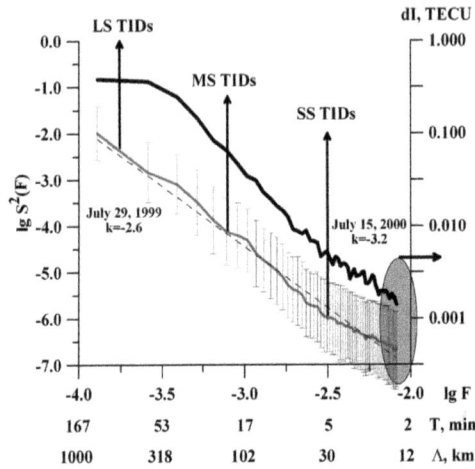

Figure 14. Characteristic spectrum of ionospheric disturbances under quiet and disturbed geomagnetic conditions (Afraimovich and Perevalova, 2006). The thick gray curve stands for quiet conditions, the black curve denotes disturbed conditions, and the dashed line is approximation of gray curve. Root-mean-square deviations are represented as grey segments.

According to Figure 14, intensity of ED irregularities of all spatial-temporal scales increases under disturbed geomagnetic conditions. In this case, spectral slope varies insignificantly, which suggests a proportional energy redistribution of geomagnetic disturbance under the ionospheric disturbances at all scales. The major portion of disturbance energy corresponds to large-scale and medium-scale TIDs (LS TIDs and MS TIDs, respectively). Such disturbances introduce smooth refraction errors of ranging to the GNSS equipment.

Small-scale TIDs (SS TIDs) may cause fast fluctuations of the refraction ionospheric ranging error. However, the main contribution to degradation of SV signals is made by scintillations of the radio signal received on earth due to diffraction on small-scale ED irregularities with a size of about one radius of the first Fresnel zone. The oval and horizontal arrows in Fig. 14 depict the region of these small-scale ionospheric inhomogeneities. According to recent research, such inhomogeneities accompanying magnetic storms are generated not only in low and high but also in middle latitudes (Demyanov et al., 2012; Afraimovich et. al., 2009b).

Smooth refraction effects caused by LS and MS TIDs can be effectively removed by two-frequency measurements of radio navigation parameters or by using differential GNSS mode. As for fast refraction variations and especially ionospheric scintillations, ranging errors or signal tracking failures caused by these events can not be corrected effectively due to the two-frequency measurements or the differential mode.

The most effective way out is to develop methods and algorithms for online detection of these events and giving immediate notice to GNSS users. This requires a universal index that could relate to the quality of radio navigation measurements to the current conditions of radio wave propagation in the Earth's ionosphere, given small-scale ED irregularities.

According to the results by Bhattacharyya et al. (2000), it is possible to use the scintillation index S4 for this purpose. Bhattacharyya et al. (2000) have established functional relationship between variations in integral parameter of the propagation medium - the total electron content along the line of sight of a satellite - and corresponding variations in the scintillation index at the receiving point. Such approach to determining ionospheric scintillations of SV signal parameters is based on the phase screen method. According to this method, variations in the carrier phase are related to TEC variations along the signal propagation trajectory (x) through a thin layer, containing ED irregularities, in the following way:

$$\Delta\phi(x) = -\lambda \cdot r_e \cdot \Delta I(x) \qquad (19)$$

where λ is the wavelength, r_e is the classical electron radius. The complex amplitude of SV signals at the receiving point (x, z) results from the addition of partial waves to phases $\Delta\phi_i(x)$:

$$u(x,z) = \sqrt{\frac{j}{\lambda \cdot z}} \cdot \int_{-\infty}^{\infty} u_i(x',0) \cdot \exp\left[\frac{-j\pi(x-x')^2}{\lambda \cdot z}\right] dx'$$
$$u_i(x,0) = U_{0,i} \cdot \exp\left(-j\Delta\phi_i(x)\right) \qquad (20)$$

where $u_i(x,0)$ is the amplitude of partial wave immediately after it emerges from the ionospheric irregularity layer, z is the distance from the scattering irregularity layer to the receiving point, x' is the current point coordinate along the irregularity layer.

Using the above-mentioned conception of the multiscale structure of ionospheric disturbances, we can easily determine contribution of the SV signal diffraction on ED irregularities to ionospheric scintillations and fast refraction fluctuations of SV signal parameters. For this purpose we can apply the Fourier transform to the complex amplitude (4.2), and the complex amplitude of the SV signal at the receiving point will be as follows:

$$u_F(q,z) = U_F(q,0) \cdot \exp\left(-j \cdot \pi \cdot \lambda \cdot z \cdot q^2\right), \qquad (21)$$

where $1/q$ is the spatial scale of ED irregularities in the layer. Given only large-scale disturbances whose sizes are much more than the first Fresnel zone $1/q >> d_F = \sqrt{2\lambda \cdot z}$, there will be only slow refraction phase variations (caused by the drift of the large-scale ED inhomogeneity relative to the wave propagation trajectory) in the complex amplitude, at the receiving point $u_F(q,z)$. That is why $u_F(q, z) \approx U_F(q, 0)$.

If, along with large-scale inhomogeneities, there are smaller-scale inhomogeneities, the complex amplitude at the receiving point $u_F(q,z)$ depends on exponential factor being a part of expression (20). According to (19-21), variations in the SV signal amplitude at the receiving point can be expressed through TEC variations. Contribution of ED irregularities having different scales is integrated along the line of sight.

Besides, it must be borne in mind that the propagation trajectories of the SV signal and ED irregularities move with respect to each other. To take this into account, TEC spatial variations along the 'receiver-satellite' line-of-sight should be transformed into temporal ones.

In the phase-screen approximation, variations in the signal phase and amplitude at the receiving point are interconnected (Bhattacharyya, 1999):

$$\frac{d^2}{dx^2}\phi(x,z) = \frac{2\pi}{\lambda \cdot z}\left[1 - \frac{\varepsilon(x,z)}{\varepsilon_0}\right]. \tag{22}$$

where $\varepsilon = u(x, z)^*$ and ε_0 are wave intensities before (i.e. undisturbed intensity) and after propagation through the irregularity layer. According to (19), phase variations can be directly related to TEC variations. At the receiving point, variations of the wave phase are caused by relative motion of ED irregularities at a velocity v. Spatial variations can be thus expressed through temporal ones. With this end in view change in space coordinate x should be expressed through velocity v (Bhattacharyya, 1999)

$$\frac{d^2}{dt^2}\Delta I(t) = -\frac{2\pi}{\lambda \cdot r_e}\frac{v^2}{\lambda \cdot z}\left[1 - \frac{\varepsilon(t,z)}{\varepsilon_0}\right], \tag{23}$$

where v is the relative velocity of motion of ED irregularities and wave propagation trajectories.

The scintillation index S4 is the root-mean-square deviations of variations in normalized series $\frac{\varepsilon(t, z)}{\varepsilon_0}$. Consequently, the equation (23) determines ionospheric scintillations of the SV signal amplitude, caused by TEC variations along the line-of-sight. The scintillation index S4 can be thus considered as a valid index of current conditions of the SV signal propagation through medium, given disturbances of different scales.

4.2. Analysis of the S4 index, TEC variations and positioning quality during a magnetic storm

In this section, we analyze applicability of the S4 index to immediate estimation of navigation measurement quality under disturbed geomagnetic conditions. We used experimental measurements of coordinates of the ISTP stationary station (Institute of Solar-Terrestrial Physics, Irkutsk, 52.2° N, 104.4° E, magnetic latitude is 41°) for analysis. Measurements were made with the GPS/GLONASS single-frequency navigation receiver MRK-19L. The equipment was installed at ISTP under stationary laboratory conditions. Geodetic survey of external antenna of MRK-19L was made with ASHTECH Z-XII3T, a dual-frequency GPS receiver. To measure TEC along the 'receiver - GPS satellite' line-of-sight we used data at the IRKT station of IGS network (http://garner.ucsd.edu/). The distance between ISTP and IRKT is 5 km so the ionosphere can't be changed significantly. Therefore we assumed that LOS was practically the same.

The experimental bench consists of:

• single-frequency receiver MRK-19L,

• computer controlling the operation mode of MRK-19L (control computer),

• computer registering measurement data (registration computer - R-comp).

Special software designed for MRK-19L was installed on the control computer. The software chose an operation mode of MRK-19L. Besides, the special software was responsible for communications protocol and mode between MRK-19L and R-comp. From the receiver, measurement data came to R-Comp in binary format. We have developed software to record data and installed it on R-Comp. The registration software includes a block of primary data logging and a block of reprocessing and analysis.

The programme of primary data logging performs data conversion. These data come to COM port of the R-Comp in a standard, unified data interchange format NMEA–0183. The period of data record in the primary file is 4 s. Then counts of the time of registration and of the receiver's current geodetic coordinates are extracted from the primary data file with the use of the block of reprocessing and analysis. Afterwards these coordinates are converted to rectangular geocentric coordinates.

We chose November 9, 2004 for analysis. On this day, we observed a complicated geomagnetic disturbance of the ionosphere caused by the solar flares of November 3 (01.24-01.44 UT) and November 5 (11.26-11.41 UT). On November 7, 2004, an abrupt drop in dynamics of the horizontal component H of the geomagnetic field was recorded at 22.00-24.00 UT. On November 9, 2004, moderate disturbances of the geomagnetic field were observed. In the time interval 20.00- 21.00 UT, there was a strong disturbance of the geomagnetic field once again; it was caused by the solar flare of November 5, 2004. Consequently, the day we analysed is characterised by a whole gamut of perturbing factors which affected near-Earth space and positioning accuracy.

There was increase in ED in the F layer from 14.00 to 20.00 UT on November 9, 2004. On the next day, however, such ED disturbances were not observed. The main contribution to ionospheric scintillations is made by disturbances in the F region. Consequently, we may assume that there were favourable conditions for formation of such scintillations on November 9.

The decrease in the average (of all satellites under observation) signal level at the receiving point can serve as a manifestation of intense ionospheric scintillations. To prove this, we analysed series of measurements of the SV signal-to-noise for each hour of observation. In

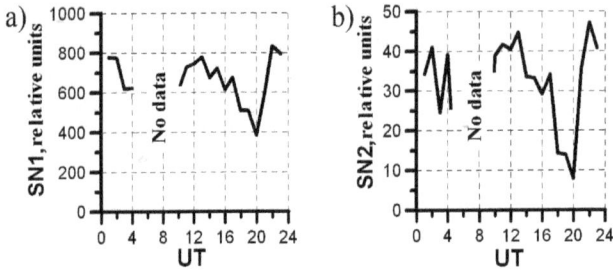

Figure 15. Averaged values of the SV signal level with elevation angles θ of more than 60° on November 9, 2004.

order to provide reliable detection of the effect, we considered measurements for which elevation angle θ was more than 60°. Because of this, the strength of effects at low elevation angles (whereat the signal attenuation can be related not only to ionospheric disturbances) decreases.

Results are presented in Fig. 15. SN1 (a) is signal-to-noise at L1 and SN2 (b) is the same at L2. From 4.30 to 9.50 UT, there were no SV with elevation angles θ of more than 60° over the observation site. A steady decrease in the signal level of both operating frequencies of SV is observed at the observation site from 17.00 to 21.00 UT. It coincides in time with occurrence of intense ionospheric disturbances caused by the magnetic storm.

The decrease in signal-to-noise ratio at the receiving point due to ionospheric scintillations may lead to deterioration in positioning quality. Fig. 16a presents series of errors in the positioning of ISTP - $\Delta S = \sqrt{\Delta X^2 + \Delta Y^2 + \Delta Z^2}$. Measurements were made with MRK-19L, on the basis of GPS data only. The time step was 4 s. A sharp increase in positioning errors was observed from 15.00 to 20.30 UT. This period corresponds to the local night (23.00-04.30 LT) and is marked by rectangle in Fig. 16a.

Notice that the said deterioration in the positioning precision correlates well (according to the occurrence time) with appearance of regions with increased ED. Besides, the period of increase in positioning errors coincides with the period of decrease in the SV signal level at both GPS frequencies (Fig. 15).

To extract effects caused by deterioration in geometry of constellation of satellites, we calculated Position Dilution of Precision (PDOP). Fig. 16b presents PDOP values for ISTP. The time step was 4 s. The maximum PDOP values were observed at 01.10 and 02.20 UT. This was due to the local peculiarity of the observation site. Note that the increase in positioning errors did not coincide with the maximum PDOP values. From 15.00 to 20.30 UT, we observed only short-term increases (> 3.5) in PDOP values. Thus, ionospheric scintillations of the SV signal amplitude were most likely to make the most substantial contribution to the increase in positioning errors during the period under consideration.

Fig. 17 exemplifies TEC observations made by GPS satellites PRN 06 and PRN 20 on November 9, 2004 at 00.00-03.00 UT. The figures also present the S4 index dynamics (see Fig. 17c,f). Time

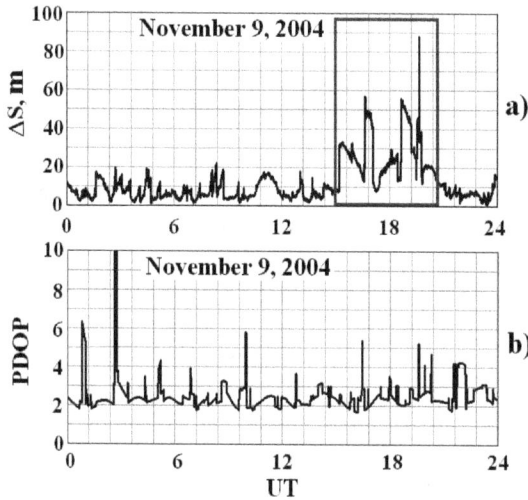

Figure 16. Positioning errors (a) and Position Dilution of Precision (b) during the magnetic storm on November 9, 2004. Data from MRK-19L were used for positioning.

series of TEC variations with eliminated trend are given in Fig. 17a,d. The method to eliminate the trend is described in (Afraimovich and Perevalova, 2006). We calculated TEC values with the use of two-frequency phase measurements. The values were measured with ASHTECH Z-XII3T (IRKT). The time step was 30 s. Fig. 17b,e illustrate series of the time derivative of TEC. The scintillation index was calculated using MRK-19L measurements of signal-to-noise ratio at the GPS L1 frequency.

$$S4^2 = \frac{\left[\langle P^2 \rangle - \langle P \rangle^2\right]}{\langle P \rangle^2}, \tag{24}$$

where P is the current value of signal-to-noise ratio measured in relative units and the time step is 4 s. When calculating the scintillation index, averaging of measurements of signal P amplitude was made in the time interval of 28 s in order to coordinate TEC measurements and estimate S4 values in time.

According to Fig. 17a-c, short-period TEC disturbances corresponding to small-scale ED irregularities are quite insignificant in this case. Smooth TEC variations are more significant. This is also the case for small values of the S4 index ($S4 \leq 0.1$).

Fig. 17d-f presents data from GPS PRN 06. As can be seen, fast TEC variations caused by small-scale ED irregularities are dominant. Consequently, high values of the S4 index (≥ 0.8) are observed (Fig. 17 f). We can draw a conclusion that, in the conditions under consideration, the S4 index reflects disturbances of propagation medium of SV signals quite reliably. The analysis substantiates the validity of using the S4 index under disturbed geomagnetic conditions. In this case, we observed superposition of effects of several geomagnetic disturbances of radio wave propagation medium. Thus, the S4 index can be applied to making online control of the quality of measurements of radio navigation parameters during magnetic storms.

There are, however, a number of restrictions that have not been taken into consideration. These restrictions should be considered individually. Noteworthy is the fact that velocity v and integration time of the right-hand and left-hand sides of equation (23) are undetermined parameters. These parameters may reliably reflect the current conditions of radio wave propagation.

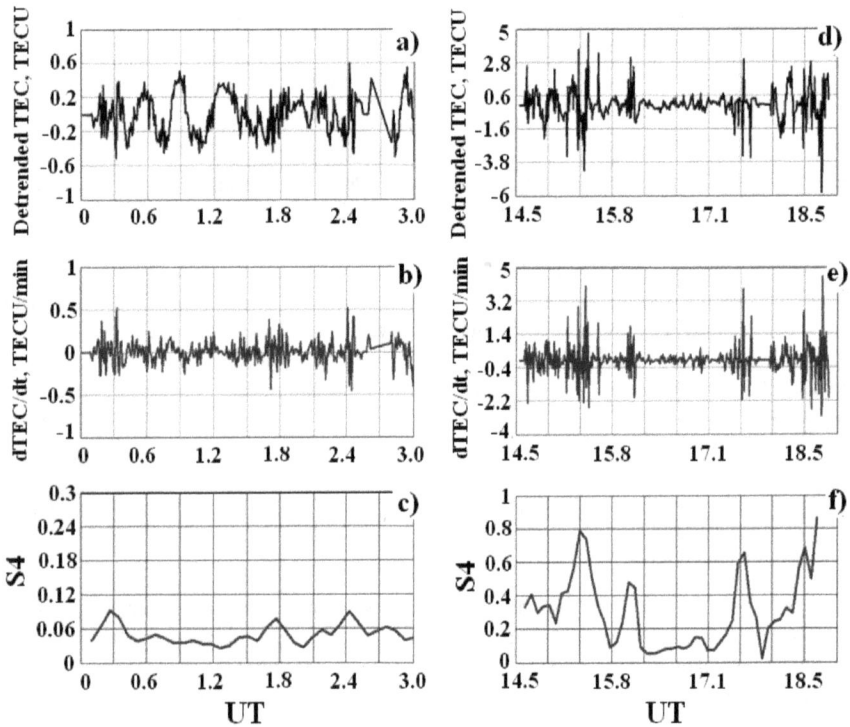

Figure 17. TEC variations (a, d), TEC derivative (b, e), and the S4 index (c, f). Data from the GPS satellite PRN 20 (a-c) and PRN 06 (d-e) were used.

The article deals mainly with ionospheric scintillations caused by the SV signal diffraction. The main contribution is made by ED irregularities with the scale of about the first Fresnel zone $d_F \approx \sqrt{2\lambda \cdot z}$. In this case, the integration time can be determined from relative velocity v. The

Fresnel frequency v_F (related to velocity v) corresponds to the magnitude inversely proportional to the characteristic scintillation period.

$$v_F = \frac{v}{\sqrt{2\lambda \cdot z}} \tag{25}$$

However, it is a common knowledge that the Fresnel frequency can vary over a wide range. For operating GPS frequencies, typical sizes of inhomogeneities located at 110-400 km that make the main contribution to amplitude scintillations are in the range 145 to 310 m. The GPS satellite moves in a circular orbit at 20200 km. The revolution period is 12 hours. If the line of sight is zenith-directed, the horizontal component of the rate of motion of the SV signal trajectory is 21 m/s for the E region and 77 m/s for the F region. The horizontal drift velocity of small-scale ED irregularities in middle latitudes is usually more than 150 m/s (Afraimovich et al., 2004). Consequently, the Fresnel frequency value can vary from 0.08 to 0.36 Hz. In high latitudes, the drift velocity can be up to 1 km/s. In this case, the Fresnel frequency can be significantly higher (Bhattacharyya et al., 2000; Afraimovich et al., 2004). Consequently, we should also analyse time of averaging of observational data that enter into the left-hand and right-hand sides of equation (23) to estimate adequately the current conditions of scintillation formation.

The restriction can be insignificant, if the phase-screen approximation is adequate. According to (Bhattacharyya et al., 2000), equation (23) can be replaced by the following empirical expressions:

$$DROTI = \alpha \cdot v_F \cdot S4$$
$$DROTI = \sqrt{\left\langle \left(\frac{d^2}{dt^2} \Delta I(t)\right)^2 \right\rangle - \left\langle \frac{d^2}{dt^2} \Delta I(t)\right\rangle^2}, \tag{26}$$

where α is the empirical parameter chosen for current observational conditions. In (Bhattacharyya et al., 2000, Bhattacharyya, 1999), it was found that $\alpha = 1.6 \cdot 10^3 - 8.5 \cdot 10^3$.

The phase screen approximation is not applicable when waves pass through the layer of inhomogeneities having high intensity and thickness. This is the case for the low-latitude and equatorial ionosphere. In the case under consideration, we have observed saturation of amplitude scintillations. If the irregular layer is thick, amplitude fluctuations are developed inside of it. As a result, the wave becomes subject to both phase and amplitude disturbances in the ionospheric layer output (Yeh and Liu, 1982). The well-known observational results show that amplitude scintillations were usually registered in such cases. Scintillations of the phase, angle of arrival and polarisation of received radio emission were observed much less often (Gunze, 1982). In this case, the S4 index determined only intensity of amplitude scintillations. These data are not enough to give immediate notice to GNSS consumers of a sharp deterioration in conditions of radio wave propagation. There are known cases of tracking loss of SV signals due to abrupt changes in the carrier phase. Noteworthy is the fact that variations in the signal amplitude are relatively low (Afraimovich et al., 2009b).

Additional research is needed to find new indices of signal propagation medium under different conditions. These indices should reflect variations in other parameters of the SV signal. The analysis should include influence of fluctuations of different radio-signal parameters on stability of system operation.

Acknowledgements

The authors are grateful a lot to Prof. E.L. Afraimovich, who will never see this text, sad to say. These results couldn't be obtained without him. Authors express profound gratitude to Prof. G. Y. Smolkov for his support and interest in this investigation. We are thankful to N.S. Gavriluk, A.B. Ishin, Prof. M.V. Tinin for their assistance in the preparing some parts. We are obliged to Dr. V. V. Grechnev for his help in using the 1 GHz and 2 GHz data of the Nobeyama Radio Polarimeters and colleagues from the Nobeyama Radio Observatory for solar radio emission data on December 13, 2006, as well as the International GNSS service and Geographical Survey Institute of Japan for RINEX data. The study was partially supported by RFBR (under grant No. 12-05-33032 a), by the Ministry of Education and Science of the Russian Federation (under agreement Nos. 8699, 8388, and 14.518.11.7065) and the Russian Federation President Grant MK-2194.2011.5.

Author details

Vladislav V. Demyanov[1*], Yury V. Yasyukevich[2] and Shuanggen Jin[3]

*Address all correspondence to: sword1971@yandex.ru

1 Irkutsk State Railway University, Russia

2 Institute of Solar-Terrestrial Physics, the Russian Academy of Sciences, the Siberian Branch, Irkutsk, Russia

3 Shanghai Astronomical Observatory, Chinese Academy of Sciences, Shanghai, China

References

[1] Afraimovich, E. L. (2000). GPS global detection of the ionospheric response to solar flares // Radio Sci., N 6, , 35, 1417-1424.

[2]   Afraimovich, E. L, & Perevalova, N. P. (2006). GPS monitoring of the Earth's upper atmosphere. Irkutsk. in Russian)., 480.

[3]   Afraimovich, E. L, Lesyuta, O. S, Ushakov, I. I, & Voeykov, S. V. (2002). Geomagnetic storms and the occurrence of phase slips in the reception of GPS signals // Ann Geophys, N 1, , 45, 55-71.

[4]   Afraimovich, E. L, Astafieva, E. I, Berngardt, O. I, Lesyuta, O. S, Demyanov, V. V, Kondakova, T. N, & Shpynev, B. G. (2004). Mid-latitude amplitude scintillation of GPS signals and GPS performance slips at the auroral oval boundary // Radiophysics and Quantum Electronics, N 7, , 47, 453-468.

[5]   Afraimovich, E. L, Demyanov, V. V, Ishin, A. B, & Smolkov, G. Y. (2008). Powerful solar radio bursts as a global and free tool for testing satellite broadband radio systems, including GPS-GLONASS-GALILEO // J Atmospheric and Solar-Terrestrial Phys, , 70, 1985-1994.

[6]   Afraimovich, E. L, Demyanov, V. V, & Kondakova, T. N. (2003). Degradation of performance of the navigation GPS system in geomagnetically disturbed conditions // GPS Solut., N 2, , 7, 109-119.

[7]   Afraimovich, E. L, Demyanov, V. V, & Smolkov, G. Ya. ((2009a). The total failures of GPS functioning caused by the powerful solar radio burst on December 13, 2006 // Earth, Planets and Space, , 61, 637-641.

[8]   Afraimovich, E. L, Astafieva, E. I, Demyanov, V. V, & Gamayunov, I. F. Amplitude Scintillation of GPS Signals and GPS Performance Slips // Adv. Space Res. , 43, 964-972.

[9]   Anderson, P. C, & Straus, P. R. (2005). Magnetic field orientation control of GPS occultation observations of equatorial scintillation // Geophys. Res. Lett., doi: 10.1029/2005GL023781., 32, L21107.

[10] Astafyeva, E. I, Afraimovich, E. L, & Voeykov, S. V. (2008). Generation of secondary waves due to intensive large-scale AGW traveling // Adv. Space Res., , 41, 1459-1462.

[11] Bhattacharyya, A, Beach, T. L, Basu, S, & Kintner, P. M. (2000). Nighttime Equatorial Ionosphere: GPS Scintillations and Differential Carrier Phase Fluctuations // Radio Sci., , 35, 209-224.

[12] Bhattacharyya, A. (1999). Deterministic retrieval of ionospheric phase screen from amplitude scintillations // Radio Sci., , 34, 229-240.

[13] Burke, W. J, Donatelli, D. E, Sagalyn, R, & Kelley, M. (1979). Low density regions observed at high altitudes and their connection with equatorial spread F // Planet. Space Sci., , 27, 593-601.

[14] Carrano, C. S, Groves, K. M, & Bridgwood, C. T. (2007). Effects of the December 2006 Solar Radio Bursts on the GPS Receivers of the AFRL-SCINDA Network // In: Doherty PH (ed) Proc International Beacon Satellite Symp, June , 11-15.

[15] Cerruti, A. P, Kintner, P. M, Gary, D. E, Lanzerotti, L. J, De Paula, E. R, & Vo, H. B. (2006). Observed Solar Radio Burst Effects on GPS/WAAS Carrier-to-Noise Ration. // Space Weather, doi:10.1029/2006SW000254., 4, S10006.

[16] Chen, Z, Gao, Y, & Liu, Z. (2005). Evaluation of solar radio bursts' effect on GPS receiver signal tracking within International GPS Service network // Radio Sci, doi: 10.1029/2004RS003066., 40, RS3012.

[17] Conker, R. S, Arini, M. B, Hegarty, J, & Hsiao, T. (2003). Modeling the effects of ionospheric scintillation on GPS/satellite-based augmentation system availability // Radio Sci., N 1, P. P. 1001, doi:10.1029/2000RS002604,, 38

[18] ICD-GLONASS ((2002). Global navigation satellite system-GLONASSInterface Control Document (in Russian). http://www.glonassgsm.ru/upl_instructions/-ICD-2002r.pdf

[19] Demyanov, V. V. Yasyukevich Yu.V., Ishin A.B., Astafyeva E.I.. ((2012). Effects of ionosphere super-bubble on the GPS positioning performance depending on the orientation relative to geomagnetic field // GPS solutions, N 2, DOI:s10291-011-0217-9., 16, 181-189.

[20] Doherty, P, Coster, A. J. A, & Murtagh, W. (2004). Space Weather Effects of October-November 2003 // GPS Solutions, , 8, 267-271.

[21] Gurtner, W, & Estey, L. (2009). RINEX: The Receiver Independent Exchange Format. Version 3.01. June 2009. Available from <http://igscb.jpl.nasa.gov/igscb/data/format/rinex301.pdf>.

[22] Gunze, E, & Zhaohang, L. (1982). The Ionospheric radiowave scintillations // TIIER, N.4, , 70, 5-45.

[23] Icd-gps, c. (1993). Interface Control Document. http://www.navcen.uscg.gov/pubs/gps/icd200/ICD200Cw1234.pdf

[24] Hewish, A, & Symons, M. D. (1969). Radio investigations of the solar plasma // Planet. Space Science, N 3, , 17, 313-320.

[25] Huang, C, Foster, S, & Sahai, J. C. Y. ((2007). Significant depletions of the ionospheric plasma density at middle latitudes: A possible signature of equatorial spread F bubbles near the plasmapause // J. Geophys. Res., , 112, A05315.

[26] Jin, S. G, Luo, O, & Park, P. (2008). GPS observations of the ionospheric F2-layer behavior during the 20th November 2003 geomagnetic storm over South Korea // J. Geod., N 12, doi:s00190-008-0217-x., 82, 883-892.

[27] Kaplan, E. D. ed) ((1996). Understanding GPS: Principles and applications. Artech House, 556 p.

[28] Kelley, M. C, Makela, J. J, Paxton, L. J, Kamalabadi, F, Comberiate, J. M, & Kil, H. (2003). The first coordinated ground- and space-based optical observations of equatorial plasma bubbles // Geophys. Res. Lett., N 14, doi:10.1029/2003GL017301., 30, 1766-1769.

[29] Kintner, P. M, & Ledvina, B. M. (2005). The ionosphere, radio navigation, and global navigation satellite systems // Adv. Space Res., N 5, , 35, 788-811.

[30] Kintner, P. M, Kil, H, & De Paula, E. (2001). Fading Time Scales Associated with GPS Signals and Potential Consequences // Radio Science, N 4, , 36, 731-743.

[31] Kintner, P. M, Ledvina, B. M, De Paula, E. R, & Kantor, I. J. (2004). Size, shape, orientation, speed, and duration of GPS equatorial anomaly scintillations // Radio Sci., doi: 10.1029/2003RS002878., 39, RS2012.

[32] Kintner, P. M, Humphreys, T, & Hinks, J. (2009). GNSS and ionospheric scintillation. How to Survive the Next Solar Maximum // Inside GNSS, N 4, , 4, 22-31.

[33] Klobuchar, J. A, Kunches, J. M, & Van Dierendonck, A. J. (1999). Eye on the ionosphere: Potential solar radio burst effects on GPS signal to noise // GPS Solut., N 2, , 3, 69-71.

[34] Ledvina, B. M, & Makela, J. J. (2005). First observations of SBAS/ WAAS scintillations: Using collocated scintillation measurements and all-sky images to study equatorial plasma bubbles // Geophys. Res. Lett., doi:10.1029/2004GL021954., 32, L14101.

[35] Ledvina, B. M, Makela, J. J, & Kintner, P. M. (2002). First observations of intense GPS L1 amplitude scintillations at midlatitude // Geophys. Res. Lett., N 14, DOI:GL014770., 29

[36] Ledvina, B. M, Kintner, P. M, & Makela, J. J. (2004). Temporal properties of intense GPS LI amplitude scintillations at midlatitudes // Radio Sci., doi: 10.1029/2002RS002832., 39, RS1S18.

[37] Ma, G, & Maruyama, T. (2006). A super bubble detected by dense GPS network at east Asian longitudes // Geophys. Res. Lett., DOI:GL027512., 33, L21103.

[38] Manchester, R. N, & Taylor, J. H. (1977). Pulsars. Freeman, San Francisco, 176 pp.

[39] Maurits, S. A, Gherm, V. E, Zernov, N. N, & Strangeways, H. J. (2008). Modeling of scintillation effects on high-latitude transionospheric paths using ionospheric model (UAF EPPIM) for background electron density specifications // Radio Sci., DOI: 10.1029/2006RS003539., 43, RS4001.

[40] Mcclure, J. P, Hanson, W. B, & Hoffman, J. H. (1977). Plasma bubbles and irregularities in the equatorial ionosphere // J. Geophys. Res., , 82, 2650-2656.

[41] Meggs, R. W, Cathryn, N. M, & Smith, A. M. (2006). An investigation into the relationship between ionospheric scintillation and loss of lock in GNSS receivers // Proceedings of the Meeting RTO-MP-IST-056 Characterising the Ionospere. Paper 5. Alaska, Fairbanks, US, June , 12-16.

[42] Obara TOya H. ((1994). EXOS-B (Jikiken) observations of the field aligned plasma density depletion at high altitude region // in Low-Latitude Ionospheric Physics COSPAR Colloquia Sen, 7, edited by F.-S. Kuo, Elsevier, New York, 1994, 275 pp.

[43] Pätzold, M, Neubauer, F. M, & Bird, M. K. (1995). Radio occultation studies with Solar Corona Sounders // Space Science Reviews, , 77-80.

[44] Perov, A. I, & Kharisov, V. N. (2005). GLONASS: Principles of Construction and Functioning. Radiotekhnika, Moscow, 720 pp. (in Russian).

[45] Skone, S. H. (2001). The impact of magnetic storms on GPS receiver performance // Geodesy, N 9-10, doi:10.1007/S001900100198,., 75, 457-468.

[46] Skone, S, & De Jong, M. (2001). Limitations in GPS receiver tracking performance under ionospheric scintillation // Physics and Chemistry of the Earth. Part A, N 6-8, , 26, 613-621.

[47] Tereshchenko, E. D, Milichenko, A. N, Frolov, V. L, & Yurik, R. Yu. ((2008). An observation of the magnetic zenith effect by using GPS/GLONASS satellites signals // Radiophysics and Quantum Electronics, N 11, , 51, 842-846.

[48] Tsui, J. B. (2005). Fundamentals of global positioning system receivers: a software approach.- 2nd ed. 0-47170-647-7G109.5.T85 2005.

[49] Wernik, A. W, Liu, C. H, Franke, S. J, & Gola, M. (1990). High-latitude irregularity spectra deduced from scintillation measurements // Radio Sci., N 5, , 25, 883-895.

[50] Woodman, R. F. LaHoz C. ((1976). Radar observations of F region irregularities // J. Geophys. Res., , 81, 5447-5466.

[51] Yeh, K. C, & Liu, C. H. (1982). Radio wave scintillations in the ionosphere // Proc. IEEE, N 4, , 70, 24-64.

[52] Zhuk, N. (1980). Scintillation studies of cosmic source angular structure (Review) // Radiophysics and Quantum Electronics, N 8, , 23, 597-615.

Chapter 7

Impact of Solar Forcing on Thermospheric Densities and Spacecraft Orbits from CHAMP and GRACE

Jiuhou Lei, Guangming Chen, Jiyao Xu and
Xiankang Dou

Additional information is available at the end of the chapter

1. Introduction

The thermosphere is the outer gaseous shell of a planet's atmosphere that exchanges energy with the space plasma environment. The energy deposition of solar irradiation and magneto-spheric inputs into the upper atmosphere can change the thermospheric density significantly. From a practical standpoint, unanticipated changes in the density of the thermosphere cause satellites to deviate from their anticipated paths, or ephemerides. Many studies have been pursued to investigate the variations of thermospheric densities caused by solar forcing, which includes solar irradiation and magnetospheric energy deposition [1-12]. However, the quantitative examination of the impact of thermospheric density changes associated with solar forcing on satellite orbits is rare, given that the simultaneous measurements of thermospheric density and precise tracking data of satellite are sparse.

Recently, we utilized the measurements obtained from the Challenging Minisatellite Payload (CHAMP) and the Gravity Recovery and Climate Experiment (GRACE) satellites to study the impact of solar irradiation and solar wind forcing on thermospheric density and satellite orbits as well [e.g., 13-14]. The CHAMP and GRACE satellites provided simultaneous observations for both thermospheric density and satellite orbit data. The CHAMP satellite was launched in July 2000 at 450 km altitude in a near-circular orbit with an inclination of 87.3°. Meanwhile, two identical satellites GRACE-A and GRACE-B were launched in March 2002 at approxi-mately 500 km altitude, in near-circular 89.5° inclination orbits with GRACE-B following approximately 220 km behind GRACE-A. Then the mass densities are obtained from CHAMP and GRACE accelerometer measurements using standard methods [15]. Note that only GRACE-A data are used for this study, given that the mass densities from GRACE-A and GRACE-B show very similar variations. On the other hand, the GPS receiver aboard CHAMP and GRACE satellites can provide precise tracking data of the spacecraft orbits [16]. The change of satellite altitude is caused by thermospheric drag that is proportional to thermospheric density ρ [13]:

$$\frac{dr}{dt} = -C_D \frac{A}{m} \sqrt{GMr} \rho \tag{1}$$

where r is the mean radius between the satellite and the Earth, C_D is the drag coefficient, m is the satellite mass, M is the mass of the Earth, G is the gravitation constant and A is the surface area of the satellite. Therefore, these simultaneous observations from CHAMP and GRACE for both thermospheric density and orbit tracking data at a high temporal resolution provide a good opportunity to explore the impact of thermospheric density on spacecraft orbits in a quantitative way.

2. Solar activity dependence of thermospheric density and satellite orbit

Figure 1 shows the temporal variations of daily mean thermospheric density, decay rate of satellite orbit per day of CHAMP and the corresponding NOAA hemispheric power (HP) from 2001 to 2006. It is clear that the variations of thermospheric density are well correlated with those of satellite orbit decay rate. Both thermospheric density and orbit decay rate show evident seasonal and solar cycle variations. The seasonal variation of thermospheric density and the resultant change in the orbit decay rate are explained by the thermospheirc spoon effect [17] and the seasonal variation of the lower atmospheric forcing associated with eddy mixing in the mesopause region [18]. In addition, the long term trend of thermospheric density and orbit decay rate is mainly driven by the corresponding changes of solar forcing [19], which is indicated by the F10.7 proxy and auroral hemispheric power HP [20].

There is a 27-day period variation of the densities, which is mainly caused by the periodic oscillation of solar radiation induced by solar rotation [3, 13]. The orbit decay rate of CHAMP, which is caused by the atmospheric drag, has the similar oscillations with the densities. Figure 2 gives the results for solar EUV flux, thermospheric densities observed by CHAMP and GRACE and orbital radiuses of the two satellites after a band-pass filter. The band-pass filter was centered at the period of 27 days, with half-power points at 22 and 32 days. It is obvious that both thermospheric densities and satellite orbital radiuses had a strong response to the 27 day oscillation in solar radiation. High correlations were found for the oscillations between thermospheric density, mean radius of the satellite orbit and EUV, especially when a strong quasi-27 day periodicity was present.

As seen in Figure 2, the amplitudes of the oscillations in thermosperic density and mean radius of satellite orbit of CHAMP were larger than those of GRACE. Given that the altitude of the GRACE was about 100 km higher than that of the CHAMP, the effect of thermospheric drag was weaker at the GRACE altitude. The oscillation of mean radius of the satellite orbit per revolution was about 0.1 km for CHAMP, while it was about 0.05 km for GRACE during the

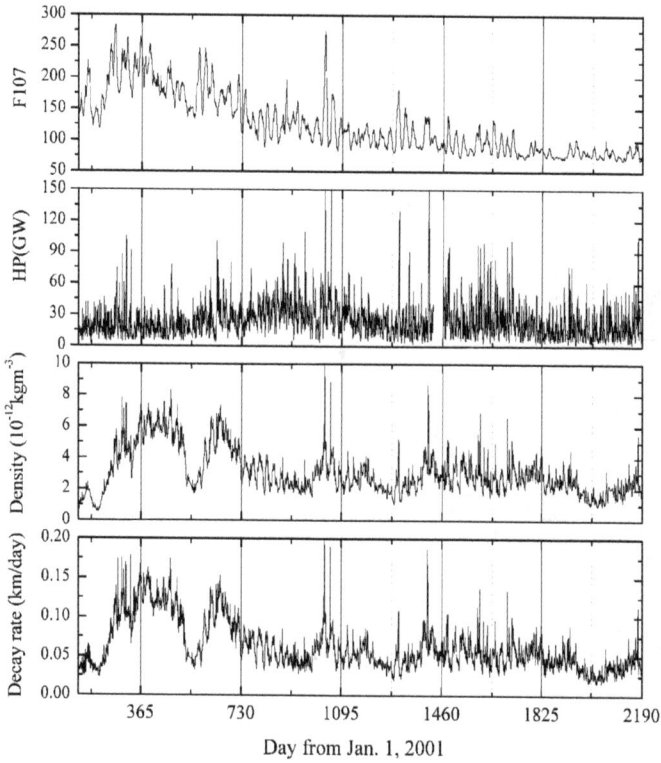

Figure 1. Variations of the F10.7, daily averaged auroral hemispheric power (HP), thermospheric density, and decay rate of the satellite orbit per day of CHAMP during 2001-2006. The day number is accounted from January 1, 2001.

first half of 2003. As the solar activity declines, the oscillations in thermospehric density and satellite orbit tend to decrease, for example in 2005.

Note that multi-day oscillations at the periods of 7 and 9 days were observed in the thermospheric densities [8-10], which are caused by the solar wind high-speed streams and the associated recurrent geomagnetic activity. The effect of the multi-day oscillations in thermospheric density is also imbedded in the satellite orbital radiuses [13]. Besides the periodic oscillations, sudden enhancements of thermospheric density and decay rate of satellite orbit, which are caused by the geomagnetic storms, are also seen simultaneously in Figure 1. For example, the enhancements during those storms in October, November in 2003 and November in 2004 are especially significant due to the large magnetospheric energy deposition. The values of the thermospheric density and orbit decay rate in the October 2003 Halloween storm reach their maximum, which are larger than 1×10^{-11} kg m^{-3} and 0.22 km/day, respectively. In the next section, we will give more details about the variations of thermosphere density and satellite orbits during the storm events.

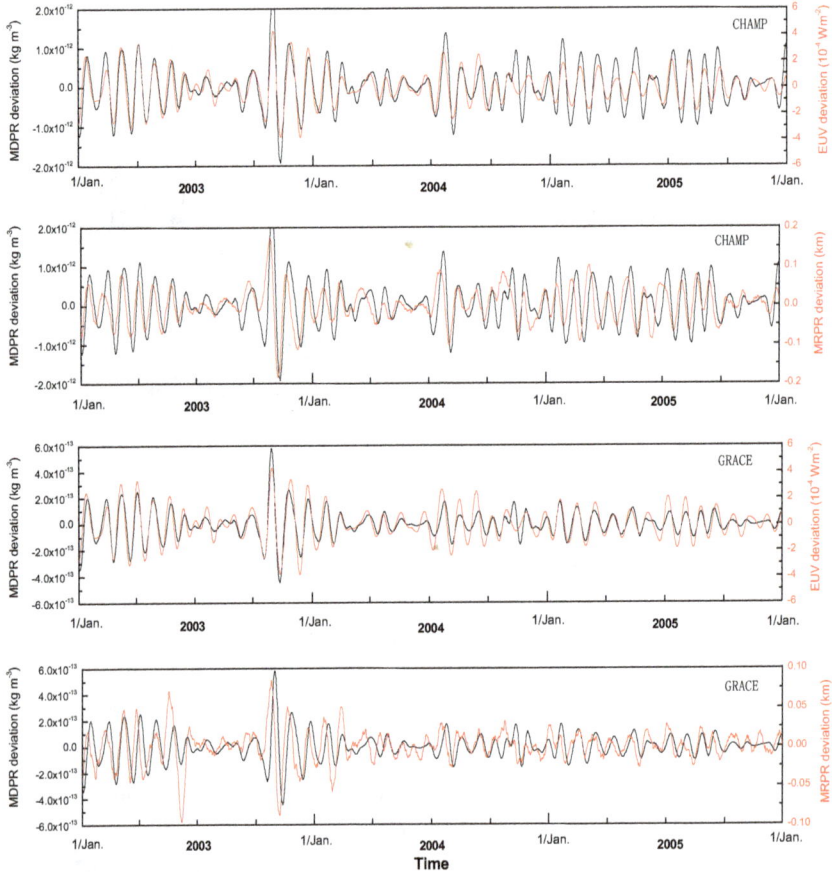

Figure 2. Band-pass filtered time series of EUV flux, mean thermospheric density per revolution (MDPR), and mean radius of satellite orbit per revolution (MRPR). From top to bottom: first panel, EUV flux and CHAMP MDPR; second panel, CHAMP MDPR and MRPR; third panel, EUV flux and GRACE MDPR; fourth panel, GRACE MDPR and MRPR. See [13].

3. Orbital variations induced by CME and CIR storms

There are significant differences between geomagnetic storms driven by coronal mass ejections (CME) and by corotating interaction regions (CIRs)/high speed solar wind streams [21, 12].

Usually, the strength of magnetospheric convection electric field of a CME-storm is stronger than that of a CIR-storm [22]. However, the duration of a CIR-storm is oftern much longer than that of CME [23, 14]. Although the rate of solar wind energy input into the magnetosphere of CIR is far less than that during coronal mass ejection (CME) magnetic storm intervals, the energy input over longer durations of time, e.g., several days or even longer, can be greater during high speed stream intervals [e.g., 24-26]. In this section, thermospheric densities and the orbit parameters from CHAMP are used to address the responses of satellite orbital altitudes to geomagnetic activity caused by CME and CIR storms.

Figure 3 shows, from top to bottom, the $F_{10.7}$ index, IMF B_z, solar wind density and velocity, Dst, ap and AE indices, auroral hemispheric power (HP) and thermospheric densities observed by CHAMP, satellite orbit decay rate and total orbit decay on days 323-326, 2003. The detail for the calculation of orbit decay rate can be found in Chen et al. [14]. The proxy F10.7 varied around 170 during this period, indicating high solar activity condition for this case. A geomagnetic storm occurred at 0728 UT on day 324, which is indicated by vertical line in Figure 3. This storm had a large IMF Bz southward component of about -46 nT during the main phase of the event. After the start of the storm, solar wind speed underwent a gradual increase from ~500 km/s to 700 km/s on day 324. The storm had a minimum Dst of -422 nT, which made it a strongest storm in the 23rd solar cycle using the *Loewe and Prölss* classification [27]. The maximum values of ap and AE reached 300 nT and 1698 nT, respectively. The auroral hemispheric power also increased significantly with a maximum of ~300 GW, indicating enhanced particle precipitation energy deposited into the ionosphere/thermosphere. At about 0100 UT on day 325, Bz turned northward and AE and HP recovered gradually to their pre-storm values. In response to geomagnetic activity and energy deposition, thermospheric density increased significantly. The averaged thermospheric densities reached a maximum of ~1.3×10^{-11} kg•m^{-3} at ~1900 UT on day 324 from its pre-storm value of ~3×10^{-12} kg•m^{-3}. Similar to the thermospheric density, the satellite orbit decay rate also reached a maximum of 279 m/day at ~2100 UT on day 324. Both thermospheric density and orbital decay rate recovered to their pre-storm values around the end of day 325.

Another geomagnetic storm (Storm 2) we are focusing on was caused by high speed solar wind streams and the resultant CIR that hit the Earth at about 1900 UT on day 258. As shown in Figure 4, this storm had a southward Bz excursion with a maximum amplitude of about 7 nT. The solar wind velocity increased from ~400 km/s to 800 km/s on day 261. The minimum of Dst for this storm was -57 nT on day 260, which denotes a moderate geomagnetic storm [27]. The maximum values of ap, AE index and HP reached 132 nT, 1228 nT and 125 GW, respectively, which are much smaller than those in the CME in Figure 3. Unlike the CME storm as shown previously (Storm 1), in which IMF Bz turned northward and AE, HP as well as thermospheric density, recovered rapidly after 0200 UT on day 325, this CIR storm had an oscillating IMF Bz that lasted for several days with high AE and HP values. Thermospheric density also stayed an elevated level till day 270. This is related to Alfvenic fluctuations in IMF and high speed streams that followed the CIR interval as discussed by *Tsurutani and Gonzalez* [23] and *Tsurutani et al.* [25]. The thermospheric density and orbital decay rate increased rapidly and reached the values of ~3.2×10^{-12} kg•m^{-3} and 68 m/day from their pre-storm values

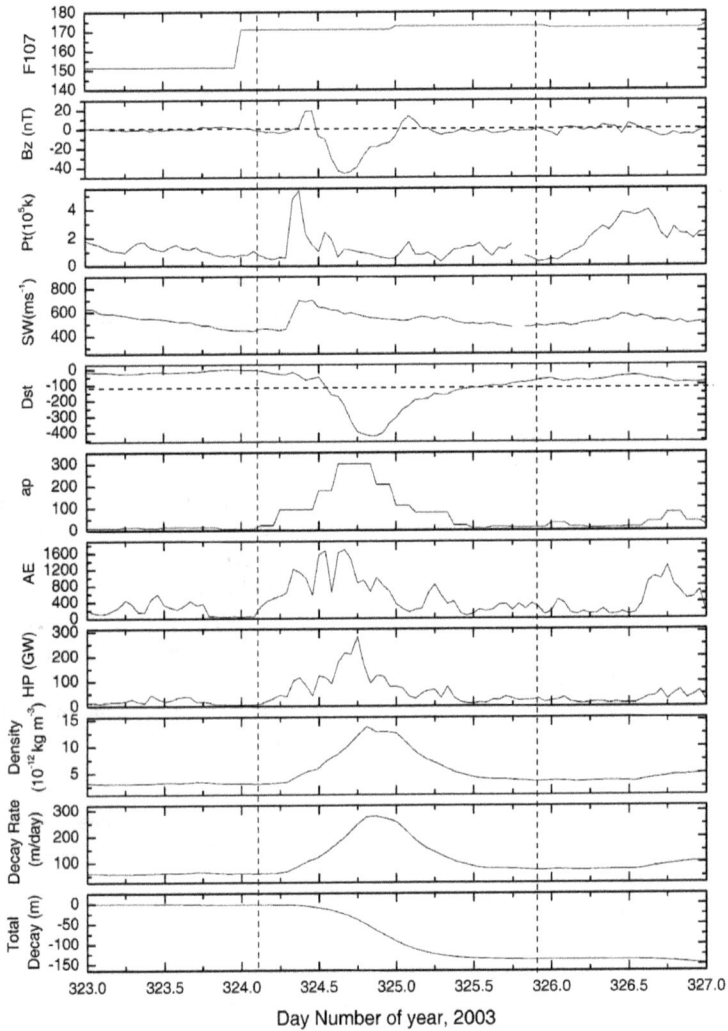

Figure 3. Variations of F10.7, hourly averaged Bz (nT), solar wind temperature (K), solar wind speed (km/s), Dst, ap, AE (nT), auroral hemispheric power (HP) and thermospheric density, CHAMP orbit decay rate and geomagnetic activities induced total orbit decays during November 19-22, 2003 (day 323-326, 2003). The dashed lines show the start and the end of the storm.

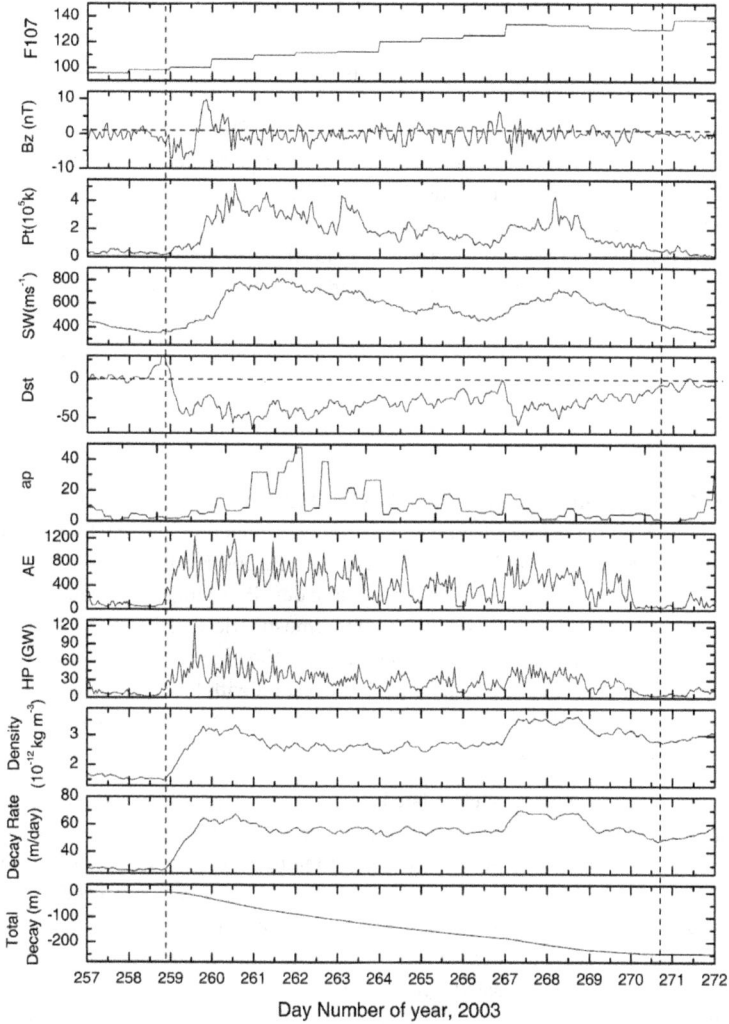

Figure 4. Same as Figure 3, but for the storm event during September 14-28, 2003 (day 257-271, 2003)

of ~1.5×10^{-12} kg•m^{-3} and 27 m/day, respectively. As indicated by the vertical lines in Figures 3-4, the duration of Storm 1 was 1.59 days, and that of Storm 2 was 11.98 days. Thus Storm 2

persisted for a much longer period of time, which consequently produced sustained perturbations to thermospheric densities and satellite orbits.

The total changes of orbit mean semi-major axis induced by these two storms are then
calculated by subtracting the observed variations in the semi-major axis from presumed semi-
major axis variations as a result of the drag by the quiet-time, background thermosphere. For
Strom 1, the storm-induced total variation of the semi-major axis was 130 m. For Storm 2, the
corresponding variation was 242 m, about a factor of 1.8 of Storm 1. Storm 1 is evidently
stronger, with deeper Dst minimum, stronger auroral activity, greater density changes and
larger orbital decay rates than Storm 2; however, it lasted a much shorter time. Thus, the
cumulative effect on thermospheric density and satellite orbit is less than that of Storm 2. As
a result, the total orbit decay caused by a strong CIR-storm can be larger than that by a severe
CME-storm. However, further comprehensive data analysis is required to explore the impact
of CME and CIR storms on the satellite orbit changes in a statistical way.

4. Summary

Thermosphere densities can be inferred from the CHAMP and GRACE accelerometer measurements with much higher temporal and spatial resolution than previous satellite drag data
in the upper thermosphere. Thus, the CHAMP and GRACE observations provide a unique
opportunity to investigate the impact of thermospheric density changes associated with the
solar forcing on satellite orbits. It is found that both thermospheric densities and the resultant
satellite orbit change vary significantly with solar activity. The oscillation amplitude of mean
radius of the satellite orbit per revolution, which is associated with the periodic oscillation of
solar radiation induced by solar rotation, can be as large as 0.1 km for the CHAMP, while it
was about 0.05 km for the GRACE.

The CHAMP and GRACE data have elucidated the thermosphere response to geomagnetic
storms in unprecedented detail. However, the effectiveness of the CME- and CIR-type storms
on satellite orbits is not well understood, albeit the differences between geomagnetic storms
driven by CME and by CIRs/high speed solar wind streams were recognized from the
interplanetary/solar wind structure viewpoint. Our case studies showed that the severe CME
storm caused larger thermosphere density disturbance and the resultant orbital decay rates
during its main phase, whereas it lasted a much shorter duration to compare with the CIR/
high speed stream event. However, the CIR storm can persist for many days and then produce
sustained perturbations to thermospheric densities and satellite orbits. As demonstrated in
our calculation, total variation of the semi-major axis was 242 m for the CIR storm during
September 15-27, 2003 in contrast to 130 m for the CME superstorm event during November
20-21, 2003. Therefore, the CIR storm can also cause significant impact on the thermospheric
density and the resultant satellite orbit change, given that it has long duration and occurs
frequently during the declining phase of a solar cycle and solar minimum.

Acknowledgements

This work is partly supported by the National Natural Science Foundation of China (41174139, 41274157, 41104098, 41004062), the Project of Chinese Academy of Sciences (KZZD-EW-01), China Postdoctoral Science foundation (20100481450, 201104799) and the Open Research Foundation of Science and Technology on Aerospace Flight Dynamics Laboratory (2012afdl1027). We also acknowledge the CEDAR data based at the National Center for Atmospheric Research (NCAR) for providing the auroral hemispheric power data used in this study. The ap and F10.7 indices were downloaded from NGDC database, and the ACE solar wind data were obtained from the GSFC/SPDF OMNIWeb interface at http://omni-web.gsfc.nasa.gov.

Author details

Jiuhou Lei[1*], Guangming Chen[2], Jiyao Xu[2] and Xiankang Dou[1]

*Address all correspondence to: leijh@ustc.edu.cn

1 CAS Key Laboratory of Geospace Environment, University of Science and Technology of China, Hefei, Anhui, China

2 State Key Laboratory for Space Weather, Center for Space Sciences and Applied Research, Chinese Academy of Sciences, Beijing, China

References

[1] Eastes, R., S. Bailey, B. Bowman, F. Marcos, J. Wise, and T. Woods (2004), The correspondence between thermospheric neutral densities and broadband measurements of the total solar soft X-ray flux, Geophys. Res. Lett., 31, L19804, doi: 10.1029/2004GL020801.

[2] Forbes, J. M., G. Lu, L. S. Bruinsma, R. S. Nerem, and X. Zhang (2005), Thermosphere density variations due to the 15–24 April 2002 solar events from CHAMP/STAR accelerometer measurements, J. Geophys. Res., 110, A12S27, doi:10.1029/2004JA010856.

[3] Forbes, J. M., S. Bruinsma, and F. G. Lemoine (2006), Solar rotation effects in the thermospheres of Mars and Earth, Science, 312, 1366– 1368.

[4] Sutton, E. K., J. M. Forbes, and R. S. Nerem (2005), Global thermospheric neutral density and wind response to the severe 2003 geomagnetic storms from CHAMP accelerometer data, J. Geophys. Res., 110, A09S40, doi:10.1029/2004JA010985.

[5] Liu, H., and H. Lühr (2005), Strong disturbance of the upper thermospheric density due to magnetic storms: CHAMP observations, J. Geophys. Res., 110, A09S29, doi: 10.1029/2004JA010908.

[6] Bruinsma, S. L., J. M. Forbes, R. S. Nerem, and X. Zhang (2006), Thermosphere density response to the 20–21 November 2003 solar and geomagnetic storm from CHAMP and GRACE accelerometer data, J. Geophys. Res., 111, A06303, doi: 10.1029/2005JA011284.

[7] Guo, J., W. Wan, J. M. Forbes, E. Sutton, R. S. Nerem, T. N. Woods, S. Bruinsma, and L. Liu (2007), Effects of solar variability on thermosphere density from CHAMP accelerometer data, J. Geophys. Res., 112, A10308, doi:10.1029/2007JA012409.

[8] Thayer, J. P., J. Lei, J. M. Forbes, E. K. Sutton, and R. S. Nerem (2008), Thermospheric density oscillations due to periodic solar wind high-speed streams, J. Geophys. Res., 113, A06307, doi:10.1029/2008JA013190.

[9] Lei, J., J. P. Thayer, J. M. Forbes, E. K. Sutton, and R. S. Nerem (2008), Rotating solar coronal holes and periodic modulation of the upper atmosphere, Geophys. Res. Lett., 35, L10109, doi:10.1029/2008GL033875.

[10] Lei, J., J. P. Thayer, J. M. Forbes, E. K. Sutton, R. S. Nerem, M. Temmer, and A. M. Veronig (2008), Global thermospheric density variations caused by high-speed solar wind streams during the declining phase of solar cycle 23, J. Geophys. Res., 113, A11303, doi:10.1029/2008JA013433.

[11] Lei, J., J. P. Thayer, G. Lu, A. G. Burns, W. Wang, E. K. Sutton, and B. A. Emery (2011), Rapid recovery of thermosphere density during the October 2003 geomagnetic storms, J. Geophys. Res., 116, A03306, doi:10.1029/2010JA016164.

[12] Lei, J., J. P. Thayer, W. Wang, and R. L. McPherron (2011), Impact of CIR storms on thermosphere density variability during the solar minimum of 2008, Sol. Phys., 274, 427-437, doi:10.1007/s11207-010-9563-y.

[13] Xu, J., W. Wang, J. Lei, E. K. Sutton, and G. Chen (2011), The effect of periodic variations of thermospheric density on CHAMP and GRACE orbits, J. Geophys. Res., 116, A02315, doi:10.1029/2010JA015995.

[14] Chen, G., J. Xu, W. Wang, J. Lei, and A. G. Burns (2012), A comparison of the effects of CIR- and CME-induced geomagnetic activity on thermospheric densities and spacecraft orbits: Case studies, J. Geophys. Res., 117, A08315, doi: 10.1029/2012JA017782.

[15] Sutton, E. K., R. S. Nerem, and J. M. Forbes (2007), Density and winds in the thermosphere deduced from accelerometer data, J. Spacecr. Rockets, 44, 1210–1219, doi: 10.2514/1.28641.

[16] Reigber, C., H. Lühr, and P. Schwintzer (2002). CHAMP mission status. Adv. Space Res., 30 (2), 129–134.

[17] Fuller-Rowell, T. J. (1998), The "thermospheric spoon": A mechanism for the semiannual density variation, J. Geophys. Res., 103, 3951– 3956.

[18] Qian, L., S. C. Solomon, and T. J. Kane (2009), Seasonal variation of thermospheric density and composition, J. Geophys. Res., 114, A01312, doi: 10.1029/2008JA013643.

[19] Knipp, D. J., W. K. Tobiska, B. A. Emery (2004), Direct and indirect thermospheric heating sources for solar cycles 21-23, Sol. Phys., 224, 495-505.

[20] Emery, B. A., I. G. Richardson, D. S. Evans, R. J. Rich, and W. Xu (2009), Solar wind structure sources and periodicities of global electron hemispheric power over three solar cycles, J. Atmos. Sol. Terr. Phys., 71, 1157–1175, doi:10.1016/j.jastp.2008.08.005.

[21] Borovsky, J. E., and M. H. Denton (2006), Differences between CME-driven storms and CIR-driven storms, J. Geophys. Res., 111, A07S08, doi:10.1029/2005JA011447.

[22] Denton, M. H., J. E. Borovsky, R. M. Skoug, M. F. Thomsen, B. Lavraud, M. G. Henderson, R. L. McPherron, J. C. Zhang, and M. W. Liemohn (2006), Geomagnetic storms driven by ICME- and CIR-dominated solar wind, J. Geophys. Res., 111, A07S07, doi:10.1029/2005JA011436.

[23] Tsurutani, B. T., and W. D. Gonzalez (1987), The cause of high-intensity long-duration continuous AE activity (HILDCAAs): Interplanetary Alfven wave trains, Planet. Space Sci., 35, 405-412.

[24] Tsurutani, B. T., W. D. Gonzalez, A. L. C. Gonzalez, F. Tang, J. K. Arballo, and M. Okada (1995), Interplanetary origin of geomagnetic activity in the declining phase of the solar cycle, J. Geophys. Res., 100(A11), 21,717– 21,733.

[25] Tsurutani, B. T., et al. (2006), Corotating solar wind streams and recurrent geomagnetic activity: A review, J. Geophys. Res., 111, A07S01, doi:10.1029/2005JA011273.

[26] Kozyra, J. U., et al. (2006), Response of the upper/middle atmosphere to coronal holes and powerful high-speed solar wind streams in 2003, in Recurrent Magnetic Storms: Corotating Solar Wind Streams, Geophys. Monogr. Ser., vol. 167, edited by B. T. Tsurutani et al., pp. 319–340, AGU, Washington, D. C.

[27] Loewe, C. A., and G. W. Pröss (1997), Classification and mean behavior of magnetic storms, J. Geophys. Res., 102 (A7), 14209-14214.

Geodetic Geophysics

Chapter 8

Earth Rotation – Basic Theory and Features

Sung-Ho Na

Additional information is available at the end of the chapter

1. Introduction

Earth rotation is, in most case, meant to be spin rotation of the Earth. Diverse seasonal variations are the result of Earth's obliquity to the ecliptic and orbital rotation of the Earth around the Sun. Although the linear motion of Earth's orbital rotation is faster than that of Earth's spin rotation, human beings cannot easily recognize it except aberration, which was first noticed by Bradley. In the Renaissance and also ancient times, scholars comprehended that orderly movements of stars in night sky are due to the Earth's spin rotation. To them, the planets, Moon and Sun moved with different periodicities in complicated way on the celestial sphere. As a matter of fact, the Earth's orbital/spin rotations are quite stable, and their stabilities far exceed human perception and most man-made instruments. However, the variations in both angular speed and direction of Earth spin have become detectable as technologies improved. Earth rotation is one of the most interesting scientific phenomena ever known. Moreover the importance of accurate knowledge of Earth rotation cannot be overestimated, because both the spatial and time systems of human civilization are referenced to the Earth and its spin rotational state.

Precession of equinox has been known since early day astronomy. The period of Earth's precession is about 26 thousand years. Nutation, which is periodic perturbation of the Earth's spin axis, is much smaller in amplitude than precession and often regarded as associated motion of precession. While precession and nutation are the motions of the Earth's spin axis (angular momentum) viewed by an observer in space outside of the Earth, the pole of Earth's spin rotation also changes with respect to observer on the Earth. Nowadays the Earth's rotational pole is usually represented by the Celestial Intermediate Pole (CIP). Slow drift of the Earth's pole in a time scale of thousand years is called polar wander. The polar wander is mostly due to the glacial isostatic adjustment and slow internal processes in the Earth. The pole of Earth's reference ellipsoid coincides with the Reference Pole, which was determined as average position of the Earth's pole between 1900 and 1905 (formerly called the Conventional International Origin). Pole position (CIP) with respect to the Reference Pole is represented by rectangular coordinates (x_p, y_p), which slowly draws a rough circle on the Earth's surface in a year or so. This pole offset is termed as polar motion.

Not only the orientation of the Earth's spin rotation but also its speed is slightly variable. Before the advent of quartz clocks, rotation of the Earth was regarded as the most reliable clock except planetary motion. Periodic variations in the length of day (LOD) have been found since the observations of star transit across meridian with accuracy better than 1 millisecond in the 1950s. Annual, semi-annual, and fortnightly perturbations are the first ones identified in the LOD variation. The Universal Time (UT), which is based on Earth rotation has been replaced by the Coordinated Universal Time (UTC) or the International Atomic Time (TAI), which are based on atomic clocks. Secular deceleration of the Earth due to tidal friction has been presumed since G. Darwin, and was confirmed later on.

Space geodetic technology since 1980s greatly enhanced precision and accuracy of measuring Earth rotation. Very Long Baseline Interferometry (VLBI) on the electromagnetic waves, which were emitted from quasar and detected at stations on Earth, has provided most important dataset, particularly for UT variation. VLBI and others as Global Positioning System (GPS), Satellite Laser Ranging (SLR) are capable of measuring pole offset with precision better than 1 milliarcsec as subdaily basis. Recently Ring Laser Gyroscope emerged as a unique and promising instrument to measure directly the Earth's spin rotational angular velocity with unprecedented accuracy.

Any position on the Earth's surface is usually denoted by its latitude and longitude in the Terrestrial Reference Frame (TRF), while direction to an astronomical object can be conveniently represented by its right ascension and declination in the Celestial Reference Frame (CRF). Transformation between TRF and CRF is attained with the necessary information; Earth rotation angle due to time passage from a certain epoch and small changes in orientation due to precession-nutation and polar motion.

In this chapter, characteristics of aforementioned variations in the Earth's rotation are compiled. Explanation of theoretical principle and observational features for each aspects are given one by one. But at times of much elaboration, more thorough treatment is avoided and proper references are recommended instead. The pre-requisite mechanics and mathematics are summarized in Appendix, where elementary vector algebra, harmonic oscillator, and basics for rotational mechanics *etc.* are covered. Old but still worth-reading monographs about Earth rotation are Munk and Macdonald [1], Lambeck [2], and Moritz and Mueller [3]. Explanations on terms and concepts of Earth rotation can be found in Seidelmann [4]. Extensive description about recent developments of Earth rotation study with emphasis on wobble and LOD was given by Gross [5]. In the same volume, Dehant and Mathews gave summary of Earth rotation theories with emphasis on nutation [6]. A technical note of International Earth Rotation and Reference Systems Service; IERS Conventions 2010 gives descriptions of new definitions in Earth rotation [7].

2. Precession and nutation

Precession of equinox has been noticed by careful observers since early days. Earth's spin rotational pole and vernal/autumnal equinoxes slowly change their positions on the celestial sphere. Like any other gyroscopic motions, the equation of motion for precession of the Earth can be simply expressed as follows.

$$\frac{d\vec{L}}{dt} = \vec{\tau} \tag{1}$$

Precessional torque comes mainly from the Moon and Sun; their gravitational pull to the Earth's equatorial bulge. But planets in the solar system also affect and contribute to the precession of the Earth in small amount. Earth's precession caused by all those effects together is called general precession, while luni-solar precession is referred to that caused by the Moon and Sun only. Recently it has been recognized that planetary perturbations lead to minute change in Earth's orbital plane, *i.e.* ecliptic. It is now recommended to use terms as precession of equator and precession of ecliptic [7]. Compared with the angular speed of the Earth's precession, lunar orbital rotation and apparent solar annual rotation around the Earth are much faster (similar comparison holds for the 18.6 years lunar orbital precession). Therefore, the luni-solar precessional torque can be evaluated by treating the masses of the Moon and Sun as hula-hoop shaped distribution (Figure 1).

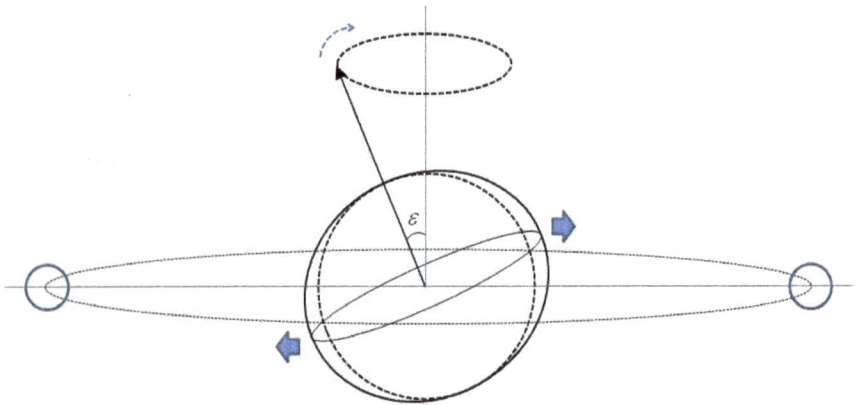

Figure 1. Due to the Earth's equatorial bulge, the Moon and Sun exert torque to Earth's spin rotation. The resulting slow precession is clockwise viewed from above the pole of ecliptic.

In fact, the precessional torque is exerted by differential gravitational force (= tidal force) on the Earth's equatorial bulge. Qualitative understanding may be directly acquired from Figure 1. If one considers both directions of the Earth's spin angular momentum $\vec{L}$ (solid arrow in the Figure) and the time averaged solar precessional torque $\vec{\tau}$ (the direction of the torque is perpendicular and into the paper), then, according to Equation; $\Delta\vec{L} = \vec{\tau}\,\Delta t$, the small change $\Delta\vec{L}$ of $\vec{L}$ in time increment Δt should be $\vec{\tau}\,\Delta t$. Average solar torque for the Earth's precession can be written as

$$\tau_s = \frac{3GM_s}{2r_s^3}(C - A)\sin\varepsilon\cos\varepsilon \tag{2}$$

where G, M_s, r_s, and ε are the constant of gravitation, mass of the Sun, mean solar distance, and the Earth's obliquity angle to the ecliptic plane. C and A are the principal moments of inertia of the Earth. Average lunar torque can be written similarly. Since the magnitude of Earth's spin angular momentum is $L = C\omega_0$, and the precession locus is shortened by a factor of $\sin\varepsilon$, the total angular speed of the Earth's precession due to both of the lunar and solar torques is given as the following.

$$\omega_{L.S.} = \frac{3G}{2\omega_E}\frac{C - A}{C}(\frac{M_m}{r_m^3} + \frac{M_s}{r_s^3})\cos\varepsilon \tag{3}$$

Derivation of Equations 2-3 can be found in [8-10], or other equivalent material. The angular velocity of general precession was recently found as $\omega_{G.P.} = 50.287946" / yr$.

Due to its precession, orientation of the Earth's spin axis ceaselessly changes in the celestial sphere. This change can be represented by the three angles ζ, θ, and z as illustrated in Figure 2. The transformation matrix corresponding to three successive rotations with these angles is represented as $R_{prec} = R_3(-z)R_2(\theta)R_3(-\zeta)$. Followings are the three angles of IAU2000A in terms of T, which is the time of given epoch in Julian century since J2000.0 - Greenwich noon on January 1st of year 2000 [11].

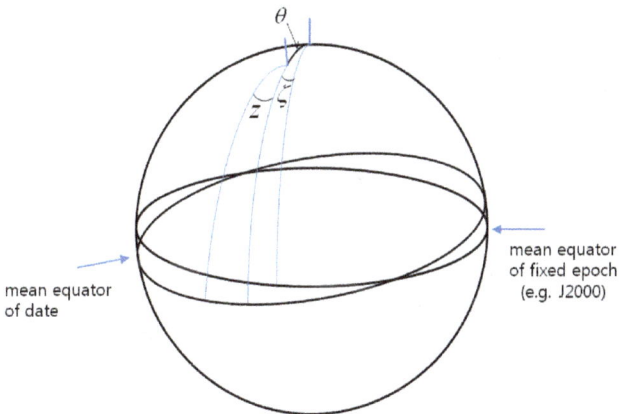

Figure 2. Representation of the Earth's precession by three angles ζ, θ, and z. The associated coordinate transformation matrix is given as $R_3(-z)R_2(\theta)R_3(-\zeta)$.

ζ = 2.5976176" + 2306.0809506" T + 0.3019015" T^2 + 0.0179663" T^3 − 0.0000327" T^4 − 0.0000002" T^5

θ = 2004.1917476" T − 0.4269353" T^2 − 0.0418251" T^3 − 0.0000601" T^4 − 0.0000001" T^5 (4)

z = − 2.5976176" + 2306.0803226" T + 1.0947790" T^2 + 0.0182273" T^3 + 0.0000470" T^4 − 0.0000003" T^5

Change in the right ascension and declination of any fixed position on the celestial sphere can readily be acquired with known precession matrix.

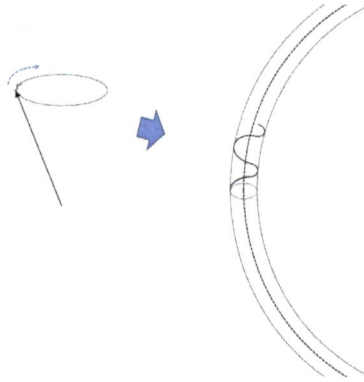

Figure 3. Schematic illustration of 18.6 year nutation superposed on the precession of the Earth. The locus of angular momentum vector of the Earth during about one and half nutation period is drawn, which is enlarged on the right.

Besides precession, there exist small periodic oscillations superposed on it, and these are called nutations. The cause of nutation is the luni-solar gravitational pull to Earth's equatorial bulge, same as for precession but only with different frequency. In Figure 3, the main nutation (18.6 year period) is illustrated as a superposition on the precession. There are numerous different periodic components in the Earth's nutation. Treating the lunar and solar masses as circularly distributed rings was only an approximation to consider precession, but the true motions of the Moon and Sun relative to the Earth definitely are periodic. Accordingly effect of the periodic lunar or solar torque exist as oscillatory perturbations, *i.e.*, nutations. Like the response of a simple harmonic oscillator to periodic driving force (away from resonance), which is proportional to inverse square of the driving frequency, amplitude of longer period nutation should be larger.

The nutational torque vector and resultant change in angular momentum vector can be briefly described as follows (here, precession itself is neglected for the time being. for more detailed treatment, see [3, 6]). Denote the nutational angular momentum and torque as $\vec{L}_n$ and $\vec{\tau}_n$. Write the nutational torque vector as elliptically rotating in time;

$\vec{\tau}_n = \hat{e}_1 A_1 \cos\omega_n t \mp \hat{e}_2 A_2 \sin\omega_n t$ ('−' for retrograde, '+' for prograde). Then steady state solution

for $\vec{L}_n$ of the equation of motion; $\dfrac{d\vec{L}_n}{dt} = \vec{\tau}_n$, readily follows as $\vec{L}_n = \hat{e}_1 \dfrac{A_1}{\omega_n} \sin\omega_n t \pm \hat{e}_2 \dfrac{A_2}{\omega_n} \cos\omega_n t$

('+ ' for retrograde, '–' for prograde). These are illustrated in Figure 4. Viewed from above, the motions of each set of retrograde/prograde nutational torque and angular momentum are clockwise/anticlockwise rotations, and nutational torque is ahead of phase $\pi/2$ for both cases. The 18.6 year nutation is retrograde. Other major nutational components, such as fortnightly and semiannual, are mostly prograde, since the orbital rotations of the Moon and Sun (apparent motion relative to the Earth) are prograde. The number 1, 2, and 3 in Figure 4 represent successive times separated by a time increment Δt, and angular momentum change is given as $\Delta \overline{L}_n \simeq \overline{\tau}_n \Delta t$. The nutational angle is given as $\arctan(L_n/C\omega_e) \simeq L_n/C\omega_e$. While torque itself is the same for different nutational components of same origin (lunar or solar), lower frequency nutation evidently has larger amplitude due to the inverse frequency dependence of nutation angular momentum; $L_n = |\overline{L}_n| = |\tau_n|/\omega_n$. This is the reason why 18.6 year nutation has much larger amplitude than other short period ones. Likewise semiannual nutation is roughly six times larger than fortnightly one (remember lunar tide is twice larger than solar tide). However, semi-annual nutation is of larger amplitude than annual one (particularly for the case of $\Delta\varepsilon$). Similarly fortnightly nutation amplitude is larger than monthly one. These are due to symmetric nature of tidal force. If the eccentricities of lunar/solar orbits were smaller, the monthly and annual nutations would have been virtually nil.

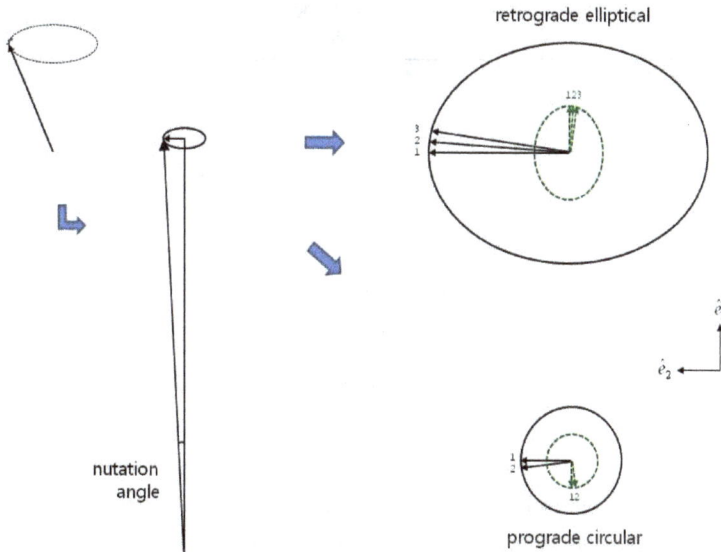

Figure 4. Schematic illustrations of precessional/nutational angular momentum and nutational torque. On the left, total angular momentum is shown. In the middle, nutation angle is described by two angular momentum vectors. On the right, two cases of nutation. (above: retrograde elliptical, below: prograde circular) are shown. Nutational torque vectors are drawn in olive. The two unit vectors $\hat{e}_1$ and $\hat{e}_2$ are in the direction of longitude decrease and obliquity increase respectively. For convenience, half cycle phase is added to vectors of prograde case.

Two angles used to specify nutation are denoted as $\Delta\varepsilon$ and $\Delta\psi$, which are illustrated in Figure 5.

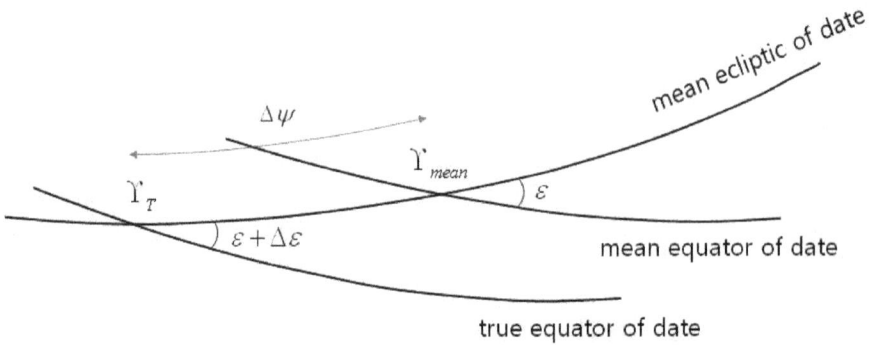

Figure 5. Nutation described by two angles $\Delta\varepsilon$ and $\Delta\psi$.

In Table 1, several nutation components of largest amplitude are listed.

Period (day)	Amplitude of obliquity $\Delta\varepsilon$	Amplitude of longitude $\Delta\psi\sin\varepsilon$	Remark
6798.4	9203	6858	lunar orbital precession
365.3	5	57	annual
182.6	574	526	semi annual
27.6	1	28	monthly
13.7	98	91	fortnightly
9.1	13	12	modulated

Table 1. Amplitudes of six largest nutation components [unit: milliarcsec]

The amplitude ellipses of four major components of Earth's nutation are shown in Figure 6.

The transformation matrix corresponding to the nutation described by the two angles $\Delta\varepsilon$ and $\Delta\psi$ is given as $R_{nut} = R_1(-\varepsilon - \Delta\varepsilon)R_3(-\Delta\psi)R_1(\varepsilon)$. The period of the largest nutational component is 18.6 years, which corresponds to the retrograde precession of the lunar orbital plane. Next largest components are semiannual, fortnightly, annual and monthly nutations. The most recent model for the precession and nutation has been reported by Mathews et al. [12], which is adopted by International Astronomical Union (IAU) as current standard model [11-12]. Although out-dated, nutation calculated by using rigid Earth approximation of former studies [13-14] show close match with the most recent one [12]. This is due to the fact that the Earth behaves as an almost perfectly rigid body at such slow variations as the precession and nutations. It is noted that the model of Mathews *et al.* was preceded by several other nutation models for nonrigid Earth. For example, see [15-18].

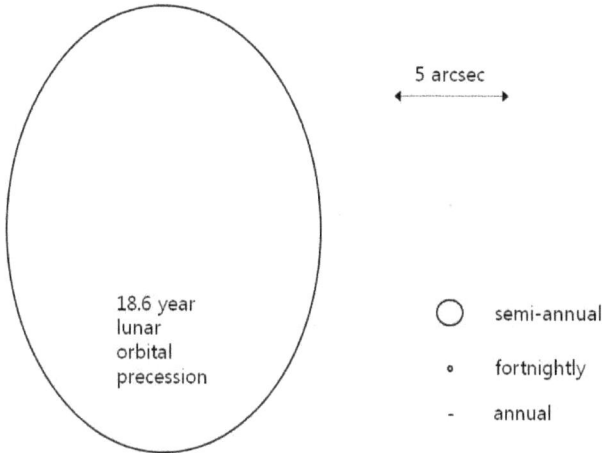

Figure 6. Comparison on size and shape of the nutation ellipses of four major components; 18.6 year, semiannual, fortnightly, and annual. A scale arrow of 5 arcsec is drawn for reference.

There are other ways to represent the coordinate transform associated with the Earth's precession and nutation. One of them is to use the position of pole projected onto the ICRF plane, of which orientation to quasars is invariable. IAU 2006 recommended to use a set of associated new terms and definitions as described below. For details, see [7] and references therein.

The transformation matrix, which incorporates both precession and nutation together, can be written as follows.

$$R_{prec+nut} = \begin{pmatrix} 1-bX^2 & -bXY & -X \\ -bXY & 1-bY^2 & -Y \\ X & Y & 1-b(X^2+Y^2) \end{pmatrix}$$

where b is defined as $b = 1/(1+Z)$ with $Z = \sqrt{1-X^2-Y^2}$. Followings are IAU 2006 expressions for X and Y without explicitly showing all the oscillatory terms of nutation.

$X = -0.016617" + 2004.191898" \, T - 0.4297829" \, T^2 - 0.19861834" \, T^3 + 0.000007578" \, T^4$
$+ 0.0000059285" \, T^5 +$ oscillatory terms

$Y = -0.006951" - 0.025896" \, T - 22.4072747" \, T^2 + 0.00190059" \, T^3 + 0.001112526" \, T^4$
$+ 0.0000001358" \, T^5 +$ oscillatory terms

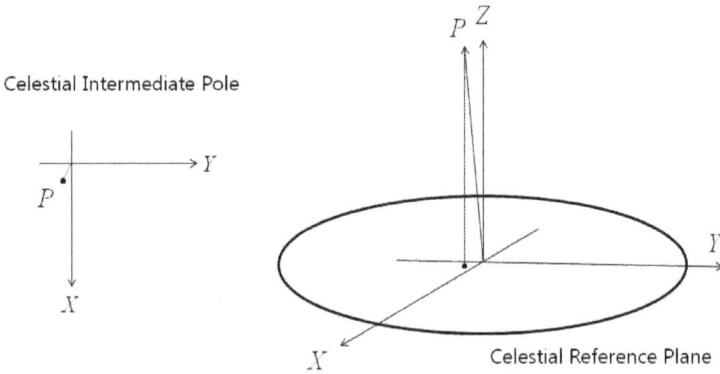

Figure 7. Celestial Intermediate Pole referenced to the CRF equator.

3. Secular deceleration

Although variation in the magnitude of Earth's spin rotation had not been detected easily in their days, some investigators, such as I. Kant and G. Darwin, suspected the Earth's deceleration. With careful reasoning only, they correctly concluded that the Earth should be secularly decelerated due to tidal friction in the oceans and solid Earth and that the lunar orbit should be modified accordingly. A schematic illustration for this secular interaction is given in Figure 7. In the figure, the two identical Earth tidal bulges exist with minute phase delay due to tidal friction. Amplitude of body tide in the Earth exceeds 20 cm in most area over the world, and that of ocean tide is usually larger. Phase lag of body tide is found to be a few degrees, and differs for each tidal constituent. Phase lag of ocean tide is known to vary largely at places. The associated energy dissipation in ocean and solid Earth exceeds 3.0 Terra Watt. Tidal torque, which is due to the gravitational pull from the tide raising body (either the Moon or the Sun) to the misaligned tidal bulges, reduces spin angular momentum of the Earth. Equal and opposite torque should exist and increase the angular momentum of orbital rotation. Total angular momentum of the Earth-Moon system can be approximately expressed as

$$L = (M_e r_e^2 + M_m r_m^2)\omega_m = \frac{M_e M_m}{M_e + M_m} r^2 \omega_m,$$ where r is Earth-Moon distance, r_e and r_m are two each

distances from the center of mass (approximately, $\frac{1}{81}r$ and $\frac{80}{81}r$), and M_e and M_m are masses of the Earth and Moon respectively. Since lunar orbital parameters satisfy Kepler's third law, increase of lunar angular momentum is accompanied with increase of the Earth-Moon distance and decrease of lunar angular velocity; from $r^3 \omega_m^2 = G(M_e + M_m)$, relation between the two increments dr and $d\omega_m$ is acquired as $3\,dr/r = -2\,d\omega_m/\omega_m$. Tide raised in the Moon by the Earth should be larger in amplitude than Earth tide, but its effect is quite smaller in the secular deceleration due to the spin-orbit synchronous rotation of the Moon. It should be noted that the above arguments are based on the approximation neglecting the Earth's obliquity and lunar orbital inclination to the ecliptic.

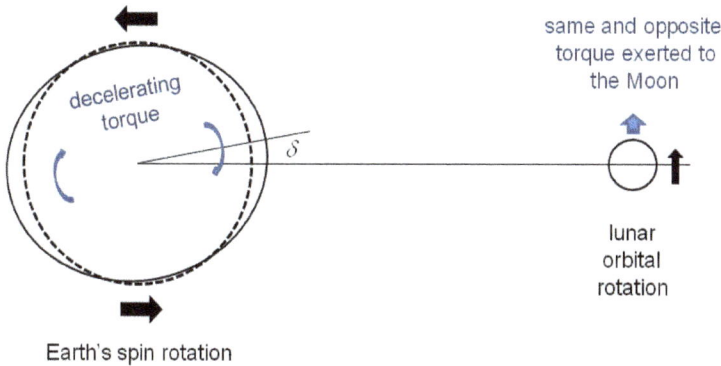

Figure 8. Schematic illustration for tidal deceleration of the Earth and Earth-Moon distance increase due to their tidal interaction.

Solar tide in the Earth is about half of lunar tide in amplitude, and it significantly contributes to Earth's deceleration (about 20%). However, unlike the lunar orbital change, the change in the Earth-Sun distance or Earth's orbital rotation period due to the Earth-Sun tidal interaction is too small to be accurately observed.

G. Darwin carried elaborate formulation and calculation for changes in the dynamical state of the Earth Moon system, and extended further arguments [19]. Faster rotation of the Earth in the geological past was later identified from paleontological evidences. Historical eclipse records also revealed positive indications about the Earth's deceleration. Direct confirmations of the secular changes in the Earth's rotational state and lunar orbit have become available after operation of accurate atomic clocks and lunar laser ranging. Lunar laser ranging started since Apollo 11 landing in 1969, and enabled direct estimate of the present lunar recession as 3.82 cm/yr. Other related studies are satellite orbit analysis and ocean tide modeling. The fact that the Moon was closer to the Earth in the past, led G. Darwin to the fission hypothesis for lunar origin. Other hypotheses suggested for lunar origin were capture, binary accretion, and impact theories. This interesting topic has been occasionally re-visited [20].

In these days, estimates about the Earth's deceleration rate by different approaches seem to converge, so that the LOD is now generally believed to increase with a rate of 1.8 millisecond / cy. Without glacial isostatic adjustment, this rate would be 2.3 millisecond /cy, however, there is large uncertainty in the variation of LOD mainly due to slow and complicated flows in the deep interior of the Earth. Recent fast retreat of glaciers and associated changes in the Earth's moments of inertia contribute to the variation of LOD by a certain amount, which is not accurately known.

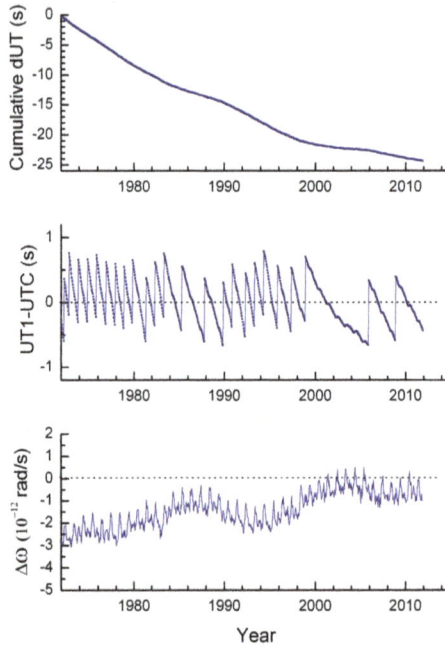

Figure 9. Recent 40 years data of delay in the universal time (UT) and Earth's angular velocity; cumulative delay in UT1 (top), difference between UT1 and UTC with leap seconds seen as steps (middle), and the excessive Earth's angular velocity deduced from UT1 delay (bottom).

In Figure 9, recent variation of UT1 is illustrated by three different ways. Above, cumulative delay in UT1 is shown, and then, in the middle, difference between UT1 and UTC (UT1-UTC) is shown. Including the last leap second introduced at the midnight of June 30th of 2012, there were total 25 leap seconds since 1972. The bottom graph shows the excessive amount of Earth's spin angular speed $\Delta\omega$ from its nominal value $\omega_0 = 7.292115 \times 10^{-5} (rad/s)$. Recent Earth's spin angular velocity, as can be seen in Figure 9(bottom), has been increasing in minute amount. This tendency can also be noticed from reduced number of leap seconds during the last twelve years or so.

Energy dissipation and angular momentum transfer via the tidal interaction process can be expressed simply in terms of dr(infinitesimal increase in the Earth-Moon distance) for isolated Earth-Moon system with neglecting Earth's obliquity and lunar inclination. The work involved in the process of angular momentum transfer is given as $dW = -d(\frac{1}{2}I_e\omega_e^2) = -I_e\omega_e d\omega_e = -\tau_e\omega_e dt$. After a little algebra, this can re-written as $dW = (\frac{M_e M_m r}{2(M+M)} - \frac{3I_m}{2r})\omega_m\omega_e dr$. The mechanical energy increase dE is given as

$dE = [-\frac{1}{2}(\frac{M_e M_m}{M_e + M_m})r - \frac{3I_m}{2r}]\omega_m^2 dr + \frac{GM_e M_m}{r^2}dr$. The difference between dW and dE, i.e., $dQ = dW - dE$ is the dissipated energy during the process. In fact, the majority of the work (96.3%) is dissipated into heat. This argument is valid regardless of the places where the dissipation occur. The estimates for the work rate $\frac{dW}{dt}$ and the tidal torque τ of present days are found as 3.30×10^{12} $Watt$ and 4.52×10^{16} $N \cdot m$. More elaborate calculation with including Earth's obliquity, lunar inclination, and solar tidal dissipation leads to slightly different estimates; 96.6%, 3.56×10^{12} $Watt$, and 4.88×10^{16} $N \cdot m$. The corresponding values for present Earth deceleration and lunar orbital retardation are $\frac{d\omega_e}{dt} = -6.08 \times 10^{-22}$ rad / s^2 and $\frac{d\omega_m}{dt} = -1.26 \times 10^{-23}$ $rad / s^2 = -25.9" / cy^2$.

Analysis on satellite orbit can yield estimate of the tidal torque, which decelerates the Earth. If there were not any other force exerting on a satellite rather than Earth's central gravitational attraction, its orbit would be a perfect ellipse. Due to luni-solar gravitational attraction, solar radiation pressure, and others, satellites undergo certain changes in their orbital configurations. Although the Earth's gravity field perturbation due to its tidal deformation is not quite large, satellite tracking has been precise enough to detect such effect on satellite orbits since late 70s [21-23]. By extending this kind approach to the Moon, the only natural satellite of the Earth, they estimated lunar orbital retardation as 27.4, 25.3, and 24.9 " / cy^2 [21-23]. Christodoulidis et al. accordingly estimated the present Earth deceleration rate as -5.98×10^{-22} rad / s^2 [22].

Since the late 60s, ocean tide modeling has been attempted, and global ocean tide models were acquired as numerical solution of Laplace's tidal equation with grid spacing larger than 2 ×2 . By using acquired global ocean tide model, it is possible to calculate the tidal torque that decelerate the Earth. Estimates for Earth's secular deceleration based on the first-generation ocean tide models were summarized by Lambeck [24]. With fast development of computing devices, much more extensive modelings have been acquired lately, such as NAOJ99, FES2004, EOT11, etc. In fact, better performance of recent models is feasible due also to sea surface height data derived from satellite altimetry (TOPEX/POSEIDON). Ray et al. combined the satellite altimetry and orbit analysis and estimated the Earth's deceleration rate as $-1304"/cy^2$, which is -6.35×10^{-22} rad/s^2 and corresponds to LOD increase rate of 2.37 ms/cy [25]. Table 2 is based on their results.

Tidal constituent	M_2	K_1	S_2	O_1	N_2	P_1	K_2	Q_1	Total
Acceleration	-919	-120	-73	-71	-40	-13	-10	-3	-1304

Table 2. Estimate of Earth's secular deceleration by each ocean tide constituents. [unit: arcsec/cy²] [25].

Accumulation of universal time delay ΔT due to a constant LOD increase of 2.0 millisecond/cy for a century corresponds to 36.525s; $\Delta T = \frac{1}{2} \times 36525 \times 0.002s = 36.525s$. A thousand year accumulation (ten times longer) by the same rate would result in $\Delta T = 3652.5$ s (hundred times) delay, which is about one hour. Similarly two thousand year accumulation would result in about four hours delay. In fact, a correction term linear to time span should be added unless the time span is measured from the reference epoch, which is the year AD 1820 [26]. Astronomical evidences about the Earth's deceleration and lunar orbital retardation can be categorized into two different kinds of records; (i) 'telescopic observation,' which has been carried for a few hundred years, and (ii) 'solar eclipse record,' which can be found from thousands year old Chinese literature or Babylonian inscriptions on clay tablets. Figure 10 is a redrawing of ΔT curve based on the observations of transit of Mercury and lunar occultation [27].

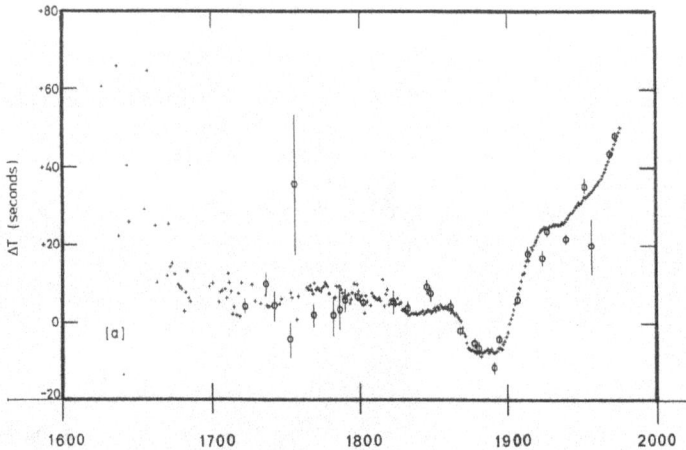

Figure 10. Values of ΔT between year 1627 and 1975 based on the observations of lunar occultation and transit of Mercury [27].

The parabolic trend of ΔT curve is due to the Earth's secular deceleration. As shown in Figure 10, the secular deceleration is superposed with large fluctuations, which are ascribed to processes in the Earth's core and mantle. From these data, Morrison determined lunar orbital retardation as $\dfrac{d\omega_m}{dt} = -26'' \Big/ cy^2$ [27], which is quite close to recent and reliable estimate by lunar laser ranging and others. From vast amount of historical records of solar eclipse and others, an empirical formula was found for ΔT as follows;

$$\Delta T = -20 + 31t^2 \tag{5}$$

where ΔT is in second, and t is time from AD 1820 in century [26]. Later the quadratic coefficient was adjusted from 31 to 32 [28]. Two corresponding rates of LOD increase are 1.70 and 1.75 millisecond/cy. It is noted that there were other estimates of same sorts, which largely differ from those introduced above. For example, see [29-30]. In Figure 11, curve of ΔT determined from observations between 500 BC and AD 1300 [26], and an approximate curve of ΔT based on Morrison's table [28] are shown.

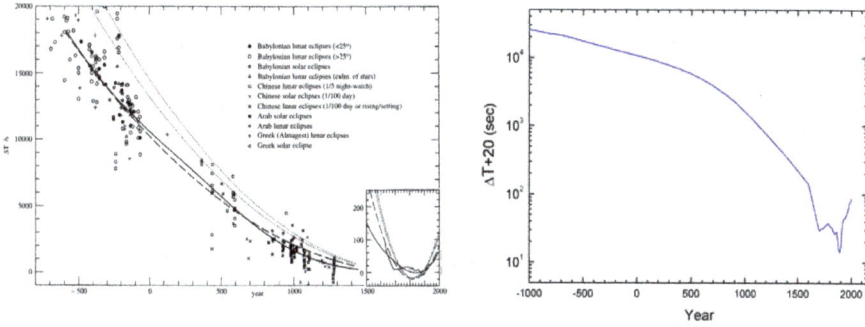

Figure 11. Two sets of ΔT data and curves: ΔT determined from observations between 500 BC and AD 1300 and their parabola fittings. Inset is set of data and fittings for later times [26] (left), and an approximate curve of ΔT based on tabulated data in [28] (right, $\Delta T + 20s$ is drawn in logarithmic scale).

Paleontological evidences exist for faster rotation of the Earth in the Mesozoic and Paleozoic era. As trees retain yearly growth rings, certain organisms, such as coral and shell, can record high and low ocean tides in their hard parts due to differential growth. Careful counting of these growth lines in fossil specimen yielded the numbers of days per month and year at geological past [31-33]. Lambeck estimated Earth's deceleration rate and lunar orbital retardation rate from those numbers [33]. His estimates based on fossil bivalve data were $\dfrac{d\omega_e}{dt} = -5.9 \times 10^{-22}\, rad\big/s^2$ and $\dfrac{d\omega_m}{dt} = -1.3 \times 10^{-23}\, rad\big/s^2 \approx 27"\big/cy^2$. These estimates were proved to be quite reasonable by later studies. Therefore, tidal dissipation and associated lunar orbital change must have occurred rather consistently during hundreds of millions years regardless of decadal fluctuations due to core/mantle processes in the Earth or other perturbations. Followings are the formulation developed by Lambeck for the analysis of paleontological data.

$$N_1 = \frac{\omega_e - \omega_s}{\omega_s}\ ,\quad N_2 = \frac{\omega_e - \omega_s}{\omega_m - \omega_s}\ ,\quad N_3 = \frac{\omega_m - \omega_s}{\omega_s} \tag{a}$$

$$\frac{dN_1}{dt} = \frac{1}{\omega_s}\frac{d\omega_e}{dt}\ ,\quad \frac{dN_2}{dt} \approx \frac{1}{\omega_m - \omega_s}\frac{d\omega_e}{dt} - \frac{\omega_e - \omega_s}{(\omega_m - \omega_s)^2}\frac{d\omega_m}{dt}\ ,\quad \frac{dN_3}{dt} = \frac{1}{\omega_s}\frac{d\omega_m}{dt} \tag{b}$$

$$\frac{1}{\omega_s}\begin{bmatrix} 1 & 0 & t & 0 \\ 1 & -\beta & t & -\beta t \\ 0 & 1 & 0 & t \end{bmatrix}\begin{bmatrix} \Delta\omega_e \\ \Delta\omega_m \\ d\omega_e/dt \\ d\omega_m/dt \end{bmatrix} = \begin{bmatrix} N_1 + 1 - \omega_e(t_0)/\omega_s \\ \dfrac{\omega_m(t_0) - \omega_s}{\omega_s}N_2 + 1 - \omega_e(t_0)/\omega_s \\ N_3 + 1 - \omega_m(t_0)/\omega_s \end{bmatrix} \tag{c}$$

(6)

N_1, N_2, N_3 are the numbers of solar days per year and synodic month, and the number of synodic month per year respectively. Time derivatives of these three numbers are expressed as in equation (6b)

Write $\omega_e(t) = \omega_e(t_0) + \Delta\omega_e + (d\omega_e/dt)\,t$ and $\omega_m(t) = \omega_m(t_0) + \Delta\omega_m + (d\omega_m/dt)\,t$, then the condition equation for the unknowns $\dfrac{d\omega_e}{dt}$, $\dfrac{d\omega_m}{dt}$, $\Delta\omega_e$, and $\Delta\omega_m$ is written as in equation (6c) with

$\beta = \dfrac{\omega_e(t_0) - \omega_s}{\omega_m(t_0) - \omega_s}$. While the estimates of different studies based on the historical eclipse records

varied widely on two quantities $\dfrac{d\omega_e}{dt}$ and $\dfrac{d\omega_m}{dt}$, the reliability of estimate based on paleontological record is noticeable.

Numerical extrapolation for the past/future lunar orbit was first attempted by Darwin [19], and later by many other investigators including Goldreich, Migard, Hansen, and Webb [34-37]. Goldreich calculated the past state of the Earth-Moon system by using three time steps; year, period of lunar orbital precession, and secular [34]. Goldreich and Mignard avoided time scale problem by using the Earth-Moon distance as the independent variable. Calculations of Hansen and Webb were based on each idealized ocean tide modeling. Hansen specified two kinds of paleo-ocean configurations for his calculation; one - circular continent at polar region and the other - circular continent at equator [37]. Webb used the orientation averaged ocean with one continental cap. By extending Kaula's satellite orbit theory, Lambeck derived differential equations for lunar orbital evolution. By using those Lambeck's formulae with modification, the extrapolation was recently repeated [38], and its set of the calculated past Earth-Moon distance for different tidal friction parameter is shown in Figure 12. According to this result, the Moon could have been at close approach to the Earth at different times, from 1.6 - 4.0 billion years ago or even earlier depending on the amount of reduction in tidal friction in the past.

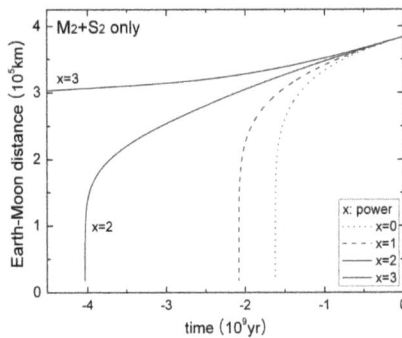

Figure 12. Calculated Earth-Moon distance in the past. According to the modified Lambeck's formulae with assuming power law behavior for tidal phase lag angle. M_2 and S_2 tidal constituents only are considered [38].

Calculated values of five parameters (eccentricity, obliquity, lunar orbital inclination, ω_e, and ω_m) in the future are shown in Figure 13. The scale factor is ratio of Earth-Moon distance to its present value, *i.e.*, r / r_0. As the ratio increases from 1.0 (present) to 1.2-1.3, Earth's spin angular velocity decreases almost linearly, and obliquity angle increases abruptly up to and over 60 degrees, while other parameters of lunar orbit gradually change in small amounts.

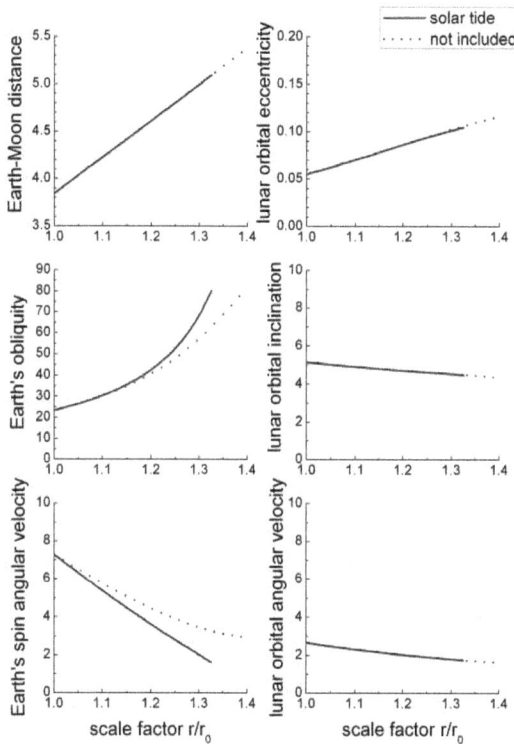

Figure 13. Calculated Earth-Moon system configuration in future by modified Lambeck's formulae. The horizontal axis is the ratio r / r_0. Solid lines correspond to those including solar tidal dissipation, while dotted lines correspond to those not including it. The units are [10^8m], [none], [deg], [deg], [10^{-5} rad/s], and [10^{-6} rad/s] from left top.

There exist four hypotheses suggested for the lunar origin; (i) 'fission', (ii) 'capture', (iii) 'binary accretion', and (iv) 'impact.' It is difficult to verify/trace such a far distant past event like the Earth-Moon system formation, and one may consider only feasibilities of those hypotheses. A brief sketch of these four hypotheses is given here. Fission hypothesis, suggested by Darwin [19], is an explanation of lunar origin as a result of mechanical resonance of the early Earth, which should have been rotating very fast. Since the natural period of foot ball mode of Earth's free oscillation is 54 minutes, the mechanical resonance to eject a part of mantle might have been possible, provided the early Earth had been rotating with a period about two hours or so. If the Earth would gain the angular momentum, which was lost due to both lunar and solar tidal interactions during the whole past of billion years, this fast rotation could be possible. Fission theory is also compatible with the fact that the lunar mass density is quite close to that of the Earth's mantle. Problem with the fission theory is that direct calculation of the Earth-Moon system back to the past should lead to high inclination angle of the lunar orbit. One has to imagine a certain scenario, such as close approach of Venus to perturb lunar orbit, to reconcile the imperfection. Capture hypothesis states that the Moon was initially formed independently and later captured by the Earth's gravity field during its passage near the Earth. Suppose it approached by following a hyperbolic/parabolic orbit with respect to the Earth, there should have been a mechanism to explain the change of lunar orbit. Gerstenkorn event is one such explanation claiming the Moon had approached in a retrograde orbit and then undergone its orbital change from hyperbolic to elliptic due to severe lunar tidal dissipation. Problem with the Gerstenkorn event is that, even at closest distance, tidal interaction cannot perform with such an extreme efficiency. Could there be other alternative stopping mechanism, then the capture hypothesis may remain feasible. Binary accretion hypothesis states that the Moon was formed with the Earth together as two isolated bodies by planetesimal accumulation from the beginning stage of the Earth-Moon system. Had the Moon been formed as an isolated body from the Earth, the low density of the Moon cannot be explained. Impact hypothesis emerged in the late 70s and was numerically simulated later. Suppose an object of 0.1 Earth's mass or so had collided with the early Earth, the Moon could have been formed afterwards by accretion of the remaining fragments. Impact hypothesis has been favored, because the Moon is depleted of volatile elements.

Like other major natural satellites, such as Galilean satellites, the Moon is already in synchronous rotation so that it always shows the same face toward the Earth. This synchronous rotation of the Moon can be maintained due also to another tidal interaction - dissipation of tide in the Moon raised by the Earth. Unless other impact with large third body or comparable perturbations, the Earth-Moon system will undergo secular changes as described above; the Earth's spin will slow down with its obliquity increase, and the lunar orbit will become larger with small decrease in its inclination angle.

4. Liouville equation and excitation function

In this section, the equation of motion for Earth's angular velocity perturbation is derived in the Earth fixed reference frame by neglecting elastic deformation and other complicated features of the real Earth. First, the general equation of motion for rigid body rotation and Euler's equation for wobble are derived. Further considerations will be followed after then.

Spin angular momentum of a rotating body can be written as $\vec{L} = \int \vec{r} \times \vec{v} \, dm$, where the center of its mass can be taken as the coordinate origin. Equation of motion states that the time rate of angular momentum equals to the exerting torque; $\dfrac{d\vec{L}}{dt} = \vec{\tau}$. In rotating coordinate frame, time derivative should be considered with retaining terms due to change in the direction of reference frame in space (in Appendix), so that the equation of rotational motion in body frame is rewritten as follows.

$$\frac{d\vec{L}}{dt} + \vec{\omega} \times \vec{L} = \vec{\tau} \tag{7}$$

Taking the principal axes of body as reference frame axes, then $\vec{L}$ and $\vec{\omega} \times \vec{L}$ are simply expressed as follows.

$$L_i = \sum_{j=1}^{3} I_{ij}\omega_j \Rightarrow I_i\omega_i \ , i.e., \vec{L} = \omega_1 I_1 \hat{e}_1 + \omega_2 I_2 \hat{e}_2 + \omega_3 I_3 \hat{e}_3 \qquad \text{a}$$

$$\vec{\omega} \times \vec{L} = (\omega_1 \hat{e}_1 + \omega_2 \hat{e}_2 + \omega_3 \hat{e}_3) \times (\omega_1 I_1 \hat{e}_1 + \omega_2 I_2 \hat{e}_2 + \omega_3 I_3 \hat{e}_3) \tag{8}$$
$$= \omega_2\omega_3(I_3 - I_2)\hat{e}_1 + \omega_3\omega_1(I_1 - I_3)\hat{e}_2 + \omega_1\omega_2(I_2 - I_1)\hat{e}_3 \qquad \text{b}$$

where the principal moments of inertia I_i replace inertia tensor I_{ij}(underlying basic summarized in Appendix).

Then Equation 7 is rewritten as

$$I_1 \frac{d\omega_1}{dt} + \omega_2\omega_3(I_3 - I_2) = \tau_1 \qquad \text{a}$$

$$I_2 \frac{d\omega_2}{dt} + \omega_3\omega_1(I_1 - I_3) = \tau_2 \qquad \text{b} \tag{9}$$

$$I_3 \frac{d\omega_3}{dt} + \omega_1\omega_2(I_2 - I_1) = \tau_3 \qquad \text{c}$$

In case two values I_1 and I_2 are same ($I_1 = I_2 \Rightarrow I$), then Equation 9c becomes simple; $I_3 \frac{d\omega_3}{dt} = \tau_3$. Its solution can be expressed as $\omega_3(t) = \omega_0 + \frac{1}{I_3}\int_{t_0}^{t} \tau(t')\,dt'$. The solution $\omega_3(t)$ for sinusoidal torque of frequency ω_d as $\tau(t) = \tau_0 \cos\omega_d t$ is readily acquired as $\omega_3(t) = \text{constant} + \frac{\tau_0}{I_3\omega_d}\sin\omega_d t$. As a further restriction, if the external torque does not exist ($\vec{\tau} = 0$), then $\omega_3(t)$ should remain constant as $\omega_3 = \omega_0$ and the two Equations 9a-b become as follows.

$$I\frac{d\omega_1}{dt} + \omega_2\omega_0(I_3 - I) = 0, I\frac{d\omega_2}{dt} + \omega_0\omega_1(I - I_3) = 0 \tag{10}$$

The coupled solution for ω_1 and ω_2 of Equation 10 is found as a circular motion as follows.

$$\omega_1 = m\omega_0 \cos\Omega t, \omega_2 = m\omega_0 \sin\Omega t \tag{11}$$

where $\Omega = \dfrac{I_3-I}{I}\omega_0$. The argument Ωt may be added with a phase angle; $\Omega t \Rightarrow \Omega t + \phi$. Amplitude $m\omega_0$ and phase angle ϕ can be specified as initial condition. This rotational motion is often called Eulerian free nutation or wobble. Suppose a body initially rotating along its principal axis is agitated by a small perturbation, then the rotating body will undergo a wobbling motion as it rotates.

If the Earth were perfectly rigid, the frequency of its wobbling motion should be specified as $\Omega = \dfrac{C-A}{A}\omega_0$, of which corresponding period is 304.5 sidereal days, *i.e.*, 303.6 days. As rigid Earth rotates daily, an observer on it would see that both the angular momentum and velocity vectors slowly encircle the x_3-axis (Figure 14). The three vectors stay on a same rotating plane (not shown in the Figure), and the angles between them remain unchanged. Due mainly to finite elasticity of the Earth's mantle, the period of real Earth's wobble is increased by roughly twenty percent. The estimate according to recent observation is about 433 ~ 434 days. To honor the one who first observed this motion, it is called Chandler wobble.

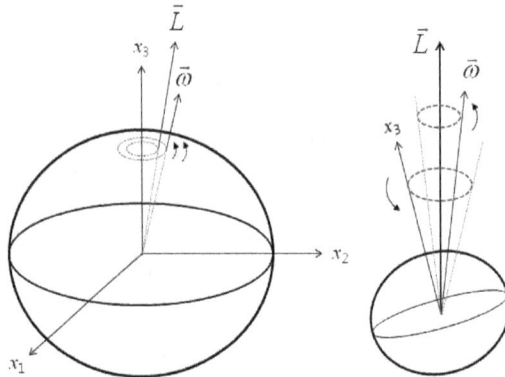

Figure 14. Two illustrations of wobble (free Eulerian nutation). Precession of the angular velocity and momentum vectors around the principal axis (x_3-axis) seen by an observer on the body (left). To an observer in space, the angular momentum vector remains unchanged, while the angular velocity vector and x_3-axis rotate around it (right).

Denote small variations in inertia tensor, angular velocity, and angular momentum as ΔI_{ij}, $\omega_0\vec{m} = \omega_0(m_1, m_2, m_3)$, and $\vec{h} = (h_1, h_2, h_3)$, then the terms for angular momentum and Equation 7 are rewritten as follows.

First, inertia tensor I_{ij} is written as

$$\begin{bmatrix} I_{11} & I_{12} & I_{13} \\ I_{21} & I_{22} & I_{23} \\ I_{31} & I_{32} & I_{33} \end{bmatrix} = \begin{bmatrix} A & 0 & 0 \\ 0 & A & 0 \\ 0 & 0 & C \end{bmatrix} + \begin{bmatrix} \Delta I_{11} & \Delta I_{12} & \Delta I_{13} \\ \Delta I_{21} & \Delta I_{22} & \Delta I_{23} \\ \Delta I_{31} & \Delta I_{32} & \Delta I_{33} \end{bmatrix}$$

It is noted that $I_{ij} = I_{ji}$, i.e., inertia tensor is symmetric.

Angular velocity $\vec{\omega}$ is written as

$$\begin{bmatrix} \omega_1 \\ \omega_2 \\ \omega_3 \end{bmatrix} = \omega_0 \begin{bmatrix} 0 \\ 0 \\ 1 \end{bmatrix} + \omega_0 \begin{bmatrix} m_1 \\ m_2 \\ m_3 \end{bmatrix} = \omega_0 \begin{bmatrix} m_1 \\ m_2 \\ 1+m_3 \end{bmatrix}$$

Angular momentum $\vec{L}$ in tensor form is given as $L_i = \sum_j I_{ij}\omega_j + h_i$, and its matrix representation

is given as

$$\begin{bmatrix} L_1 \\ L_2 \\ L_3 \end{bmatrix} = \begin{bmatrix} A+\Delta I_{11} & \Delta I_{12} & \Delta I_{13} \\ \Delta I_{21} & A+\Delta I_{22} & \Delta I_{23} \\ \Delta I_{31} & \Delta I_{32} & C+\Delta I_{33} \end{bmatrix} \omega_0 \begin{bmatrix} m_1 \\ m_2 \\ 1+m_3 \end{bmatrix} + \begin{bmatrix} h_1 \\ h_2 \\ h_3 \end{bmatrix} \simeq \omega_0 \begin{bmatrix} Am_1+\Delta I_{13} \\ Am_2+\Delta I_{23} \\ C(1+m_3)+\Delta I_{33} \end{bmatrix} + \begin{bmatrix} h_1 \\ h_2 \\ h_3 \end{bmatrix}$$

where first order terms only are retained with neglecting much smaller higher order terms. This equation has been called Liouville's equation.

After a little algebra, three components of the equation; $\dfrac{d\vec{L}}{dt} + \vec{\omega} \times \vec{L} = 0$ are found as follows.

$$\frac{dh_1}{dt} + (\frac{d\Delta I_{13}}{dt} + A\frac{dm_1}{dt})\omega_0 - h_2\omega_0 + [(C-A)m_2 - \Delta I_{23}]\omega_0^2 = 0$$

$$\frac{dh_2}{dt} + (\frac{d\Delta I_{23}}{dt} + A\frac{dm_2}{dt})\omega_0 + h_1\omega_0 + [(A-C)m_1 + \Delta I_{13}]\omega_0^2 = 0$$

$$\frac{dh_3}{dt} + (\frac{d\Delta I_{33}}{dt} + C\frac{dm_3}{dt})\omega_0 = 0$$

Divide the first and second equations by $\omega_0(C-A)$ and rearrange terms, then the following two equations are found.

$$\frac{1}{\Omega}\frac{dm_1}{dt} + m_2 = \phi_2 - \frac{1}{\omega_0}\frac{d\phi_1}{dt} \qquad \text{a}$$

$$\frac{1}{\Omega}\frac{dm_2}{dt} - m_1 = -\phi_1 - \frac{1}{\omega_0}\frac{d\phi_2}{dt} \qquad \text{b} \qquad (12)$$

$$\frac{dm_3}{dt} = -\frac{d}{dt}(\frac{h_3}{C\omega_0} + \frac{\Delta I_{33}}{C}) \qquad \text{c}$$

where the two excitation functions ϕ_1 and ϕ_2 are defined as $\phi_1 = \dfrac{h_1}{(C-A)\omega_0} + \dfrac{\Delta I_{13}}{C-A}$ and $\phi_2 = \dfrac{h_2}{(C-A)\omega_0} + \dfrac{\Delta I_{23}}{C-A}$.

The third equation can be rewritten as in equation (12c)

In case with $h_3 = 0$, this is rewritten as $m_3 = -\dfrac{\Delta I_{33}}{C}$. The other case with $\Delta I_{33} = 0$ is also obvious;

$m_3 = -\dfrac{h_3}{C\omega_0}$. Suppose a periodic variation in h_3 or ΔI_{33} exists, then corresponding variation of same periodicity in m_3 should exist in an amount divided by $C\omega_0$ or C. Equation 12c can be related with changes in LOD and UT1 as follows.

$$-\frac{d}{dt}(UT1 - UTC) = \frac{\Delta LOD}{LOD} = -m_3 = \frac{h_3}{C\omega_0} + \frac{\Delta I_{33}}{C} \tag{13}$$

If the two excitation functions ϕ_1 and ϕ_2 do not exist, Equations 12a-b are the same as Equation 10, i.e., the body will show pure wobble. In case there exists any periodic excitation, then a perturbation in polar motion of the same periodicity should be accompanied. This can be conveniently shown by complex notation. Write $m_c = m_1 + im_2$ and $\phi_c = \phi_1 + i\phi_2$, then we can rewrite Equation 12a-b simply as

$$\frac{1}{\Omega}\frac{dm_c}{dt} - im_c = -i\phi_c - \frac{1}{\omega_0}\frac{d\phi_c}{dt} \tag{14}$$

Write a periodic excitation as $\phi_c \propto e^{i\omega_d t}$, then its corresponding solution for m_c follows as

$$m_c = -\frac{\omega_0 + \omega_d}{\omega_0}\frac{\Omega}{\omega_d - \Omega}\phi_c \tag{15}$$

5. Variation in LOD and polar motion

Length of day (LOD), which is meant by the length of a solar day, is about 86400 second and slightly variable. The length of a sidereal day is about 86164 second. Besides the secular increase described in a former section, there exist periodic variations in LOD. Even before atomic clocks, precise quartz clocks provided measurement of seasonal perturbation in star transit time; being behind in spring and ahead in late summer by 20-30 millisecond, which should be accumulation of LOD variation of the same periodicity. Along with the seasonal perturbation, fortnightly and monthly variations in LOD exist. Amplitudes of LOD variations of these different periodic components are in the order of one millisecond. Some amounts of these periodic perturbations are associated with body/ocean tides in the Earth. However, there is strong atmospheric effect on LOD variation. There also exist large quasiperiodic variations of much longer period range, called decadal fluctuation.

Body tidal variation of LOD is briefly described below. Even though Earth's angular momentum remains unchanged (ignoring the small secular deceleration), angular velocity of the Earth will change when its moment of inertia along the rotational axis changes. This situation is a slight modification from Equation 13, as follows.

$$\frac{\Delta LOD}{LOD} = -m_3 = \frac{\Delta I_{33}}{C}$$

where ΔLOD is the excessive amount of LOD, and related are z-components of Earth's spin angular velocity and inertia tensor; $\omega_3 = (1 + m_3)\omega_0$ and $I_{33} = C + \Delta I_{33}$ respectively. Accumulation of ΔLOD lead to delay of UT1, and this relation can be expressed as

$$UT1 - UTC = -\sum_{days} \Delta LOD = -\int \frac{\Delta LOD}{LOD} dt$$

As ΔLOD due to tidal deformation can be expected, tidal periodicities exist in the ΔLOD spectrum. The ΔLOD associated with the change in the moment of inertia due to the zonal body tide was early studied by Woolard [39] and later by Yoder and Merriam [40-41]. Following them, the expression for ΔLOD due to elastic body tidal deformation is written as

$$\Delta LOD = -86400s \times \frac{k_2 M R_e^5 \Delta r}{C r^4} \sqrt{\frac{5}{\pi}} (1 - 3\sin^2 \delta) \tag{16}$$

where k_2 is tidal Love number, M is the mass of tide raising body (Moon or Sun), R_e is the mean radius of the Earth, r is the distance to the body from the Earth, δ is the declination of the body, and Δr is the deviation of r from its average value. As will be described below, atmospheric and oceanic contributions to ΔLOD exist in large amount, so that calculated body tidal ΔLOD alone does not suffice to explain observational data.

In Figure 15, the excessive LOD time series (IERS EOP 08 C04 – simply C04) since 1981 is shown (top). The curve of variation with period longer than 2000 days is superposed with the data (top). Below three different period band components are separately shown (period between 500 and 2000 days, between 100 and 500 days, and less than 100 days). The amplitude of interannual variation is about 0.3 millisecond or less. Decomposition of the time series of Figure 15, was done by simple spectral windowing. It is interesting to note that the present overall trend of LOD shown in this figure is decreasing, which is the reverse of secular increase of LOD due to tidal dissipation. This is ascribed to certain geophysical processes in the Earth's core and mantle, such as geodynamo. Recent fast retreat of glaciers might be related as well.

The ΔLOD time series show clear annual, semiannual, monthly signals and other minor ones distributed over wide spectral range.

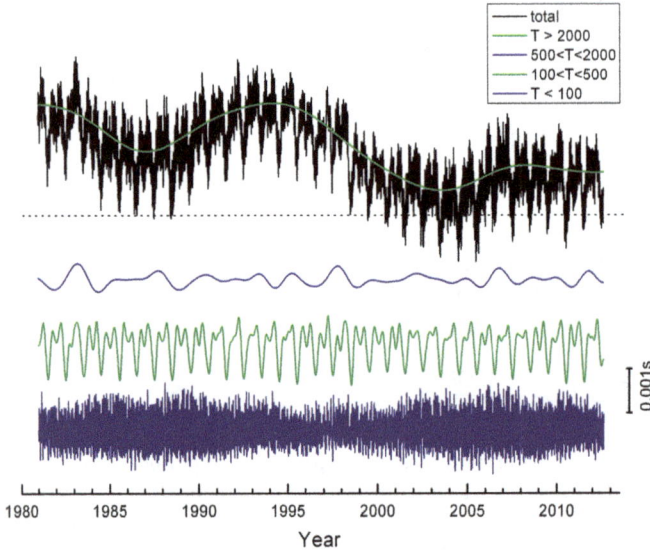

Figure 15. Variation of LOD since 1981. From the top, excessive LOD time series with its decadal trend, components of period between 500 and 2000 days, between 100 and 500 days, and less than 100 days. The dotted line is zero level. Three lower graphs are shifted for convenience, but all are in same vertical scale.

Fourier spectra of the LOD time series are shown in Figure 16. The two graphs are equivalent but differ in representation only by the axis (frequency/period). To identify the long period component accurately, calculation for these spectra was done through integration of the time series multiplied by sine/cosine sinusoid of each frequency not by using fast Fourier transform. Four largest peaks are labeled as f1, f2, f3, and f4. The two peaks f2 and f3 are annual and semi-annual ones. The period of f4 is 13.7 days. As shown in lower spectrum, the period of peak f1 is spread over a wide range. Spectral peak of f1 is split by a few minor peaks. The main period of f1 peak is about 6905 days. Including the second largest peak at 4650 days, total width of half amplitude is about 5290 days between 4150 and 9440 days. This broad peak of f1 is generally believed to be associated with geomagnetic field generation in the Earth's core and 18.6 year precession of lunar orbital plane. Peak sequence in power is follows; f1 (decadal), f4 (fortnightly), f2 (annual), f3 (semiannual). Two spectral peaks of which periods are 27.6 and 9.13 days are noticed as next large ones. There exist other smaller peaks, including 14.8 day period one.

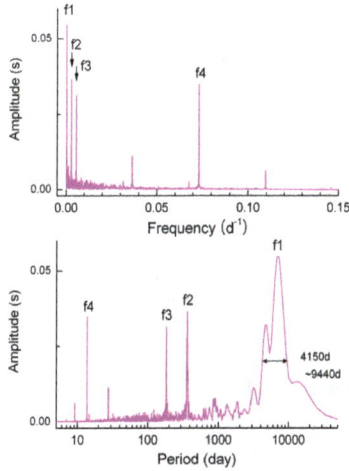

Figure 16. Fourier amplitude spectra of the LOD time series shown in Figure 15.

In Figure 17, two short period band components of ΔLOD time series of Figure 15 are shown again for three year time span between 2000 and 2003 with the body tidal ΔLOD calculated by using Equation 16. Both the data and calculated body tidal ΔLOD in Figure 17 well show annual, semiannual, monthly and fortnightly periodic components. However, there are certain differences between the two. First, the fortnightly periodicity is not quite certain in the calculated body tidal ΔLOD, and neither is the semiannual one. Moreover, annual and semiannual signals are much stronger in the data than the calculated ones. Evidently this discrepancy should be ascribed to other effects than body tidal perturbation.

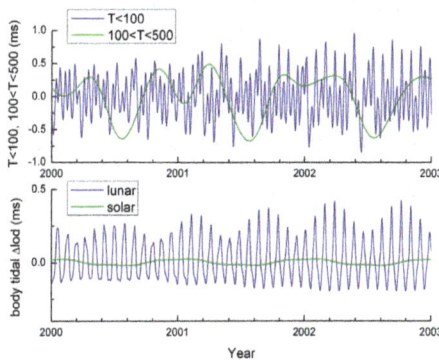

Figure 17. Observed LOD variation of period less than 500 days (above) and calculated LOD variation due to body tidal perturbation (below).

The effects of global zonal wind pattern fluctuation and ocean tidal angular momentum variation on UT and LOD had been suspected [1-2]. These effects were confirmed later [42-47], and the zonal wind pattern change was found to do the dominant role. Hydrological mass transport is also identified to have effect on annual and semiannual variation. Ocean tidal effect is found to be the main cause for diurnal and semidiurnal variation in LOD. These investigations became possible due to recent accurate ocean tide modeling and atmospheric angular momentum data. In Figure 18, both observed data and modeled LOD time series are shown after Chen *et al.* [46].

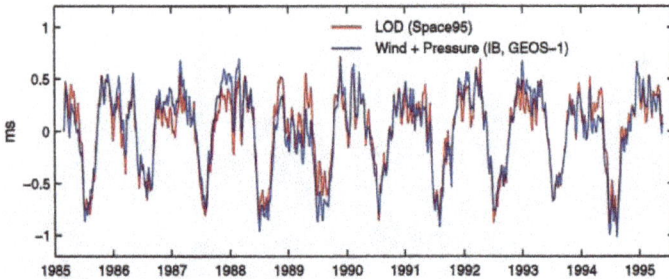

Figure 18. Observation and model for LOD; Space95 time series and calculated LOD excitation by atmospheric flow and pressure (after Chen *et al.* [46])

Position of Earth's rotational pole on the Earth's surface is slowly but ceaselessly changing. Average position of pole between 1900 and 1905 was taken as the reference point of geodetic latitude and longitude. This position was formerly called Conventional International Origin, and now is renamed as Reference Pole. Polar motion is referred to the offset of Earth's rotational pole with respect to the Reference Pole. The main feature of polar motion during the past century can be summarized as; (i) slow drift along the direction between West 70 and 80°, (ii) Chandler wobble of amplitude about 210 (150 – 280) milliarcsec, and (iii) annual wobble of amplitude about 120 (90-150) milliarcsec. There also exist various components of smaller amplitude, among which is semiannual component.

In the former section, free and forced wobble of rotating rigid Earth was considered. It is rather elaborate to calculate Chandler wobble period for more realistic Earth model. Chandler wobble frequency of elastic and oceanless Earth was acquired as $\Omega = \dfrac{C-A-D}{A+D}\omega_0$, where $D = kR_e^5\omega_0^2/3G$ [48]. Corresponding period is 446.2 days. Assuming equilibrium pole tide, the estimate was adjusted into 425.5 day [48]. Its recent theoretical estimate is 423.5 days [49]. Among many observational estimates for Chandler period and Q value, 433.0 days and 179 are the most representative ones [50].

Three other modes of Earth rotation exist due to the core of the Earth as followings; free core nutation, free inner core nutation, and inner core wobble. For an observer on the Earth's surface, free core nutation and free inner core nutation are retrograde motions having their periods approximately one day, therefore these two are called nearly diurnal free wobbles [50-52]. For an observer in space, the periods of these two are much longer; about 430 and 1000 days. Inner core wobble period has been estimated as 900-2500 days [53-54], however its existence has not yet been reported [55].

Formerly, perturbation in the Earth's angular velocity and pole offset were considered to be the same with only difference in their directional notation, i.e., $(m_1, m_2) = (x_p, -y_p)$. In fact, the relation $(m_1, m_2) \simeq (x_p, -y_p)$ approximately holds for long period components. Their exact relation is the following; $m_c = p_c - \dfrac{i}{\omega_0} \dfrac{dp_c}{dt}$, where $m_c = m_1 + im_2$ and $p_c = x_p - iy_p$. Celestial Intermediate Pole (CIP), denoted by x_p and y_p, is the one to be used for TRF-CRF conversion. For more detail, see [5] or related references cited therein. It is noted that, for dynamical effects such as pole tide, values of (m_1, m_2) should be referenced to x_3-principal axis of the Earth. Since both the x_p and y_p are small, the transformation matrix for polar motion necessary for CRF to TRF transformation can be written as $R_{pm} \simeq R_2(-x_p)R_1(-y_p) \simeq R_1(-y_p)R_2(-x_p)$ as first order approximation (see Appendix for more detail).

Recent polar motion is illustrated in Figure 19. The graphs are based on the polar motion dataset of IERS EOP 08 C04.

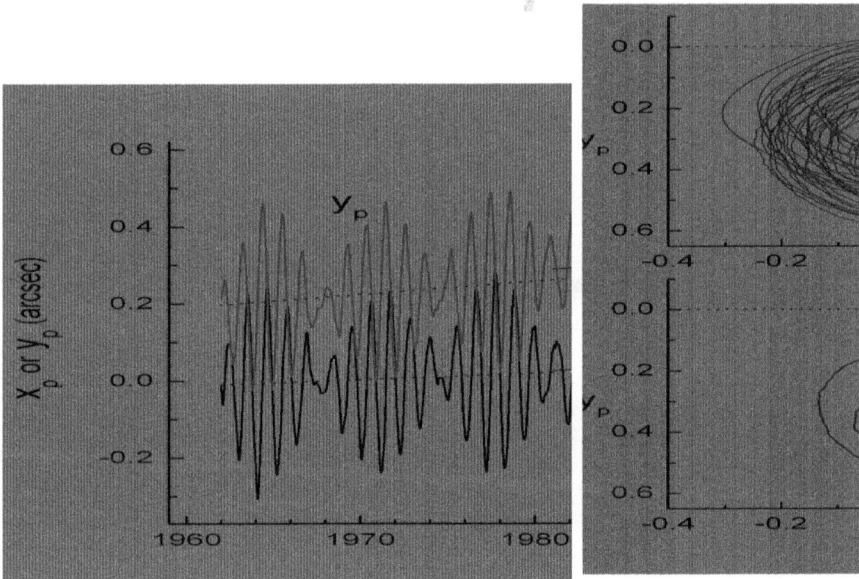

Figure 19. Polar motion; two components x_p and y_p of polar motion since 1962. Least square fit lines are drawn for whole time span and latter time span since 1981 (left). Two each (long/short) locus of pole position with respect to Reference Pole are drawn (right).

From the lines fitted with least square error in Figure 19, the linear trend of recent polar motion is read as 8.1 cm/yr along W 59° for short time span since 1981 and 12 cm/yr along W 64° for long time span since 1962. Some former estimates of the linear trend in polar motion were 10.3 cm/yr along W 75° [56], 10.9 cm/yr along W 79°[57], and 10.3 cm/yr along W 76°[58]. Therefore, recent pole drift is comparatively slower and tilted by several degrees to the East. This might be associated with the recent rapid glacier melting in Greenland.

Two main components of polar motion are Chandler and annual wobbles. Other known minor components are semi-annual, semi-Chandler, Markowitz, *etc.* There have been investigations to identify different components of polar motion and their characteristics (see, for example, [59-61]). For the polar motion time series as shown in Figure 19, two spectra were acquired by using fast Fourier transform and maximum entropy method, and are illustrated in Figure 20. Minor peaks p1, p2, and p3 in the spectrum are previously known components; semi-annual, semi-Chandler, and 300-day period ones. The 300-day period component is regarded to be associated with atmospheric phenomena [60]. Existence of 490-day period component (peak p4) in polar motion was suggested after numerical experiment and evidence of same periodicity in other geophysical phenomena [62].

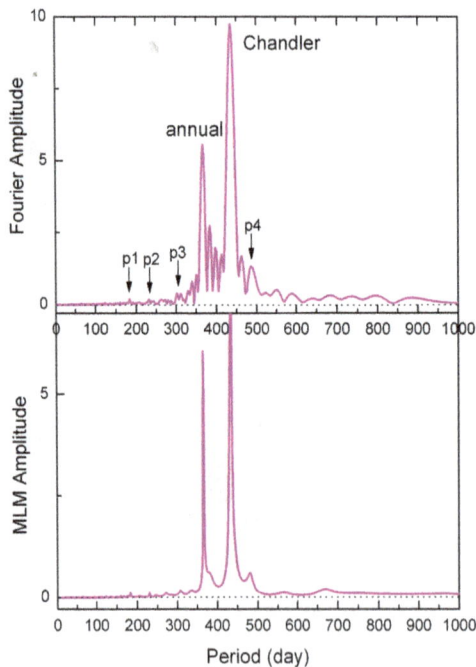

Figure 20. Fourier and Maximum likelihood amplitude spectra of recent polar motion shown in Figure 19.

In Figure 21, different period band polar motion components are shown. The each separate time series are: x_p with long time trend, long period, Chandler wobble, annual wobble, and short period components, which were acquired by spectral windowings.

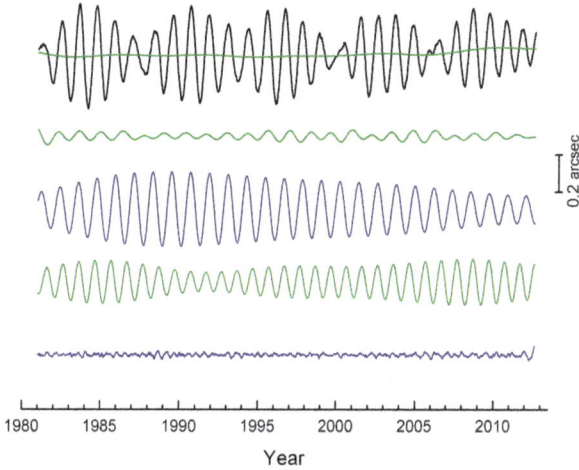

0.2 arcsec

1980   1985   1990   1995   2000   2005   2010

Year

Figure 21. Polar motion and its components of different period ranges. From above, original total data with its trend, long period, Chandler, annual, short period components. (x_p of C04 between 1981 and 2012 summer)

Variability of the Chandler wobble period has been assumed, because of observationally derived period of Chandlerian motion showed such instability. However, it was asserted that the period of Chandler wobble should be a constant, which is determined by the whole mechanical structure of the Earth [63]. The apparent variation of Chandler period is caused by variable excitation.

While Equation 14 relates excitation function and polar motion of rigid Earth, similar equation for the real Earth is acquired by replacing the Chandler frequency with the following one;

$\Omega = \dfrac{\omega_0}{433}(1 + \dfrac{i}{2Q})$. Considering the relation $m_c = p_c - \dfrac{i}{\omega_0}\dfrac{d\,p_c}{dt}$ together, the excitation function for the real Earth can be conveniently expressed in frequency domain as follows.

$$\Phi(\omega) = \frac{\omega_0}{\omega_0 + \omega}\frac{\Omega - \omega}{\Omega} M(\omega) = \frac{\Omega - \omega}{\Omega} P(\omega) \tag{17}$$

where $\Phi(\omega)$, $M(\omega)$, and $P(\omega)$ are the Fourier transform pairs of $\phi_c(t)$, $m_c(t)$ and $p_c(t)$ respectively. In Figure 22, both real and imaginary parts of the calculated excitation function $\phi_c(t)$ are illustrated (Q value is taken as 180) together with polar motion.

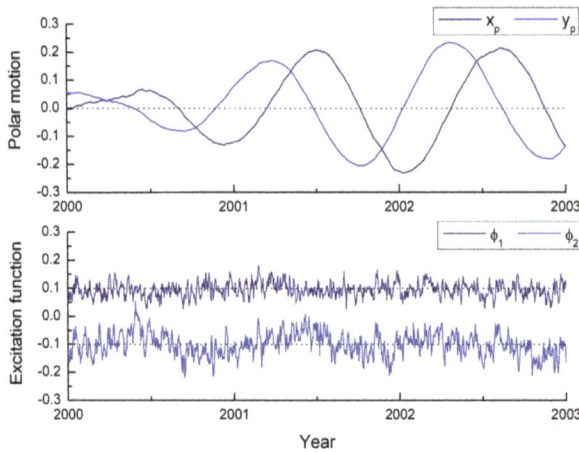

Figure 22. Polar motion and excitation function between year 2000 and 2003 [unit: arcsec]. Polar motion components are delineated, and excitation function is calculated by using Equation 17. Two components of excitation function are shifted by ±0.1 arcsec for convenience.

Excitation mechanism of Chandler wobble has been investigated for a long time, and, nowadays, fluid sphere forcing at the Earth's surface is regarded to be the main source [64-71]. Annual wobble exists with amplitude slightly smaller than Chandler wobble. Much smaller semi-annual wobble also exists. Major part of annual and semiannual wobble should be due to atmospheric excitation. Eurasian continent, North Atlantic, North Pacific, and southern oceans were found to be large sources of the atmospheric excitation for annual wobble. Explanation for polar motion excitation is sought by simultaneously considering effects from wind, atmospheric pressure, ocean current, ocean bottom pressure. In Table 3, annual and semiannual excitation components are listed after Gross *et al.* [67]. Table 3 is consisted of prograde components only, however, retrograde ones were reported of the same order of magnitude.

Excitation	Annual		Semiannual		Terannual	
	Amp (m.a.s)	Phase (deg)	Amp (m.a.s.)	Phase (deg)	Amp (m.a.s.)	Phase (deg)
Wind	2.97	-35	0.36	72	0.14	125
Atmo. pressure	15.12	-102	2.60	47	1.29	144
Ocean current	2.31	40	1.32	176	0.81	146
O.B. pressure	3.45	63	0.77	134	0.71	120
Total	12.23	-78	2.90	90	2.90	138
Observed	14.52	-63	5.67	108	2.22	109

Table 3. Calculated and observed polar motion excitation of annual, semiannual, and terannual components (prograde components only) after Gross *et al.* [67] (m.a.s. = milliarcsec, Atmo. = Atmospheric, O.B. = Ocean Bottom).

Oceanic excitation of periods between daily to seasonal has been recognized. Atmospheric excitation is found more important for LOD, and both oceanic/atmospheric effects are found important for polar motion [72-73]. For subdaily polar motion, ocean effect is known to dominate [72]. Zhou *et al.* found better assessment of the atmospheric excitation by considering the Earth's surface topography [74]. Nowadays, daily and sub-daily variations of LOD and polar motion are observed and modeled in submicrosecond and microarcsec levels; for example, see [75-76]. However, observation and model do not match completely in all spectral range both for LOD and polar motion. This is due to insufficient coverage of observational data for atmospheric/oceanic/hydrologic excitations [77-79]. Jin *et al.* analyzed hydrologic/oceanic excitation to polar motion by analyzing GRACE data [77, 79]. Better explanation for annual and semiannual LOD variation was found by GRACE+SLR analysis for the Earth's principal moment of inertia [78].

6. Time system and coordinate transformation between TRF and CRF

Universal Time (UT) is the hour angle of the apparent Sun from the Greenwich meridian. Greenwich Sidereal Time is the hour angle of the vernal equinox from the Greenwich meridian. UT1 is corrected of tiny daily oscillation of UT, which is due to the pole offset. Terrestrial Time (TT), also called as Terrestrial Dynamical Time (TDT), is uniform time in Earth based coordinate frame. International Atomic Time (TAI) is attained by atomic clocks, and has a constant time difference with TT. Coordinated Universal Time (UTC) is based on TAI and maintained close to UT1 within 0.9 second difference by assigning leap seconds. The difference between UT1 from UTC is dUT1.

$$TT = TAI + 32.184s$$
$$UTC = TAI - \left(\text{sum of leap seconds}\right) \qquad (18)$$
$$UT1 = UTC + dUT1$$

Two Earth centered coordinate frames in common use are celestial reference frame (CRF) and terrestrial reference frame (TRF). For most civilian purposes, TRF is the one to use, while CRF is convenient in astronomy. Due to the spin rotation of the Earth, transformation between TRF and CRF is needed. As summarized in former sections, not only the simple rotation angle specified as Greenwich sidereal time, but also corrections due to the precession, nutation and polar motion are necessary.

Coordinate transformation from CRF to TRF can be expressed as follows (IAU 2000A).

$$\begin{bmatrix} x_1 \\ x_2 \\ x_3 \end{bmatrix}_{Terrestrial} = R_{pm}R_{spin}R_{nut}R_{prec} \begin{bmatrix} x_1 \\ x_2 \\ x_3 \end{bmatrix}_{Celestial} \qquad (19)$$

where R_{prec}, R_{nut}, R_{spin}, and R_{pm} represent the rotation matrices corresponding to the precession, nutation, and spin with time passage, and polar motion.

$$R_{prec} = R_3(-z)R_2(\theta)R_3(-\zeta)$$
$$R_{nut} = R_1(-\varepsilon - \Delta\varepsilon)R_3(-\Delta\psi)R_1(\varepsilon)$$
$$R_{spin} = R_3(GAST)$$
$$R_{pm} = R_2(-x_p)R_1(-y_p)$$

Explanations for GAST (Greenwich Apparent Sidereal Time) and Time System are given below.

$$GAST = GMST + \Delta\psi\cos\varepsilon \text{ (in radian)}$$

where GMST is the Greenwich Mean Sidereal Time in radian, and $\Delta\psi\cos\varepsilon$ is the nutation of right ascension.

$$GMST = GMST_0 + \alpha UT1 \text{(in second)}$$

$GMST_0$ is given as

$$GMST_0 = 24110.54841 + 8640184.812866T + 0.093104T^2 - 6.2\times10^{-6}T^3 \text{(in second)}$$

where T is time in Julian century (36525 days) from J2000.0 to 0h UT1 of the day. Julian day conversion from calendar date is preferred to calculate T (for the conversion, see [4]). α is the factor for sidereal time conversion.

$$\alpha = 1.002737909350795 + 5.9006\times10^{-11}T - 5.9\times10^{-15}T^2$$

Brief description for the transformation between TRF and CRF adopted by the IAU 2006 is given here.

$$\begin{bmatrix} x_1 \\ x_2 \\ x_3 \end{bmatrix}_{Terrestrial} = R_{pm}R_{spin}R_{prec+nut} \begin{bmatrix} x_1 \\ x_2 \\ x_3 \end{bmatrix}_{Celestial} \tag{20}$$

where $R_{prec+nut}$, R_{spin}, and R_{pm} are defined as follows.

$$R_{prec+nut} = \begin{pmatrix} 1-bX^2 & -bXY & -X \\ -bXY & 1-bY^2 & -Y \\ X & Y & 1-b(X^2+Y^2) \end{pmatrix}$$

$$R_{spin} = R_3(\theta - s + s')$$

$$R_{pm} = R_1(-y_p)R_2(-x_p)R_3(\frac{x_p y_p}{2})$$

For definitions of θ, s, and s', see [7] or other references cited therein.

7. Conclusion

Different kinds of variations in the Earth's spin rotation are classified and explained. With emphasis on basic principles, each aspects of Earth rotation; precession, nutation, secular deceleration, LOD variation, and polar motion are described concisely. Euler equation and its formal solution for rigid Earth are included with their modifications for real Earth. Transformation between CRF and TRF is summarized. For convenience, underlying mechanics and mathematics are briefly summarized in Appendix.

Appendix: Summary of pre-requisite mechanics and mathematics

In this appendix, certain basic physical and mathematical concepts needed for understanding the content of this chapter are explained.

A1. Vector algebra

A vector in three dimension can be written as $\vec{A} = \hat{i}A_1 + \hat{j}A_2 + \hat{k}A_3$, where $\hat{i}$, $\hat{j}$, and $\hat{k}$ are the unit vectors in each directions. The unit vectors are taken to be cyclic in right handed coordinate system. Magnitude of a vector is defined as $A = |\vec{A}| = \sqrt{A_1^2 + A_2^2 + A_3^2}$. Inner product of two vectors is defined as $\vec{A} \cdot \vec{B} = A_1B_1 + A_2B_2 + A_3B_3$. Cross product of two vectors is defined as $\vec{A} \times \vec{B} = \hat{i}(A_2B_3 - A_3B_2) + \hat{j}(A_3B_1 - A_1B_3) + \hat{k}(A_1B_2 - A_2B_1)$. The magnitudes of the two products are $|\vec{A} \cdot \vec{B}| = |\vec{A}| |\vec{B}| \cos\theta$ and $|\vec{A} \times \vec{B}| = |\vec{A}| |\vec{B}| \sin\theta$, where θ is the angle between $\vec{A}$ and $\vec{B}$.

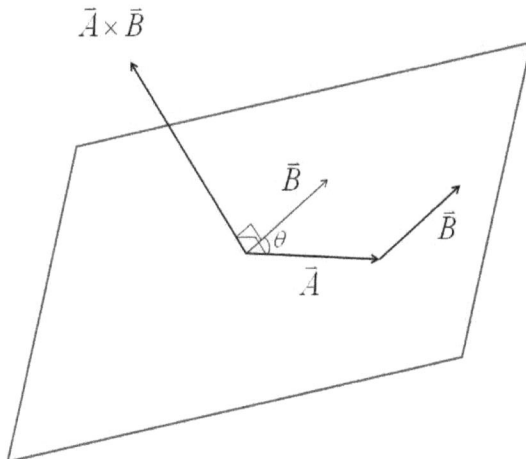

One immediately following fact is that cross product of a vector with itself or any other parallel ones vanish; $\vec{A} \times \vec{A} = 0$. A triple cross product can be composed as; $\vec{A} \times (\vec{B} \times \vec{C}) = (\vec{A} \cdot \vec{C})\vec{B} - (\vec{A} \cdot \vec{B})\vec{C}$.

A2. Harmonic oscillator

Oscillation and wave are two phenomena of fundamental importance in science and technology. Wave can be regarded as succession of harmonic oscillation in space. Mechanical oscillation exists in a system, where restoring force is accompanied to a deformation from equilibrium. Theory of one-dimensional harmonic oscillator is summarized below.

Linear restoring force and harmonic motion

For a displacement x of a spring from its equilibrium position, restoring force is exerted by the spring and can be written as $f = -kx$. Equation of motion for such system can be written as $m \dfrac{d^2 x}{dt^2} = f = -kx$. The solution for this equation is found as $x = A\cos\omega_1 t + B\sin\omega_1 t$, where the angular frequency and period are given as $\omega_1 = \sqrt{\dfrac{k}{m}}$ and $T_1 = 2\pi / \omega_1$.

Slightly damped harmonic motion

Assuming existence of a small viscous dragging force, which varies linearly with the velocity, the equation of motion is given as $m \dfrac{d^2 x}{dt^2} = -kx - c \dfrac{dx}{dt}$. The solution is found as exponentially decaying oscillation.

$x(t) = Ae^{-\gamma t}\cos(\omega_2 t - \theta)$, where $\gamma = \dfrac{c}{2m}$ and $\omega_2 = \sqrt{\omega_1^2 - \gamma^2}$.

Forced harmonic motion

When a damped harmonic oscillator is driven by external periodic force of angular frequency ω_d, its equation of motion can be written as $m \dfrac{d^2 x}{dt^2} = -kx - c \dfrac{dx}{dt} + Be^{i\omega_d t}$, and the solution is found as $x(t) = Ae^{i(\omega_d t - \theta)}$, where the amplitude and phase angle are given as $A = \dfrac{B/m}{[(\omega_1^2 - \omega_d^2)^2 + c^2 \omega_d^2 / m^2]^{1/2}}$ and $\tan\theta = \dfrac{c\omega_d / m}{\omega_1^2 - \omega_d^2}$.

Euler formula

The convenient complex notation $e^{i\alpha} = \cos\alpha + i\sin\alpha$, so called Euler formula, can be verified by using three series expansions as follows.

$$\cos\alpha = 1 - \frac{\alpha^2}{2!} + \frac{\alpha^4}{4!} - \frac{\alpha^6}{6!} + \ldots, \quad \sin\alpha = \alpha - \frac{\alpha^3}{3!} + \frac{\alpha^5}{5!} - \frac{\alpha^7}{7!} + \ldots, \quad e^{\beta} = 1 + \beta + \frac{\beta^2}{2!} + \frac{\beta^3}{3!} + \frac{\beta^4}{4!} + \ldots$$

Substitution $\beta = i\alpha$ into the third expression leads to Euler formula.

$$e^{i\alpha} = 1 + i\alpha + \frac{(i\alpha)^2}{2!} + \frac{(i\alpha)^3}{3!} + \frac{(i\alpha)^4}{4!} + \ldots = 1 - \frac{\alpha^2}{2!} + \frac{\alpha^4}{4!} + \ldots + i(\alpha - \frac{\alpha^3}{3!} + \frac{\alpha^5}{5!} - \ldots) = \cos\alpha + i\sin\alpha$$

A3. Rotational mechanics

Torque and angular momentum

Torque $\vec{\tau}$ exerted by a force $\vec{f}$ acting at position $\vec{r}$ is defined as $\vec{\tau} = \vec{r} \times \vec{f}$. Angular momentum $\vec{l}$ of a single moving body of mass m and velocity $\vec{v}$ at position $\vec{r}$ is defined as $\vec{l} = \vec{r} \times m\vec{v}$. Torque applied to a system results in change in its angular momentum. This relation is stated as

$\vec{\tau} = \frac{d\vec{l}}{dt}$, which can be readily verified by time differentiation on $\vec{l}$;

$$\frac{d\vec{l}}{dt} = \frac{d}{dt}(\vec{r} \times m\vec{v}) = \vec{v} \times m\vec{v} + \vec{r} \times m\frac{d\vec{v}}{dt} = \vec{r} \times m\vec{a} = \vec{r} \times \vec{f} = \vec{\tau}.$$

Angular velocity

For a simple rotation occurring in a plane, angular velocity is defined as time rate of rotation angle $\omega = \frac{d\theta}{dt}$. The speed due to rotation is given as $v = \omega r$, where r is the radius from the rotation axis. It should be kept in mind that rotational motion (velocity) is perpendicular to the rotating axis (angular velocity vector). For a three dimensional rotation, angular velocity vector $\vec{\omega}$ is defined by its three components; $\vec{\omega} = \omega_1\hat{e}_1 + \omega_2\hat{e}_2 + \omega_3\hat{e}_3$. Velocity $\vec{v}$ of any point $\vec{r}$ in the rotating body is given as $\vec{v} = \vec{\omega} \times \vec{r}$.

Inertia tensor

Consider the angular momentum of a rotating rigid body. Write mass element dm at position $\vec{r}$, then its linear momentum is given as $\vec{v}dm = \vec{\omega} \times \vec{r}dm$. Total angular momentum is expressed as $\vec{L} = \int \vec{r} \times \vec{v}dm = \int \vec{r} \times (\vec{\omega} \times \vec{r})dm = \int (r^2\vec{\omega} - (\vec{r} \cdot \vec{\omega})\vec{r})dm$. Three components of the total angular momentum are given as follows.

$$L_1 = \int (r^2\omega_1 - (x_1\omega_1 + x_2\omega_2 + x_3\omega_3)x_1)dm$$

$$L_2 = \int (r^2\omega_2 - (x_1\omega_1 + x_2\omega_2 + x_3\omega_3)x_2)dm$$

$$L_3 = \int (r^2\omega_3 - (x_1\omega_1 + x_2\omega_2 + x_3\omega_3)x_3)dm$$

These can be rewritten as follows.

$$L_1 = I_{11}\omega_1 + I_{12}\omega_2 + I_{13}\omega_3, \quad L_2 = I_{21}\omega_1 + I_{22}\omega_2 + I_{23}\omega_3, \quad L_3 = I_{31}\omega_1 + I_{32}\omega_2 + I_{33}\omega_3$$

where components of inertia tensor I_{ij} are defined as elements of the following matrix.

$$[I_{ij}] = \begin{bmatrix} I_{11} & I_{12} & I_{13} \\ I_{21} & I_{22} & I_{23} \\ I_{31} & I_{32} & I_{33} \end{bmatrix} = \begin{bmatrix} \int (x_2{}^2 + x_3{}^2) dm & -\int x_1 x_2 dm & -\int x_1 x_3 dm \\ -\int x_2 x_1 dm & \int (x_3{}^2 + x_1{}^2) dm & -\int x_2 x_3 dm \\ -\int x_3 x_1 dm & -\int x_3 x_2 dm & \int (x_1{}^2 + x_2{}^2) dm \end{bmatrix}$$

Rotational kinetic energy

Kinetic energy of any moving body is defined as $T = \int \frac{1}{2} v^2 dm$. The kinetic energy T of a rotating

body is given as; $\int \frac{1}{2} v^2 dm = \frac{1}{2} \int \vec{v} \cdot (\vec{\omega} \times \vec{r}) dm = \frac{1}{2} \vec{\omega} \cdot \int (\vec{r} \times \vec{v}) dm = \frac{1}{2} \vec{\omega} \cdot \vec{L}$. Therefore, $T = \int \frac{1}{2} v^2 dm$

$= \frac{1}{2} \vec{\omega} \cdot \vec{L}$ or $T = \frac{1}{2} \sum_{i=1}^{3} \sum_{j=1}^{3} I_{ij} \omega_i \omega_j$.

Principal axis

For any rigid body, set of three particular orthogonal axes (called principal axes) exists. If body frame of reference coincides with those principal axes, only diagonal component of inertia tensor remain as nonzero.

$$[I_{ij}] = \begin{bmatrix} I_{11} & 0 & 0 \\ 0 & I_{22} & 0 \\ 0 & 0 & I_{33} \end{bmatrix}$$

When a body rotates along its principal axis, its motion may be regarded as a simple planar rotation. As an example, suppose a rotation of a body along its principal axis x_3, then, the angular momentum and rotational kinetic energy are simply given as $L = L_3 = I_3 \omega_3$ and

$T = \frac{1}{2} I_3 \omega_3{}^2$. It is noted that set of principal axis and principal moment of inertia can be deter-

mined from the following condition; $L_i = \sum_{j=1}^{3} I_{ij} \omega_j = I_\lambda \omega_i$.

Rotating coordinate frame

Two coordinate frames of common origin, one stationary in space and the other rotating with angular velocity $\vec{\omega}$, are described in the figure (left). On the right figure, the unit vector $\hat{e}_2{}'$ at time t and $t + \Delta t$ are shown. The change $\Delta \hat{e}_2{}'$ in the unit vector $\hat{e}_2{}'$ is $\Delta \hat{e}_2{}' = \hat{e}_2{}'(t + \Delta t) - \hat{e}_2{}'(t) \simeq \sin\phi \, \Delta\theta \, \hat{n} = \sin\phi \, \omega \Delta t \, \hat{n}$. Here the direction of $\Delta \hat{e}_2{}'$ is denoted by a unit vector $\hat{n}$, which is parallel to $\vec{\omega} \times \hat{e}_2{}'$. Therefore the time derivative of $\hat{e}_2{}'$ is expressed as $\frac{d\hat{e}_2{}'}{dt} = \vec{\omega} \times \hat{e}_2{}'$. Likewise time derivatives of $\hat{e}_3{}'$ and $\hat{e}_1{}'$ are found as $\frac{d\hat{e}_3{}'}{dt} = \vec{\omega} \times \hat{e}_3{}'$ and $\frac{d\hat{e}_1{}'}{dt} = \vec{\omega} \times \hat{e}_1{}'$. Write the coordinates of a position vector $\vec{r}$ in the two reference frame as (x_1, x_2, x_3) and $(x_1{}', x_2{}', x_3{}')$. Then we have $\vec{r} = x_1 \hat{e}_1 + x_2 \hat{e}_2 + x_3 \hat{e}_3 = x_1{}' \hat{e}_1{}' + x_2{}' \hat{e}_2{}' + x_3{}' \hat{e}_3{}' = \vec{r}{}'$.

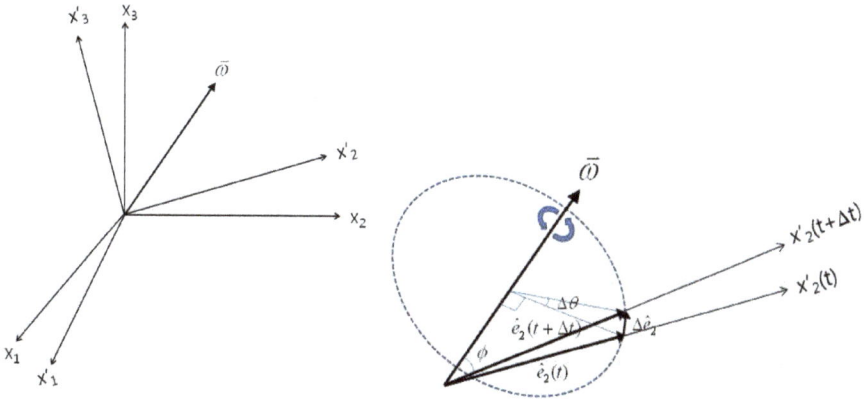

The velocity vector can be written as $\dfrac{dx_1}{dt}\hat{e}_1+\dfrac{dx_2}{dt}\hat{e}_2+\dfrac{dx_3}{dt}\hat{e}_3$ or equivalently as

$\dfrac{dx_1}{dt}\hat{e}_1'+\dfrac{dx_2}{dt}\hat{e}_2'+\dfrac{dx_3}{dt}\hat{e}_3'+x_1'\dfrac{d\hat{e}_1'}{dt}+x_2'\dfrac{d\hat{e}_2'}{dt}+x_3'\dfrac{d\hat{e}_3'}{dt}$. Since the time derivatives of unit

vectors are $\dfrac{d\hat{e}_i'}{dt}=\vec{\omega}\times\hat{e}_i'$, the three terms $x_1'\dfrac{d\hat{e}_1'}{dt}+x_2'\dfrac{d\hat{e}_2'}{dt}+x_3'\dfrac{d\hat{e}_3'}{dt}$ can be rewritten as

$\vec{\omega}\times(x_1'\hat{e}_1'+x_2'\hat{e}_2'+x_3\hat{e}_3')=\vec{\omega}\times\vec{r}'$. Then the relation is expressed as

$$\vec{v}=\vec{v}'+\vec{\omega}\times\vec{r}'$$

This time derivative expression derived for position vector in rotating frame can be extended to other vectors.

$$\dfrac{d}{dt}\bigg|_{inertial}=\dfrac{d}{dt}\bigg|_{body}+\vec{\omega}\times$$

A4. Fourier series and fourier transform

A periodic function $f(t)$ of period T can be represented by its Fourier series.

$$f(t)=\dfrac{a_0}{2}+\sum_{n=1}^{\infty}\left(a_n\cos\dfrac{2\pi nt}{T}+b_n\sin\dfrac{2\pi nt}{T}\right)$$

where the coefficients are defined as follows.

$$a_n=\dfrac{2}{T}\int_{-T/2}^{T/2}f(t')\cos\dfrac{2\pi nt'}{T}dt' \text{ and } b_n=\dfrac{2}{T}\int_{-T/2}^{T/2}f(t')\sin\dfrac{2\pi nt'}{T}dt'$$

Equivalent series for $f(t)$ can be written as

$$f(t) = \sum_{n=-\infty}^{\infty} c_n \exp(i\frac{2\pi n}{T}t) \text{ with coefficient } c_n = \frac{1}{T} \int_{-T/2}^{T/2} f(t')\exp(-i\frac{2\pi n}{T}t')dt'$$

As can be derived from above expression by taking period T as infinity, Fourier transform $F(\omega)$ of arbitrary well-behaved function $f(t)$ is defined as follows.

$$F(\omega) = \frac{1}{\sqrt{2\pi}} \int_{-\infty}^{\infty} f(t)\exp(-i\omega t)dt$$

The inverse transform is given as

$$f(t) = \frac{1}{\sqrt{2\pi}} \int_{-\infty}^{\infty} F(\omega)\exp(i\omega t)d\omega$$

Both Fourier series and Fourier transform are widely used. While Fourier transform is more convenient in theoretical development, quite often discrete Fourier transform is used for calculation in practice. For a given sequence $f[n]$, its discrete Fourier transform is defined as follows.

$$F[k] = \frac{1}{\sqrt{N}} \sum_{n=0}^{N-1} f[n]\exp(-i\frac{2\pi}{N}nk)$$

The inverse transform is given as

$$f[n] = \frac{1}{\sqrt{N}} \sum_{k=0}^{N-1} F[k]\exp(i\frac{2\pi}{N}nk)$$

Both $f[n]$ and $F[k]$ are periodic sequences such that $f[n] = f[N+n]$ and $F[k] = F[N+k]$. Fast Fourier transform is an algorithm to evaluate discrete Fourier transform by minimum number of calculations.

A5. Rotation matrix

Coordinate transformation due to a rotation along one of reference axis by an angle θ is represented as follows.

$$R_1(\theta) = \begin{bmatrix} 1 & 0 & 0 \\ 0 & \cos\theta & \sin\theta \\ 0 & -\sin\theta & \cos\theta \end{bmatrix}, \ R_2(\theta) = \begin{bmatrix} \cos\theta & 0 & -\sin\theta \\ 0 & 1 & 0 \\ \sin\theta & 0 & \cos\theta \end{bmatrix}, \ R_3(\theta) = \begin{bmatrix} \cos\theta & \sin\theta & 0 \\ -\sin\theta & \cos\theta & 0 \\ 0 & 0 & 1 \end{bmatrix}$$

Any two different axis rotations of finite angles do not commute, unless the angles are infinitesimal. For polar motion, the transformation matrix is given as $R_{pm} \simeq R_2(-x_p)R_1(-y_p) \simeq R_1(-y_p)R_2(-x_p)$ for first order approximation. Retaining second order terms, the transformation matrix is given as

$$R_{pm} \simeq R_2(-x_p)R_1(-y_p)R_3(-\frac{x_p y_p}{2}) \simeq R_1(-y_p)R_2(-x_p)R_3(\frac{x_p y_p}{2}).$$

Acknowledgements

This chapter was written by the author during his stay at Korea Astronomy and Space Science Institute (KASI) under support of KASI Basic Core Technology Development Programs. He thanks Dr. Gross R. and Dr. Jin S. for their kind influence on his works about Earth rotation.

Author details

Sung-Ho Na

Korea Astronomy and Space Science Institute, Korea

References

[1] Munk WH., Macdonald GJF. The Rotation of the Earth. Cambridge University Press; 1960.

[2] Lambeck K. The Earth's Variable Rotation: Geophysical Causes and Consequences. Cambridge University Press; 1980.

[3] Moritz H., Mueller II. Earth Rotation Theory and Observation. Ungar; 1986.

[4] Seidelmann PK., editor. Explanatory Supplement to the Astronomical Almanac. University Science Books; 1992.

[5] Gross RS. Earth Rotation Variations – Long Period. In: Herring T., editor. Geodesy, Treatise on Geophysics vol. 3. Elsevier; 2009. p239-294.

[6] Dehant V., Mathews PM. Earth Rotation Variations. In: Herring T., editor. Geodesy, Treatise on Geophysics vol. 3. Elsevier; 2009. p295-349.

[7] Petit G., Luzum B., editor. IERS Conventions (2010), IERS Technical Note No. 36, International Earth Rotation and Reference Systems Service (IERS). Verlag des Bundesamts für Kartographie und Geodäsie; 2010.

[8] Danby JMA. Fundamentals of Celestial Mechanics 2nd ed. Willman-Bell Inc; 1988.

[9] Melchior P. The Tides of the Planet Earth. Pergamon Press; 1978.

[10] Stacey FD., Davis PM. Physics of the Earth 4th ed., Cambridge University Press; 2008.

[11] Capitaine N., Wallace PT., Chapront J. Expressions for IAU 2000 precession quantities, Astronomy and Astrophysics 2003; 412, 567-586.

[12] Mathews PM., Herring TA., Buffett BA. Modeling of nutation and precession: New nutation series for nonrigid Earth and insights into the Earth's interior, Journal of Geophysical Research 2002; 107, B4, 2068.

[13] Woolard EW. Theory of the Rotation of the Earth around its Center of Mass, Astronomical Papers 1953; XV, Pt. I.

[14] Kinoshita H. Theory of the Rotation of the Rigid Earth, Celestial Mechanics 1977; 15, 277-326.

[15] Wahr J. The forced nutations of an elliptical, rotating, elastic and oceanless earth, Geophysical Journal Royal Astronomical Society 1981; 64, 705-727.

[16] Dehant V. On the nutation of a more realistic Earth model, Geophysical Journal International 1990; 100, 477-483.

[17] Schatok J. A new nutation series for a more realistic model earth, Geophysical Journal International 1997; 130, 137-150.

[18] Huang CL., Jin WJ., Liao XH. A new nutation model of a non-rigid earth with ocean and atmosphere, Geophysical Journal International 2001; 146, 126-133.

[19] Darwin GH. Scientific Papers I and II. Cambridge University Press; 1908.

[20] Brosche P, Sündermann J, editors. Tidal Friction and the Earth's Rotation. Springer-Verlag; 1978 and its continuing volume published in 1982.

[21] Goad C., Douglas B. Lunar tidal acceleration obtained from satellite-derived ocean tide parameters, Journal of Geophysical Research 1978; 83, B5, 2306-2310.

[22] Christodoulidis D., Smith D. Observed Tidal Braking in the Earth/Moon/Sun System, Journal of Geophysical Research 1988; 93, B6, 6216-6236.

[23] Lerch FJ., Nerem RS., Putney BH., Felsentreger TL., Sanchez BV., Marshall JA., Klosko SM., Patel GB., Williamson RG., Chinn DS., Chan JC., Rachlin KE., Chandler NL., MaCarthy JJ., Kuthecke SB., Pavlis NK., Pavlis DE., Robbins JW., Kapoor S., Pavlis EC. A geopotential model from satellite tracking, altimeter, and surface gravity data: GEM-T3, Journal of Geophysical Research 1994; 99, B2, 2815-2839.

[24] Lambeck K. Effects of Tidal Dissipation in the Oceans on the Moon's Orbit and the Earth's Rotation, Journal of Geophysical Research 1975; 80, 20, 2917-2925.

[25] Ray RD., Bills BG., Chao BF. Lunar and solar torques on the ocean tides, Journal of Geophysical Research 1999; 104, B8, 17653-17659.

[26] Stephenson FR., Morrison LV. Long-Term Fluctuations in the Earth's Rotation: 700 BC to AD 1990, Philosophical Transactions of the Royal Society A 1995; 351, 165-202.

[27] Morrison LV. Tidal deceleration of the Earth's Rotation deduced from astronomical observations in the period A.D. 1600 to the present. In Brosche P, Sündermann J, editors. Tidal Friction and the Earth's Rotation. Springer-Verlag; 1978. p22-27. Original

data and illustration exist in the following article: Morrison LV., Ward CG. An Analysis of the Transits of Mercury: 1677-1973, Monthly Notice of Royal Astronomical Society 1975; 173, 183-206.

[28] Morrison LV., Stephenson LV. Historical Values of the Earth's Clock Error DT and the Calculation of Eclipses, Journal for the History of Astronomy 2004; 25, 327-336.

[29] Spencer Jones H. The Rotation of the Earth and the Secular Acceleration of the Sun, Moon and Planets, Monthly Notice of Royal Astronomical Society 1939; 99, 541-548.

[30] Newton R. Secular Accelerations of the Earth and Moon, Science 1969; 166, 825-831.

[31] Rosenberg GD., Runcorn SK., editor. Growth Rhythms and the History of the Earth's Rotation. Wiley; 1975.

[32] Scrutton CT. Periodic Growth Features in Fossil Organisms and the Length of the Day and Month. In Brosche P, Sündermann J, editors. Tidal Friction and the Earth's Rotation. Springer-Verlag; 1978. p154-196.

[33] Lambeck K. The Earth's Paleorotation. In Brosche P., Sündermann J., editors. Tidal Friction and the Earth's Rotation. Springer-Verlag; 1978. p145-153.

[34] Goldreich P. History of the lunar orbit, Reviews of Geophysics 1966; 4, 411-439.

[35] Mignard F. Long Time Integration of the Moon's Orbit and references therein. In Brosche P, Sündermann J, editors. Tidal Friction and the Earth's Rotation II. Springer-Verlag; 1982. p67-91.

[36] Hansen KS. Secular Effects of Oceanic Tidal Dissipation on the Moon's Orbit and the Earth's Rotation, Review of Geophysics and Space Physics 1982; 20, 457-480.

[37] Webb DJ. On the Reduction in Tidal Dissipation Produced by Increases in the Earth's Rotation Rate and Its Effect on the Long-Term History of the Moon's Orbit. In Brosche P., Sündermann J., editors. Tidal Friction and the Earth's Rotation II. Springer-Verlag; 1982. p210-221.

[38] Na S. Tidal Evolution of Lunar Orbit and Earth Rotation, Journal of the Korean Astronomical Society 2012; 45, 49-57.

[39] Woolard EW. Inequalities in mean Solar Time from Tidal Variations in the Rotation of the Earth, Astronomical Journal 1959; 64, 140-142.

[40] Yoder CF., Williams JG., Parke ME. Tidal Variations of Earth Rotation, Journal of Geophysical Research 1981; 86, B2, 881-891.

[41] Merriam JB. Tidal Terms in Universal Time: Effects of Zonal Winds and Mantle Q, Journal of Geophysical Research 1984; 89, B12, 10109-10114.

[42] Hide R., Dickey JO. Earth's Variable Rotation, Science 1991; 253, 629-637.

[43] McCarthy DD., Luzum BJ. An analysis of tidal variations in the length of day, Geophysical Journal International 1993; 114, 341-346.

[44] Gross RS. The Effect of Ocean Tides on the Earth's Rotation as Predicted by the Results of an Ocean Tide Model, Geophysical Research Letter 1993; 20, 293-296.

[45] Chao BF., Merriam JB., Tamura Y. Geophysical analysis of zonal tidal signals in length of day, Geophysical Journal International 1995; 122, 765-775.

[46] Chen JL., Wilson CR., Chao BF., Shum CK., Tapley BD. Hydrological and oceanic excitations to polar motion and length-of-day variation, Geophysical Journal International 2000; 141, 149-156.

[47] Höpfner J. Atmospheric, oceanic and hydrological contributions to seasonal variations in length of day, Journal of Geodesy 2001; 75, 137-150.

[48] Smith ML., Dahlen FA. The period and Q of the Chandler wobble, Geophysical Journal Royal Astronomical Society 1981; 64, 223-281.

[49] Wahr J. Polar Motion Models: Angular Momentum Approach. In: Plag HP., Chao B., Gross R., van Dam T., editors. Forcing of Polar Motion in the Chandler Frequency Band: a Contribution to Understanding Interannual Climate Variation: proceedings, 21-23 April 2004. Luxemburg. European Center for Geodynamics and Seismology; 2005, 1-7.

[50] Smith ML. Wobble and nutation of the Earth, Geophysical Journal Royal Astronomical Society 1977; 50, 103-140.

[51] Sasao T., Okubo S., Saito M. A Simple Theory on the Dynamical Effects of a Stratified Fluid Core upon Nutational Motion of the Earth. In: Fedorov EP., Smith ML., Bender PL., editors. Nutation and the Earth's Rotation, IAU Symposium No. 78. D Reidel Publishing Co; 1978. p165-183.

[52] Sasao T., Wahr J. An excitation mechanism for the free core nutation, Geophysical Journal Royal Astronomical Society 1981; 64, 729-746.

[53] Dehant V., Hinderer J., Legros H., Leffz M. Analytical approach to the computation of the Earth, the outer core and the inner core rotational motions, Physics of the Earth and Planetary Interiors 1993; 76, 259-282.

[54] Rochester MG., Crossley DJ. Earth's long-period wobbles: a Lagrangian description of the Liouville equations, Geophysical Journal International 2009; 176, 40-62.

[55] Guo JY., Greiner-Mai H., Ballani L. A spectral search for the inner core wobble in Earth's polar motion, Journal of Geophysical Research 2005; 110, B10402.

[56] Wilson CR, Vicente RO. Maximum likelihood estimates of polar motion parameters. In: McCarthy DD., Carter We., editors. Variations in Earth Rotation, AGU Monograph Series vol. 59; 1990. p151-155.

[57] McCarthy DD., Luzum B. Path of the mean rotational pole from 1899 to 1994, Geophysical Journal International 1996; 125, 623-629.

[58] Gross RS., Vondrák J. Astrometric and space-geodetic observations of polar wander, Geophysical Research Letter 1999; 26, 2085-2088.

[59] Schuh H., Nagel S., Seitz T. Linear drift and periodic variations observed in long time series of polar motion, Journal of Geodesy 2001; 74, 701-710.

[60] Höpfner J. Low-frequency variations, Chandler and annual wobbles of polar motion as observed over one century, Surveys in Geophysics 2004; 25, 1-54.

[61] Höpfner J. Parameter variability of the observed periodic oscillations of polar motion with smaller amplitudes, Journal of Geodesy 2003; 77, 388-401.

[62] Na, S., Cho, J., Baek J., Kwak Y., Yoo SM. Spectral Analysis on Earth's Spin Rotation for the Recent 30 years, Journal of the Korean Physical Society 2012; 61, 152-157.

[63] Jochmann H. Period variations of the Chandler wobble, Journal of Geodesy 2003; 77, 454-458.

[64] Gross RS. The excitation of the Chandler wobble, Geophysical Research Letter 2000; 27, 2329-2332.

[65] Aoyama Y., Naito I. Atmospheric excitation of the Chandler wobble, 1983-1998, Journal of Geophysical Research 2001; 106, B5, 8941-8954.

[66] Brzezinski A., Nastula J. Oceanic excitation of Chandler wobble, Advances in Space Research 2002; 30, 195-200.

[67] Gross RS., Fukumori I., Menemenlis D. Atmospheric and oceanic excitation of the Earth's wobbles during 1980-2000, Journal of Geophysical Research 2003; 108, B8, 2370.

[68] Seitz F., Schmidt M. Atmospheric and oceanic contributions to Chandler wobble excitation determined by wavelet filtering, Journal of Geophysical Research 2005; 110, B11406.

[69] Salstein DA., Rosen RD. Regional Contributions to the Atmospheric Excitation of Rapid Polar Motions, Journal of Geophysical Research 1989; 94, D7, 9971-9978.

[70] Chao BF., Au AY. Atmospheric Excitation of the Earth's Annual Wobble: 1980-1988, Journal of Geophysical Research 1991; 96, B4, 6577-6582.

[71] Nastula J., Kolaczek B. Seasonal Oscillations in Regional and Global Atmospheric Excitation of Polar Motion, Advances in Space Research 2002; 30, 381-386.

[72] Ponte RM., Oceanic excitation of daily to seasonal signals in Earth rotation: results from a constant-density numerical model, Geophysical Journal International 1997; 130, 469-474.

[73] Chao BF., Ray RD., Gipson JM., Egbert GD., Ma C. Diurnal/semidiurnal polar motion excited by oceanic tidal angular momentum, Journal of Geophysical research 1996; 101, B9, 20151-20163.

[74] Zhou YH., Salstein DA., Chen JL. Revised atmospheric excitation function series related to Earth's variable rotation under consideration of surface topography, Journal of Geophysical Research 2006; 111, D12108.

[75] Nilsson T., Böhm J., Böhm S., Schindelegger M., Schuh H., Schreiber U., Gebauer A., Klügel T. High frequency Earth rotation variations from CONT11, EGU General Assembly 2012; Session G5.1.

[76] Brzezinski A., Dobslaw H., Thomas M., Slusarczyk L. Subdiurnal atmospheric and oceanic excitation of Earth rotation estimated from 3-hourly AAM and OAM data, EGU General Assembly 2012; Session G5.1.

[77] Jin S., Chambers DP., Tapley BD. Hydrological and oceanic effects on polar motion from GRACE and models, Journal of Geophysical Research 2010; 115, B02403.

[78] Jin S., Zhang LJ., Tapley BD. The understanding of length-of-day variations from satellite gravity and laser ranging measurements, Geophysical Journal International 2011; 184, 651-660.

[79] Jin S., Hassan A., Feng G. Assessment of terrestrial water contributions to polar motion from GRACE and hydrological models, Journal of Geodynamics 2012; in press.

Chapter 9

GNSS Observations of Crustal Deformation: A Case Study in East Asia

Shuanggen Jin

Additional information is available at the end of the chapter

1. Introduction

The East Asia is located in a complex convergent region with several plates, e.g., Pacific, North American, Eurasian, and Philippine Sea plates. Subduction of the Philippine Sea and Pacific plates and expulsion of Eurasian plate with Indian plate collision [22, 36, 15, 18, 11] make the East Asia as one of the most active seismic and deforming regions (Fig. 1). Current deformation in East Asia is distributed over a broad area extending from the Tibet in the south to the Baikal Rift zone in the north and the Kuril-Japan trench in the east, with some ambiguous blocks, such as South China (Yangtze), Ordos and North China blocks, and possibly the Amurian plate, embedded in the deforming zone. The inter-plate deformation and interaction between the blocks are very complicated an active, such as the 1978 Mw=7.8 Tangshan earthquake located between North China and Amurian blocks. Since the 1960s when the theory and models of seafloor spreading and plate tectonics were established, the large Eurasian plate has been considered as an independent rigid plate, such as RM2, P071 and NNR-NUVEL1A [6], and even the present-day global plate motion models of ITRF sequences based on the space geodetic data [10]. In fact, East Asia is very complicated deformation zone with several possible rigid blocks [22]. [36] proposed that East Asia consists of several micro-plates. However, because of low seismicity and there being no clear geographical boundary except for the Kuril-Japan trench and the Baikal rift zone, it is difficult to accurately determine the geometry and boundary of sub-plates in these areas. The borderlines of the micro-plates are not visible, especially in the complicated geological tectonic regions in North China far away from the Qinghai-Tibet plateau, and convergent belts of Eurasia, North America and the Pacific plate. Therefore, it is difficult to confirm the tectonic features and evolution of the deformation belts in East Asia.

In the northern part of East Asia, [36] first proposed the existence of the Amurian plate based on the clear geographical boundary of the Kuril-Japan trench and the Baikal rift zone, but it becomes diffuse throughout continental East Asia. The proposed Amurian plate (AM) in East Asia is of special interest to constrain the relative motion of the major and minor plate in East Asia and provides a rigorous framework for interpreting seismicity and the kinematics, especially for seismically active Japan. However, the southern boundary of the suggested Amurian plate is poorly understood. Figure 1 shows some recent proposed boundaries. The upper blue solid line is from [4], the middle black dash boundary line is from [34], and the bottom red dash line is from [8].

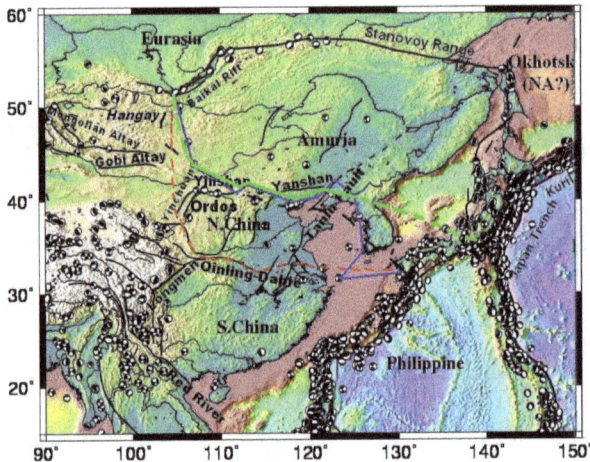

Figure 1. Tectonics setting in East Asia. The un-continuous lines are the main fault lines and the dash lines are the undefined plate boundary. Circles are the earthquakes from Harvard CMT catalogue (1976-2005, Mw >5.0). The upper blue solid line is from [4], the middle black dash boundary line is from [34], the bottom red dash line is from [8], and the upper green solid+dash lines is this study.

Therefore, the existence of the Amurian plate and its boundary geometry remain controversial. Over the last two decades a number of investigations of the micro-plates tectonics in East Asia have been conducted using geologic, seismological and geodetic data [36, 34, 8, 27, 5, 2]. However all of these studies have suffered from limited data quality and quantity, resulting in ambiguous conclusions. For instance, [34] estimated the AM motion with earthquake slip vectors and found that the spreading rates of the Baikal Rift are with the order of < 1mm/yr. [8] assumed Amuria and North China as an independent Amurian plate, and estimated spreading rates at the Baikal Rift were about 10 mm/yr. [29] used only four GPS sites in or around the AM plate and concluded that AM can neither be resolved nor excluded as a separate plate.[5] postulated that North China (including the possible AM) and South China could be a single rigid block using data from only 9 GPS sites. [2] claimed the existence of the Amurian plate, but still could not determine the location of the southern Amurian plate boundary. These investigations of the tectonics in this region are not conclusive because of the sparse and limited data that were used. Hence, there is much debate surrounding the nature of microplate and its boundaries in East Asia.

With more GPS observations in East Asia, it provides a new ways to clearly monitor the large-scale crustal motion and to distinguish the possible blocks, e.g., the national projects "Crustal Movement Observation Network of China (CMONC) with more than 1000 GPS sites and the Japanese GPS Earth Observation Network (GEONET) with more than 1000 continuous GPS sites. These dense GPS observations will obtain a more accurate estimate of plate geometry and its interior crustal deformation in East Asia. In this Chapter, we present new dense geodetic results for East Asia from about 1000 GPS sites in China, South Korea and Japan for the period 1998-2005, as well as combining recently published velocities for the Bailkal Rift and Mongolia [5]. The possibility of microplate motion independent of the Eurasian plate is tested using GPS derived velocities and its boundary and kinematics are further discussed [13].

In addition, accurate measurements of crustal strain accumulated energy rates will contribute to understand tectonic features and to evaluate the earthquake potential. Now the high precision space geodesy techniques, especially the low-cost and all weather GPS, play a key role in monitoring the crustal strain state and accumulated energy variation. Meanwhile, the significant strain accumulation caused by the tectonic activities (such as earthquakes) will provide an essential constraint on the physical processing of earthquakes. Therefore, monitoring the spatial variation of the strain and comprehensive understanding of strain accumulation pattern are beneficial to reveal the physical process of crustal tectonic activities and to evaluate the earthquake risk [14]. Here, the denser GPS velocity filed is used to estimate the strain rates and strain energy density rates, in an attempt to know largely aseismic areas and to assess the future earthquake risk potential in East Asia.

2. GPS observations and results

The Crustal Movement Observation Network of China (CMONC) was constructed since 1998, which contains a nationwide fiducial network of 25 continuous GPS sites observed since August 1998, 56 survey mode sites with yearly occupations and more than 900 regional campaign stations operated in 1999, 2001, and 2004 with continuous observations for at least 4 days during each session. In addition, the Korean GPS Network (KGN) with more than 45 permanent GPS sites was established since 2000 by the Korea Astronomy and Space Science Institute (KASI), the Ministry Of Governmental Administration and Home Affairs (MOGAHA), and the National Geographic Information Institute (NGI). Japan GPS Earth Observation Network (GEONET) was established since 1996 by the Geographical Survey Institute of Japan. These denser GPS observations can investigate and study detailed crustal deformation and kinematics in East Asia. Here we collect about 1000 GPS sites (1999.1-2004.12) in East Asia with 54 core IGS sites that were used for ITRF2000 [1] and 10 permanent IGS sites located in East Asia. These GPS sites are shown in Figure 2. The pentagon stands for the permanent GPS sites (2000-2005), the dot denotes the campaign GPS sites and the triangle is the yearly observed GPS sites (1999-2005).

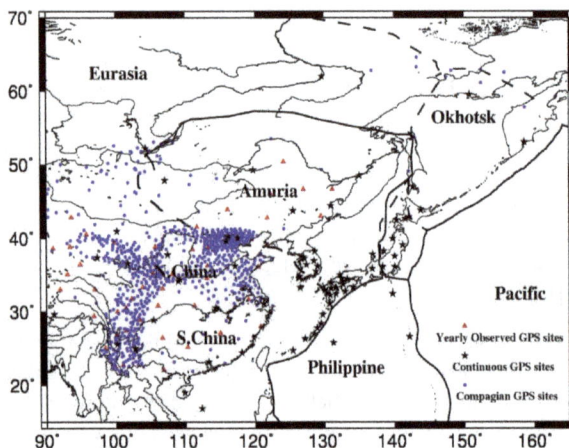

Figure 2. GPS sites distribution in this study. The triangle is the yearly observed GPS site, the pentagon is the continuous GPS site (2000-2005), and the dot is the campaign GPS site (1999-2005). The solid lines are the known plate boundaries.

All available GPS data were processed in single-day solutions using the GAMIT software [16] in a three-step approach. At the first step, loose a-priori constraints were set for all parameters and double-differenced GPS phase observations from each day were used to estimate station coordinates and the zenith tropospheric delay (ZTD) at each station every 2 hours. The IGS final orbits, IERS Earth orientation parameters, azimuth- and elevation-dependent antenna phase center models, as recommended by the IGS were used in the data processing. The 54 global IGS stations served as ties to the ITRF2000 frame [1]. At the second step, the regional daily solutions were combined with global solutions produced by the Scripps Orbital and Position Analysis Center (SOPAC, http://sopac.ucsd.edu/) using the GLOBK software [9], and the reference frame was applied to the solution by performing a seven-parameter transformation to align it to ITRF2000 (via the global 54 core stations). At the third step, the site velocities were estimated by least square linear fitting to time variations of the daily coordinates for each station [13]. For example, Figure 3 shows the North, East and Vertical GPS time series at SUWN site (South Korea). The ~1000 GPS site velocity field combining with recently published velocities at the Baikal Rift and Mongolia [5] are shown in Figure 4 with respect to the Eurasian plate.

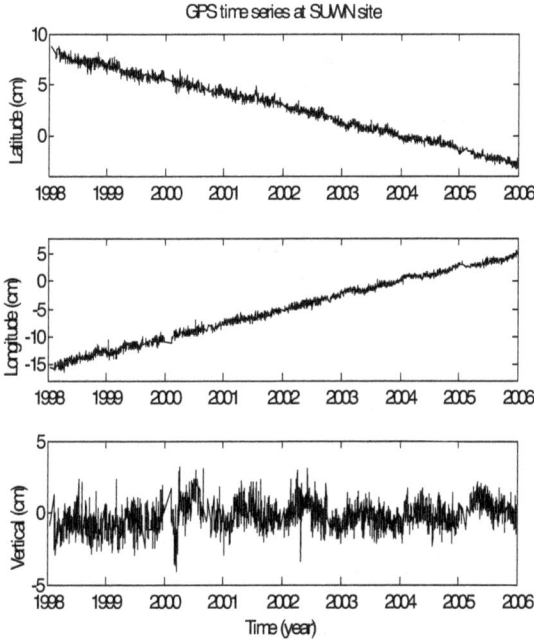

Figure 3. GPS time series at SUWN site (South Korea).

3. Sub-plates and kinematics

3.1. Block modeling method

The definition of the micro-plate geometries in East Asia is quite unclear, especially their boundaries. Denser space geodetic data can be used to define the plate boundary by testing an independent plate rotation about a best-fit Euler's rotation pole obtained by geodetic velocities. Here we assumed several plates or blocks in NE Asia whose plate boundary geometries were defined principally on the basis of seismicity and faults: EU (Eurasia), AM (Amuria), SK (South Korea), NC (North China), SC (South China), AK(AM+SK), AN (AK +NC), EA (East Asia) (see Figure 1). The boundaries are respectively the Yin Shan-Yan Shan Mountain belts for AM-NC, the Qinling-Dabie fault for NC-SC and the Tanlu fault for NC-SK blocks. As the blocks have large areas, the plates are just defined as rigid blocks motions without considering the elastic strain [32], i.e.

$$v = \Omega \times r \qquad (1)$$

where v is the velocity and r is the position vector. In order to calculate Euler's rotation parameters, we project the fault geometry and stations positions from spherical to planar geometry [20]. Then the rigid block motion are modeled to estimate the Euler ration parameters of each block (see Table 1). Here only continuously and yearly observed GPS stations are used. In addition, the angular velocity of the Eurasian plate was estimated from the velocities of 22 IGS sites located on the Eurasian plate (TROM, MADR, HERS, BRUS, KOSG, 7203, ZIMM, VILL, OBER, ONSA, WETT, POTS, GOPE, GRAS, BOR1, LAMA, KIRU, JOZE, ZWEN, IRKT, KIT3, KSTU), including the core sites in the Eurasian plate employed for the orientation and maintenance of the ITRF2000 [1] (see Table 1).

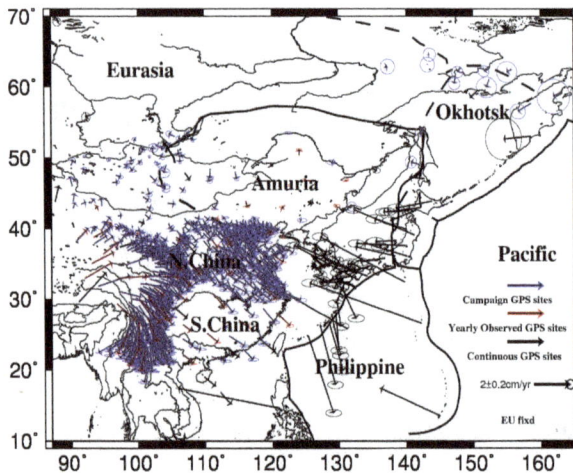

Figure 4. Crustal deformation rates in the Eurasia plate (EU) fixed reference frame with error ellipses in 95% confidence limits.

3.2. Testing results

To test whether the microblocks are independent of the Eurasian plate, we used a χ^2 test that compares how well two different models fit a set of data. χ^2 is a sum of the squares of weighted residuals defined as:

$$\chi^2 = \sum_{i=1}^{N} \frac{(v_o(i) - v_m(i))^2}{\sigma_o^2(i)} \qquad (2)$$

where $v_o(i)$ is the observation velocity of site i, $v_m(i)$ is the calculated velocity of site i from the plate rotation model, $\sigma_o^2(i)$ is the variance of observation velocity in site i, and N is the total number of observations. Table 2 shows the χ^2 for each model. The χ^2 for the model of independent AM, SK, SC, NC blocks is smaller than the Eurasian plate and a 2-block model (EU and EA), respectively. To check whether AM, SK, SC and NC are independent blocks, we perform an F-ratio statistical test.

First, we assume the one-block model in which East Asia (EA) is part of the Eurasian plate and the two-block model in which the East Asia (EA) plate rotates independently with respect to the Eurasian plate (EU). The 3-block model assumes that the EA is divided into the AN (Amuria+South Korea+North China) and South China (SC) plates, while the 4-block model contains the EU, SC (S.China), NC (N.China) and AK (Amuria+S.Korea) plates. The 5-block model is the EU, SC, NC, AM and SK plates. These blocks were defined principally on the basis of seismicity and faults (see Figure 1). We compare the misfit of each model inversion and test for significance using the F-ratio [30]:

$$F = \frac{[\chi^2(1\ block) - \chi^2(2\ block)]/3}{\chi^2(2\ block)/(N-3)} \tag{3}$$

where $\chi^2(1block)$ and $\chi^2(2block)$ stand for the sum of the squares of weighted residuals in one-block and two-block models, respectively. In Table 2 one can clearly see that for the two-block model (EU and EA) the reduced chi-squared misfit of GPS velocity observations has been greatly reduced from 6.4 (for the one-block model (EU+EA)) to 1.4 (for the two-block model (EU and EA)) and the F-ratio statistic (Eq.2) is 1086.5, which is well above the 99% confidence level of 3.8. The reduced chi-squared misfits of other independently rotating blocks for the 3-block, 4-block and 5-block models are also greatly degraded, and the calculated F-statistics between the 2-3 block and 3-4 block models are well above the 99% confidence level of 3.8 as well as between the 3-4 block and 4-5 block models. These results indicate that the AM, SK, NC and NC are independent of the Eurasian plate motion. Furthermore, it shows that the South Korea block (SK) is excluded from the Amurian plate (AM), coinciding with recent test results using fewer GPS sites [12]. Figure 5 shows residual velocities (observed minus predicted) at yearly observed and continuous GPS site in East Asia with respective to the 1 block (Eurasia), 2 blocks (Eurasia and East Asia) and 4 blocks (Eurasia, Amuria+S.Korea, North China and South China). The mean residual in 4-block model is much smaller than fewer blocks.

The estimated relative motions along the block boundaries are further obtained (Fig. 6). Comparisons of spreading or converging rates and directions along these boundaries are showing in Figures 7 and 8. The GPS-derived relative motion directions are nearly the same as the earthquake slip vector directions (Fig. 6). However there are some differences at Baikal Rift. It may be due to the larger uncertainty of earthquake slip vectors (±15). Another is possibly the Euler vector problem of the large non-rigid Eurasian plate (EU). For instance, Table 1 lists different Euler vectors of the Eurasian plate, and larger discrepancy is found between [5] and other geodetic results. The tectonic boundaries between the North China and Amuria plates, the Yin Shan-Yan Shan Mountain belts, are extending at about 2.4 mm/yr. The Qinling-Dabie fault between the North China and South China plates is moving left laterally at about 3.1 mm/yr. This difference between Amuria and South China predicted rates is about 5.5 mm/yr, almost consistent with geological results by [3] and [24]. The Tanlu fault between the North China and South Korea blocks is moving right laterally at about 3.8 mm/yr. The Amuria and South Korea blocks are extending at about 1.8 mm/yr. The convergent rates at the boundaries of the AM and Okhotsk [2] are from 9 to 17 mm/yr, similar to the seismic results of [17]. The spreading rates in the Baikal Rift zone are about 3.0 ± 1.0 mm/yr, consistent with [5] at 4 ± 1 mm/yr.

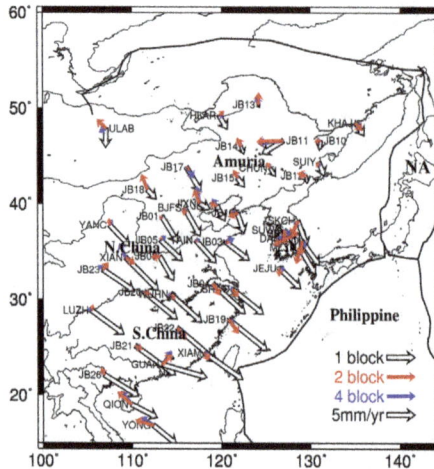

Figure 5. Residual velocities (observed minus predicted) at yearly observed and continuous GPS site in East Asia with respective to the 1 block (Eurasia), 2 blocks (Eurasia and East Asia) and 4 blocks (Eurasia, Amuria+S.Korea, North China and South China).

Plates	Longitude	Latitude	Angular rate (o /My)	Pole σ_{maj}	Error σ_{min}	Ellipse Azimuth
Eurasian plate						
This study	-100.655	56.995	0.257± 0.002	0.6	0.1	49
NNR-1A[a]	-112.3	50.6	0.234			
Altamimi et al.[2002]	-99.374	57.965	0.260± 0.005	-	-	-
Calais et al. [2003]	-107.022	52.266	0.245± 0.005	-	-	-

Plates	Longitude		Latitude	Angular rate (∘/My)	Pole σ_{maj}	Error σ_{min}	Ellipse Azimuth
ITRF2000 [b]	-99.691		57.246	0.260± 0.002	0.8	0.2	52
Sella et al. [2002]	-102.21		58.27	0.257± 0.003	1.5	0.4	34
Kreemer et al.[2003]	-97.4		56.4	0.279± 0.005	0.6	0.2	-81
				Amurian Plate			
This study	-115.285		62.474	0.291± 0.004	25.0	2.9	133
ITRF2000 [b]	-126.646		63.899	0.316± 0.021	10.9	0.9	146
Sella et al.[2002]	-133.76		63.75	0.327± 0.057	23.5	1.6	-64
Kreemer et al. [2003]	-103.8		60.0	0.302± 0.007	1.3	0.5	-20
				South Korea			
This study	177.682		64.642	0.446± 0.016	42.3	1.5	19
				North China			
This study	-123.876		64.369	0.313± 0.006	40.0	2.8	137
				South China			
This study	-109.372		57.304	0.323± 0.001	20.2	2.0	127
Sella et al. [2002]	-109.21		54.58	0.340± 0.057	16.6	1.0	-40
Shen et al. [2005]	146.70		57.92	0.22	-	-	-

Table 1. Rotation is in a clockwise direction about the pole. The error ellipses of the poles are described by the 1 σ semi-major and semi-minor axes of each error ellipse and the clockwise angle from true north of the semi-major axis. [a] No-Net-Rotation NUVEL-1A (NNR-1A) model [6] [b] Angular velocity vectors were estimated from 12 years of CGPS in ITRF2000 [25]Absolute and relative angular velocity vectors for the Eurasian, Amurian, South Korea, North China and South China plates.

Number of blocks[a]	χ^2	χ_r^2	f	F
1	5904.8	6.4	922	
2	1295.4	1.4	939	1086.5
3	1221.2	1.3	936	18.5
4	1191.5	1.3	933	7.6
5	980.5	1.1	930	65.0

Table 2. [a]1: EU (Eurasia); 2: EU and EA (East Asia); 3: EU, AN (Amuria+South Korea+North China) and SC (South China); 4: EU, AK (Amuria+S.Korea), NC, SC; 5: EU, NC, SC, AM and SK; 6: EU, NC, SC, SK, West AM and East AM. f is the number of degrees of freedom and χ_r^2 is the reduced χ^2 as the ratio of χ^2 to f .Statistic tests of different block models

4. Crustal Strain rates and Seismic Risks

Monitoring the pattern of crustal strain and comprehensive understanding of strain accumulation intensity are beneficial to reveal the physical process of crustal tectonic activities and to evaluate the earthquake risk. As the first step in the earthquake risk potential evaluation in East Asia, the strain parameters are estimated from the estimated GPS displacement rate field. In order to reduce the effects of abnormal site motions, the subnetwork with four GPS sites is used to estimate the strain parameters. Under the hypothesis that the velocity field v varies linearly inside each small sub-network covering the GPS sites, we can calculate the average horizontal velocity gradient $g = grad(v)$ over each subnetwork. Because the velocity gradient generally incorporates both deformation and rotation, this 2-D tensor is asymmetric [21]. The crustal strain rate in East Asia can be derived from GPS deformation velocities by [12, 14]:

$$v_{ei} = \frac{\partial v_{ei}}{\partial x_{ei}} x_{ei} + \frac{\partial v_{ei}}{\partial x_{ni}} x_{ni}$$

$$v_{ni} = \frac{\partial v_{ni}}{\partial x_{ei}} x_{ei} + \frac{\partial v_{ni}}{\partial x_{ni}} x_{ni}$$

(4)

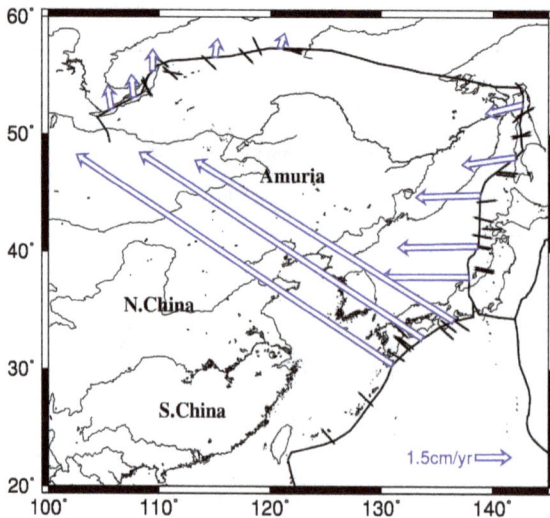

Figure 6. Relative spreading or converging motions at the plate boundaries in East Asia.

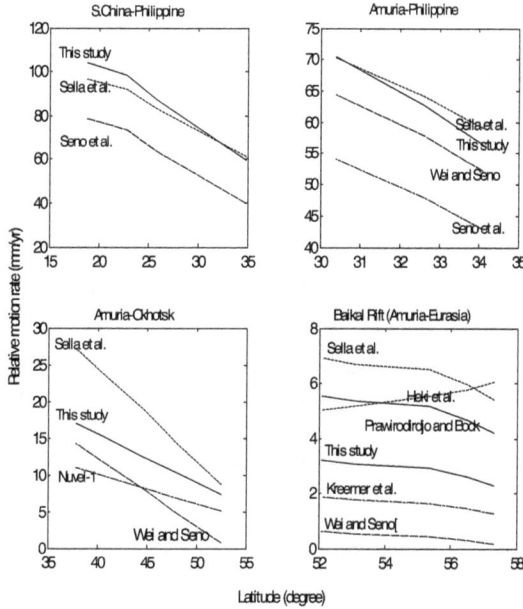

Figure 7. Relative spreading or converging rates at the plate boundaries in East Asia.

where v_{ei} and v_{ni} are the east and north component velocity at the site i located at (x_{ei}, x_{ni}).

Strain components $\dot{\varepsilon}_{ee}$, $\dot{\varepsilon}_{nn}$ and $\dot{\varepsilon}_{en}$ are expressed as $\dfrac{\partial v_e}{\partial x_e}$, $\dfrac{\partial v_n}{\partial x_n}$ and $\dfrac{1}{2}(\dfrac{\partial v_e}{\partial x_n} + \dfrac{\partial v_n}{\partial x_e})$, respectively. The dilation rates i show that East Asia is under the compressional strain regime at WNW-ENE, consistent with the focal mechanism of earthquakes in Northeast Asia. The high dilation rates appear in North China, Southwest Japan and at the boundary of Philippine Sea plate. The strong compression rates are probably caused by the extrusion force due to the subduction of the Philippine Sea and Pacific plates and the expulsion of Eurasian plate with Indian plate collision, causing frequent earthquakes in these regions. Inversely, the South Korea and South China blocks have relatively lower dilation rates, indicating a lower indirect effect of push and subduction forces or as if such forces are transmitted through South Korean peninsula and South China without causing any deformation/strain, alternatively. This may be attributed to the strong rheology or/and absence of relatively weak zones in the region.

In addition, we estimate the scalar strain rate, defined as

$$\dot{\varepsilon} = \sqrt{\dot{\varepsilon}_{ee}^2 + \dot{\varepsilon}_{nn}^2 + 2\dot{\varepsilon}_{en}^2} \tag{5}$$

where e and n are longitude and latitude directions, respectively. Figure 9 shows the contour map of scalar strain rates in Northeast Asia, implying the Korean peninsula and South China as stable blocks with low strain rates. It once again highlights that high strain rates concentrate in North China, Southwest Japan and the boundary of Philippine Sea plate with Eurasian plate, consistent with high seismicity in these areas (Figure 1).

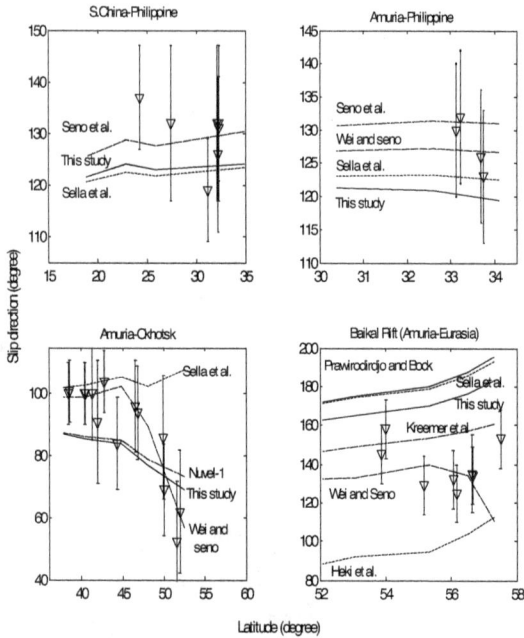

Figure 8. Relative spreading or converging directions at the plate boundaries in East Asia.

In addition, the accumulated strain energy is generally released through earthquakes until the adjacent fault blocks or plates reach a new state of equilibrium [26, 33]. Therefore, the release of tectonic strain energy stored within crustal rock is the cause of major earthquakes. The strain energy per unit volume (i.e. the strain energy density) is an important index reflecting the intensity of crustal activities, and its variation rate indicates the long-term trend of accumulated energy within the crust. The larger the variation rate of strain energy density, the higher energy accumulated in the crust, which would more probably result in earthquakes. Therefore, for the earthquake risk evaluation and prediction, it is important to estimate the strain energy density from surface displacement observations and determine the state of strain energy density within the crust and its temporal variations.

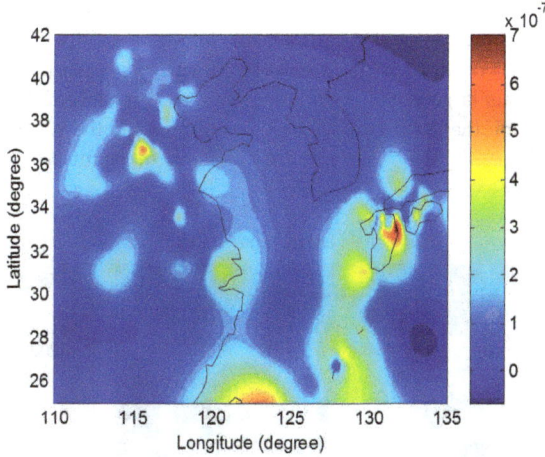

Figure 9. Map of scalar strain rates in East Asia from GPS observations.

For an elastic body, the strain energy equals the work done by external forces and its density is the strain energy per unit volume. The general tensor form for the strain energy density can be expressed in terms of strain and stress using Hooker's Law:

$$U = \frac{1}{2}\sigma_{ij}\varepsilon_{ij}$$

(6)

where U is the strain energy density (Unit: $J.m^{-3}$), σ_{ij} and ε_{ij} are the stress and strain, respectively. And the variation rate of strain energy density can be further derived from Eq. (6). The stress is obtained through the laws of elasticity theory as follows [31]:

$$\sigma_{ij} = 2\mu\varepsilon_{ij} + \delta_{ij}\lambda\Delta$$

(7)

where μ is the modulus of rigidity, λ is the Lame parameter, δ_{ij} is Kroneckerdelta, Δ is the 2-D surface dilation ($\sum_{i=1}^{2}\varepsilon_{ii}$). For Poisson's ratio v =0.25, $\lambda = \mu$, and it is assumed as standard value of 3×10^{10} Pa [7].

The stress (σ_{ij}), strain (ε_{ij}) and their rates can be derived from GPS displacements (1999-2004) and velocities, respectively. Using Eq. (6), the strain energy density variation rate in East Asia can be obtained using the derived the strain, stress and their rates, which is shown in Figure 10. The distribution of strain energy density variation rates shows that the most active areas are in North China, Southwest Japan and west margin of Philippine Sea plate, respectively, again consistent with high seismic activity zones. As the GPS measurements are made after the large historic earthquakes, the strain energy density rates derived from GPS displacement rate may include contributions from postseismic relaxation. These regions with anomalous large strain energy density rates probably indicate a high earthquake risk in future, and the lower strain energy density rates in the South Korean peninsula and South China imply that low seismicity may continue in the future.

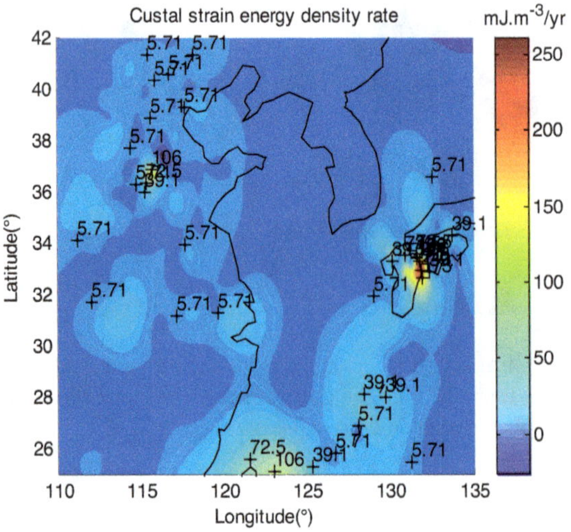

Figure 10. Variation rate of crustal strain energy density in East Asia

5. Conclusions

GPS data (1998-2005) from more than 85 continuous and about 1000 campaign stations in East Asia have been processed. The kinematics of East Asia is studied by modeling GPS-derived velocities with rigid block rotations and elastic deformation. It has been found that the deformation in East Asia can be well described by a number of rotating blocks, which are independent of the Eurasian plate motion with statistical significance above the 99% confidence level. The tectonic boundary between the North China and Amuria plates is the Yin Shan-Yan Shan Mountain belts with about 2.4 mm/yr extension. The boundary between North China and South China is the Qinling-Dabie fault, moving left laterally at about 3.1 mm/yr. The Amuria and South Korea blocks are extending at about 1.8 mm/yr. The Baikal Rift between the Amurian and Eurasian plates is spreading at about 3.0 mm/yr. The 9~17 mm/yr relative motion between the Amuria and Okhotsk blocks is accommodated at the East Sea-Japan trench zone. Furthermore, the relative motion rates and deformation types are nearly consistent with seismic and geological solutions along their boundaries. In addition, the AM, SK and SC blocks are almost rigid with residual velocities on order of 1.0~1.2 mm/yr, while the NC block has larger residual velocities on order of 1.6 mm/yr, indicating un-modeled deformation in block boundaries. Localized deformation near the Qinling-Dabie fault and Yin Shan-Yan Shan Mountain belts may be elastic strain accumulation due to interseismic locking of faults.

The strain and energy density rates in East Asia are investigated with GPS observations (1999.1-2004.12). The dilation rates show that East Asia is under the compressional strain regime at WNW-ENE, consistent with the focal mechanism of earthquakes in East Asia. The high dilation rates focus on North China, Southwest Japan and the boundary of Philippine Sea plate, probably caused by the compression force due to the subduction of the Philippine Sea and Pacific plates and the expulsion of Eurasian plate with Indian plate collision. In contrast, the South Korean Peninsula and South China blocks havbe relatively lower dilation rates, indicating a possible lower effect of push and subduction forces or that such forces are transmitted through South Korean peninsula and South China without causing any deformation/strain. This may be attribute to the strong rheology or/and absence of weak zones in the region, which leads to fewer earthquakes. Moreover, the scalar strain rates and strain energy density rates further imply the Korean peninsula and South China as a stable block with low rates, and high rates mainly concentrate on North China and Southwest Japan and the western boundary of Philippine Sea plate, consistent with highly seismic occurrences in these areas. In addition, the strain energy density rate reflects a long-term trend of strain energy accumulation and release. Therefore, North China, Southwest Japan and western boundary of Philippine Sea plate with high strain energy density rates are still highly seismic and the low seismicity in South Korea and South China with lower strain energy density rates may continue in the future.

Acknowledgements

Figures were made with the public domain software GMT [Wessel and Smith, 1998]. We are grateful to those who created the Crustal Motion Observation Network of China and made the observation data available, Korean GPS Network and Japan GPS Earth Observation Network (GEONET).

Author details

Shuanggen Jin

Shanghai Astronomical Observatory, Chinese Academy of Sciences, Shanghai, China

References

[1] Altamimi, Z., Sillard, P., & Boucher, C. (2002). ITRF2000: A New Release of the International Terrestrial Reference Frame for Earth Science Applications. *J. Geophys. Res,* 107(A10), 2214, 10.1029/2001JB000561.

[2] Apel, E. V., Burgmann, R., Steblov, G., Vasilenko, N., King, R., & Prytkov, A. (2006). Independent active microplate tectonics of northeast Asia from GPS velocities and block modeling. *Geophys. Res,* 33, L11303, 10.1029/2006GL026077.

[3] Avouac, J. P., & Tapponnier, P. (1993). Kinematic model of deformation in central Asia, . *Geophys. Res. Lett,* 20, 895-898.

[4] Bird, P. (2003). An updated digital model of plate boundaries. *Geochem. Geophys. Geosyst,* 4(3), 1027, 10.1029/2001GC000252.

[5] Calais, E., Vergnolle, M., San'kov, V., Lukhnev, A., Miroshnitchenko, A., Amarjargal, S., & Déverchère, J. (2003). GPS measurements of crustal deformation in the Baikal-Mongolia area (1994-2002): Implications for current kinematics of Asia. *J. Geophys. Res,* 108(B10), 2501, 10.1029/2002JB002373.

[6] De Mets, C., Argus, D. F., Gordon, R. G, et al. (1990). Current plate motions. *Geophys J Int,* 101, 425-478.

[7] Hanks, T. C., & Kanamori, H. (1979). A moment-magnitude scale. *J. Geophys. Res.,* 84, 2348-2350.

[8] Heki, K., Miyazaki, S., Takahashi, H., Kasahara, M., Kimata, F., Miura, S., & An, K. (1999). The Amurian plate motion and current plate kinematics in East Asia. *J. Geophys. Res,* 104, 29147-29155.

[9] Herring, A. (2002). GLOBK global Kalman filter VLBI and GPS analysis program, version 10.0, Mass. *Inst. of Technol., Cambridge Mass, USA.*

[10] Jin, S. G., & Zhu, W. (2002). Present-day spreading motion of the mid-Atlantic ridge. *Chin. Sci. Bull,* 47(18), 1551-1555, 10.1360/02tb9342.

[11] Jin, S. G., & Zhu, W. Y. (2003). Active Motion of Tectonic Blocks in Eastern Asia: Evidence from GPS Measurements. *ACTA Geological Sinica-English Edition,* 77(1), 59-63.

[12] Jin, S. G., & Park, P. (2006). Crustal stress and strain energy density rates in South Korea deduced from GPS observations. *Terr. Atmos. Ocean. Sci,* 17(1), 169-178.

[13] Jin, S. G., Park, P., & Zhu, W. (2007). Micro-plate tectonics and kinematics in Northeast Asia inferred from a dense set of GPS observations. *Earth Planet. Sci. Lett,* 257(3-4), 486-496, 10.1016/j.epsl.2007.03.011, 2007a.

[14] Jin, S. G., Park, P., & Park, J. (2007 b). Why is the South Korean peninsula largely aseismic? Geodetic evidences. *Curr. Science,* 93(2), 250-253.

[15] Kato, T., Kaotake, Y., & Nakao, S. (1998). Initial results from WING, the continuous GPS network in the western Pacific area. *Geophysical Research Letters*, 125(3), 369-372.

[16] King, R. W., & Bock, Y. (1999). Documentation for the GAMIT GPS Analysis Software. *Mass. Inst. of Technol., Cambridge Mass.*

[17] Kogan, M. G., Bürgmann, R., Vasilenko, N. F., Scholz, C. H., King, R. W., Ivashchenko, A. I., Frolov, D. I., Steblov, G. M., Kim, Ch U., & Egorov, S. G. (2003). The 2000 M w 6.8 Uglegorsk earthquake and regional plate boundary deformation of Sakhalin from geodetic data. *Geophys. Res. Lett*, 30(3), 1102, 10.1029/2002GL016399.

[18] Kogan, M. G., Steblov, G. M., King, R. W., Herring, T. A., Frolov, D. L., Egorov, S. G., Levin, V. Y., & Jones, A. (2000). Geodetic constrains on the rigidity and relative motion of Eurasian and North American. *Geophys. Res. Lett*, 27, 2041-2044.

[19] Kreemer, C., Holt, W. E., & Haines, A. (2003). An integrated global model of present-day plate motions and plate boundary deformation. *Geophys. J. Int*, 8-34.

[20] Meade, B. J., & Hager, B. H. (2005). Block models of crustal motion in southern California constrained by GPS measurements. *J. Geophys. Res*, 110, B03403, 10.1029/2004JB003209.

[21] Malvern, L. E. (1969). Introduction to the mechanics of a continuum medium. *Prentice-Hall, Englewood Cliffs, NJ.*

[22] Molnar, P., & Tapponnier, P. (1975). Cenozoic tectonic of Asia: effects of a continental collision. *Science*, 189, 419-426.

[23] Okada, Y. (1985). Surface deformation due to shear and tensile faults in a half space. *Bull. Seismol. Soc. Am*, 75, 1135-1154.

[24] Peltzer, G., & Saucier, F. (1996). Present-day kinematics of Asia derived from geologic fault rates. *J. Geophys. Res*, 101(27).

[25] Prawirodirdjo, L., & Bock, Y. (2004). Instantaneous global platemotionmodel from 12 years of continuous GPS observations. *J. Geophys. Res*, 109, B08405, 10.1029/2003JB002944.

[26] Savage, J. C., & Simpson, R. W. (1997). Surface strain accumulation and the seismic moment tensor. *Bulletin Seismic Society of America*, 87, 1345-1353.

[27] Sella, G. F., Dixon, T. H., & Mao, A. (2002). REVEL: A model for recent plate velocities from space geodesy. *J Geophys Res*, 107(B4), ETG11-1-32.

[28] Shen, Z., , K., Lü, J., Wang, M., & Bürgmann, R. (2005). Contemporary crustal deformation around the southeast borderland of the Tibetan Plateau. *J. Geophys. Res*, 110, B11409, 10.1029/2004JB003421.

[29] Steblov, G. M., Kogan, M. G., King, R. W., Scholz, C. H., Bürgmann, R., & Frolov, D. I. (2003). Imprint of the North American plate in Siberia revealed by GPS. *Geophys. Res. Lett*, 30(18), 1924, 10.1029/2003GL017805.

[30] Stein, S., & Gordon, R. (1984). Statistical tests of additional plate boundaries from plate motion inversions. *Earth Planet. Sci. lett*, 69, 401-412.

[31] Straub, C. (1996). Recent crustal defoirmation and strain accumulation in the Marmara sea region, inferred from GPS measurements. *Ist. of Geod. and Photogram. FTHZ Mitt.*, 58.

[32] Thatcher, W. (2007). Microplate model for the present-day deformation of Tibet. *J. Geophys. Res*, 112, B01401, 10.1029/2005JB004244.

[33] Weber, J., Stein, S., & Engeln, J. (1998). Estimation of intraplate strain accumulation in the New Madrid seismic zone from repeat GPS surveys. *Tectonics*, 17, 250-266.

[34] Wei, D., & Seno, T. (1998). Determination of the Amurian plate motion, in Mantle dynamics and plate interaction in East Asia. *edited by M. Flower, S. Chung, C. Lo and T. Lee*, 337-346.

[35] Wessel, P., & Smith, W. H. F. (1998). New, improved version of Generic Mapping Tools released: Eos. *Trans. Amer. Geophys. Union*, 79, 579.

[36] Zonenshain, L. P., & Savostin, L. A. (1981). Geodynamics of the Baikal rift zone and plate tectonics of Asia. *Tectonophysics*, 76, 1-45.

Chapter 10

Global Geoid Modeling and Evaluation

WenBin Shen and Jiancheng Han

Additional information is available at the end of the chapter

1. Introduction

Geoid determination with high accuracy remains a major issue in physical geodesy and attracts significant attention from the international geodetic and geophysical community. The realization of one centimeter-level geoid is still a common challenge in geodetic science today. Geoid, which is defined as the closed equi-geopotential surface nearest to the mean sea level (Listing, 1872; Grafarend, 1994), serves as height datum system and plays a significant role in different application fields.

Generally, two classical geoid modeling methods, Stokes' method (Stokes, 1849; Heiskanen and Moritz, 1967) and Molodensky's method (Molodensky et al., 1962; Heiskanen and Moritz, 1967), are used to determine a geoid or quasi-geoid via solving the corresponding boundary-value problems, namely, the Stokes boundary-value problem and the Molodensky boundary-value problem (e.g., Hofmann-Wellenhof and Moritz, 2005). By Stokes' method, one determines a geoid, while by Molodwnsky's method, one determines a quasi-geoid. Both the Stokes' method and Molodensky's method have disadvantages. Concerning the Stokes' method, to determine the geoid, the masses outside the geoid should be removed to the inside of the geoid, for instance, the masses outside the geoid should be condensed so as to form a layer right on the geoid using the Helmert's second condensation method (e.g., Heiskanen and Moritz, 1967). However, by the mass adjustment, the geoid will be changed, and corrections are needed. Concerning the Molodensky's method, though it may avoid the mass adjustment, it provides a quasi-geoid, which is unfortunately not an equi-geopotential surface and constrains its applications in practice.

In order to overcome the aforementioned disadvantages, as an alterative, Shen (2006) proposed a new method, which is different from the classical ones. In principle, the new method is based on the classical definition of geoid, and takes full information of the external gravity field model (e.g. EGM2008; Pavlis et al., 2008; 2012), digital topographic model (e.g. DTM2006.0, Shuttle Radar Topography Mission; see Pavlis et al., 2007; Farr et al., 2007), and crust density model (e.g. CRUST2.0; see Bassin et al. 2000; Tsoulis 2004; Tsoulis et al. 2009). These models are publicly available (EGM2008 and DTM2006.0 are both developed by the EGM2008 team, from the website http://earth-info.nga.mil/GandG/wgs84/gravitymod/ egm2008/egm08_wgs84.html; and CRUST2.0, from the website http://igppweb.ucsd.edu/ ~gabi/crust2. html).

This chapter introduces the main idea of the new method and the computational strategies used to determine a global geoid (section 2), describes briefly the needed datasets (section 3), provides a 30'×30' global geoid as an application example and its evaluations by comparison with the EGM2008 geoid and globally available GPS/leveling benchmarks (GPSBMs; section 4), discusses relevant problems related to this topic and concludes this chapter (section 5).

2. A new method for modeling a global geoid

A new method for determining a global geoid put forward by Shen (2006) will be introduced and technical strategies for realizing the determination of the global gravimetric geoid (Shen and Han 2012a) will be provided, including the technique in computing the terrain effects. This section is referred to Shen (2006; 2007) and Shen and Han (2012a).

2.1. Theoretical model

In this subsection we introduce the main idea of the new method, which was put forward by Shen (2006). The contents in details are referred to Shen (2007) as well as Shen and Han (2012a).

The gravitational potential $V_1(P)$ generated by the mass of a shallow layer, a layer from the Earth's surface to a depth D below the surface (see caption of Figure 1), can be determined using the following Newtonian integral

$$V_1(P) = G\int_\tau \frac{\rho(K)}{l} d\tau, \qquad P(r,\varphi,\lambda) \in \bar{\Gamma} \tag{1}$$

where $P(r, \varphi, \lambda)$ is the field point, (r, φ, λ) the spherical coordinate of the field point, G the gravitational constant (the 2006 CODATA adjustment is 6.67428×10^{-11} $m^3kg^{-1}s^{-2}$; Mohr et al., 2008), $\rho(K)$ the three-dimensional density distribution of the masses which constitute the shallow layer, where $K(r', \varphi', \lambda')$ is the moving point of the volume integral element $d\tau$, (r', φ', λ') the spherical coordinate of the moving point; l is the distance between P and K, $\bar{\Gamma}$ denotes the domain outside $\partial\Gamma$, which includes the Earth's external domain $\bar{\Omega}$ and the domain occupied by the shallow layer (Cf. Figure 1).

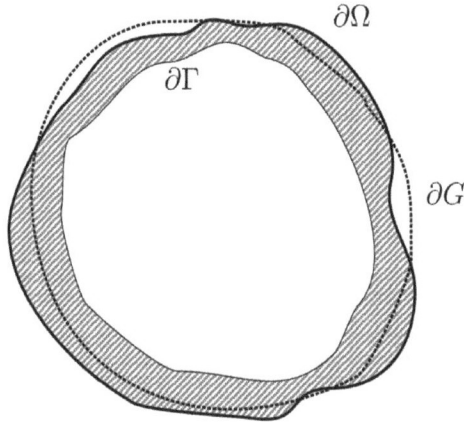

Figure 1. Definition of the shallow layer, redrawn after Shen (2006). The thick solid line denotes the Earth's surface ∂G, the dotted line denotes the geoid ∂G, and the thin solid line denotes a closed surface $\partial \Gamma$, which is below the geoid. The masses bounded by ∂G and $\partial \Gamma$, namely the shadow part, are referred to as the shallow layer

Given the external gravitational potential field $V(P)$ of the Earth, the gravitational potential $V_0(P)$ generated by the inner masses bounded by the surface $\partial \Gamma$ can be determined by the following expression

$$V_0(P) = V(P) - V_1(P), \qquad P \in \bar{\Omega} \tag{2}$$

where $V_1(P)$ is determined by Eq.(1). It should be noted that Eq.(2) is defined only in the domain $\bar{\Omega}$, as $V(P)$ is a priori given only in this region.

The potential field $V_0(P)$ given by Eq.(2) is defined, regular and harmonic in the domain $\bar{\Omega}$, and it is generated by the inner masses bounded by the surface $\partial \Gamma$. It can then be easily confirmed (Shen, 2006; 2007) that the potential field $V_0^*(P)$ defined in the region $\bar{\Gamma}$ (the region outside the surface $\partial \Gamma$) that is generated by the masses enclosed by $\partial \Gamma$ is just the natural downward continuation of the potential field $V_0(P)$.

Then, the geopotential field $W^*(P)$ generated by the Earth in the domain $\bar{\Gamma}$ can be expressed as

$$W^*(P) = V^*(P) + Q(P), \qquad P \in \bar{\Gamma}$$
$$V^*(P) = V_1(P) + V_0^*(P), \qquad P \in \bar{\Gamma}$$

(3)

where $Q(P)$ is the centrifugal potential.

Now, the position $P(P \in \partial G) \equiv P_G$ of the geoid may be determined by simply solving the following equation

$$V(P) + Q(P) = W_0, P \in \partial G$$

(4)

where ∂G denotes the geoid, W_0 is the geopotential constant, namely, the geopotential on the geoid. A rounded value W_0=62636856.0 m²/s² (Burša et al., 2007) is adopted in our application example (see section 4, and Shen and Han, 2012b). Then, the geoid undulation N may be determined based on the reference ellipsoid (e.g. WGS84) and the obtained position P_G which runs over the geoid.

2.2. Modeling the gravitational potential of the shallow layer

This subsection introduces a technique of modeling the gravitational potential of the shallow layer, referred to Shen and Han (2012a).

The gravitational potential generated by the shallow layer (masses) is computed by discretized numerical integration using elementary bodies such as right-rectangular prisms and tesseroids. The integration of Eq.(1) can be completed by using prism modeling if the mass density $\rho(K)$ of each volume integral element is homogeneous. Figure 2 demonstrates the geometry of the right-rectangular prism. The prism is bounded by planes parallel to the coordinate planes, defined by the coordinates $X_1, X_2, Y_1, Y_2, Z_1, Z_2$, respectively in the Cartesian coordinate system, and the field point P is denoted by (X_P, Y_P, Z_P).

The result of the integration is provided in the following form (Nagy et al., 2000, 2002; Heck and Seitz, 2007; Tsoulis et al., 2009)

$$V^*(P) = G\rho \Big| \Big| \Big| xy\ln(\frac{z+r}{\sqrt{x^2+y^2}}) + yz\ln(\frac{x+r}{\sqrt{y^2+z^2}}) + zx\ln(\frac{y+r}{\sqrt{x^2+z^2}})$$
$$-\frac{x^2}{2}\tan^{-1}\frac{yz}{xr} - \frac{y^2}{2}\tan^{-1}\frac{xz}{yr} - \frac{z^2}{2}\tan^{-1}\frac{xy}{zr} \Big|_{x_1}^{x_2} \Big|_{y_1}^{y_2} \Big|_{z_1}^{z_2}$$

(5)

where

$$
\begin{aligned}
x_1 &= X_1 - X_P, & x_2 &= X_2 - X_P \\
y_1 &= Y_1 - Y_P, & y_2 &= Y_2 - Y_P \\
z_1 &= Z_1 - Z_P, & z_2 &= Z_2 - Z_P \\
r &= \sqrt{x^2 + y^2 + z^2}
\end{aligned}
\tag{6}
$$

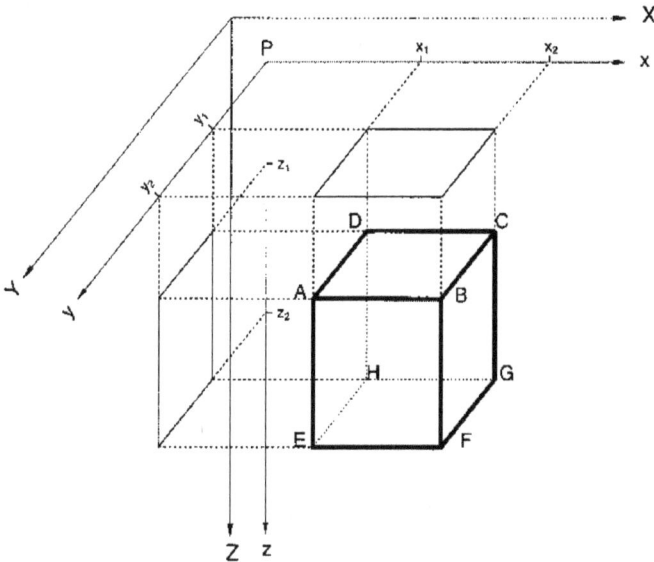

Figure 2. Sketch map of the definition of the right-rectangular prism (after Nagy et al., 2000)

Eq.(5) defines a rigorous, closed analytical expression for the computation of the gravitational potential $V(x, y, z)$ of the right-rectangular prism. Although the potential $V(x, y, z)$ is continuous in the entire domain $\mathbb{R}^3$, its solution is not defined at certain places in $\mathbb{R}^3$: 8 corners, 12 edges and 6 planes of the prism (Nagy et al., 2000; 2002). The direct computation of Eq.(5) will fail when P is located on a corner, an edge or a plane, as mentioned above, but one can calculate the corresponding limit values in a manner as given by Nagy et al. (2000, 2002) at these special positions.

The main drawback in computing the potential using Eq.(5) is the prerequisite of the repeated evaluations of several logarithmic and arctan functions for each prism. Furthermore, the formulae for computing the potential generated by prisms are given in Cartesian coordinates. This implies a planar approximation and requires a coordinate transformation for every single prism before the application of Eq.(5). One needs to perform transformations between the edge system of the prism and the local vertical system at the computation point. The explicit formulae for the transformations can be found in Heck and Seitz (2007) and Kong et al. (2001). Due to the above reason, although the prism modeling is rigorous and precise, the corresponding computation is time-consuming, especially when one needs to perform computations for a region with dense grids.

Compared to the low efficiency of the prism modeling, the tesseroid modeling is much faster. The notion "tesseroid" (see Figure 3), which was first introduced by Anderson (1976), is an elementary unit bounded by three pairs of surfaces (Heck and Seitz, 2007; Grombein and Heck, 2010): a pair of surfaces with constant ellipsoidal heights (spherical approximation is applied in practice r_1=const, r_2=const), a pair of meridional planes (λ_1=const, λ_2=const) and a pair of coaxial circular cones (φ_1=const, φ_2=const).

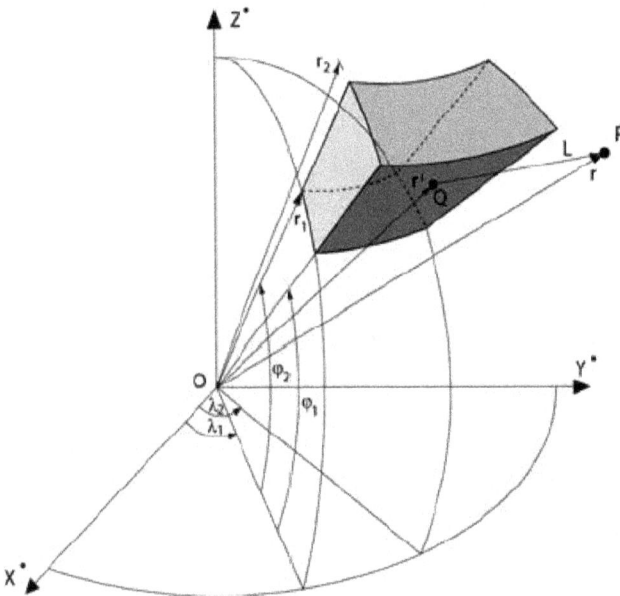

Figure 3. Geometry of a tesseroid (after Kuhn, 2003)

Based on a Taylor series expansion and choosing the geometrical center of the tesseroid as the initial point by Taylor expansion, truncated after the 3rd-order terms, the realization of Eq. (1) reads (Heck and Seitz, 2007)

$$V(r,\varphi,\lambda) = G\rho\Delta r\Delta\varphi\Delta\lambda\Big[K_{000} + \frac{1}{24}\Big(K_{200}\Delta r^2 + K_{020}\Delta\varphi^2$$
$$+K_{002}\Delta\lambda^2\Big) + O(\Delta^4)\Big] \tag{7}$$

where $\Delta r = r_2 - r_1$, $\Delta\varphi = \varphi_2 - \varphi_1$, $\Delta\lambda = \lambda_2 - \lambda_1$ denote the figuration of the tesseroid, K_{ijk} denote the trigonometric coefficients involved in the Taylor expansion, and the Landau symbol $O(\Delta^4)$ indicates that it contains only the 4th-order terms and higher ones, which could be neglected at present accuracy requirement. The trigonometric coefficients depend on the relative positions of the computation point (r, φ, λ) with respect to the geometrical center of the tesseroid $(r_0, \varphi_0, \lambda_0)$. The zero-order term of Eq.(7), which is formally equivalent to the point-mass formula, has the following form

$$K_{000} = \frac{r_0^2 \cos\varphi_0}{l_0}, \quad l_0 = \sqrt{r^2 + r_0^2 - 2rr_0\cos\Psi_0}$$
$$\Psi_0 = \sin\varphi\sin\varphi_0 + \cos\varphi\cos\varphi_0\cos\delta\lambda, \quad \delta\lambda = \lambda_0 - \lambda \tag{8}$$

The mathematical expressions of the second-order coefficients K_{200}, K_{020} and K_{002} are relatively complicated and can be found in Heck and Seitz (2007). The tesseroids are well suited to the definitions and numerical calculations of DEMs/DTMs, which are usually given on geograph-ical grids. The tesseroid modeling is also modest in terms of the computation costs versus the prism method, and it runs about ten times faster for the computation of the gravitational potential and four times faster for gravitational acceleration than those implemented by the prism method (Heck and Seitz, 2007). The numerical efficiency can be improved even further by computing the potential or acceleration along the same parallel: there is no need to re-calculate the trigonometric terms (mainly the sine, cosine functions and their squares) related to the constant latitude φ_0, and the computation load will be reduced greatly.

The prism modeling offers rigorous, analytical solution but its implementation efficiency is low and requires very demanding computations. The tesseroid modeling on the other hand shows high numerical efficiency but may provide results at a sufficient accuracy level at present, and the approximation errors due to the truncation of the Taylor series do exist but decrease very quickly with the increasing distance between the tesseroid and the computation point (Heck and Seitz, 2007). Hence, an effective way is to combine these two methods together for practical computations, which would take full advantag-es of both methods and overcome their disadvantages (Tsoulis et al., 2009). Here we in-troduce the combination method to compute the gravitational potential of the shallow layer as stated in the following strategy (Shen and Han, 2012a). After the masses of the shallow layer are partitioned into elementary units, the prism modeling is adopted to evaluate the contribution of the units which are located at the nearest vicinity surrounding the computation point, while the tesseroid modeling is employed for computing the contribution of the units located outside the mentioned vicinity area. In this case, one can maintain a manageable computation load with sufficient accuracy. This combination method is hereinafter referred to as the combination modeling method (CMM).

3. Models and datasets

In this section, we describe the needed datasets, namely the EGM2008 model, CRUST2.0 model and DTM2006.0 model (Shen and Han, 2012a).

To determine a geoid, the $5'\times5'$ resolution (~10 km) geopotential model EGM2008 (Pavlis et al., 2008; 2012) could be used, which was released by the US National Geospatial Intelligence Agency (NGA) in 2008, and it is at present the most precise global geopotential model of the Earth's external gravity field. It is complete to spherical harmonic degree and order 2159, and contains additional spherical harmonic coefficients extending to degree 2190 and order 2159. EGM2008 has been developed by combining the spaceborne GRACE satellite data, terrain and altimetry data, and the surface gravity data (Kenyon et al., 2007). Based on SRTM (Shuttle Radar Topography Mission, Farr et al., 2007) data and other altimetry datasets, the high-resolution global digital topographic model DTM2006.0 that is complete to degree/order 2160 (Pavlis et al., 2007) became publicly available at the same time.

In order to evaluate the gravitational potential of the shallow layer, according to Eq.(1), one has to know (a) its interior structure, especially the density distribution and (b) the geometry of the entire layer. The former aspect, namely, the density distribution, is usually provided by geological investigations (rock samples, deep drilling projects, etc.) and seismic methods. Dziewonski and Anderson (1981) established the preliminary reference Earth model (PREM), which has a spherical symmetric density distribution. From then on, many different models have been established with various levels of details. The best currently available global crustal model is CRUST2.0 (Bassin et al., 2000; Tsoulis, 2004). Based on seismic refraction data and a fine-tuned dataset of ice and sediment thickness, CRUST2.0 was established and released by the US Geological Survey and the Institute for Geophysics and Planetary Physics at the University of California in 2000. CRUST2.0, a significant upgrade of CRUST5.1 ($5°\times5°$, Mooney et al., 1998), offers a seven-layered density distribution and structure of the crust at a $2°\times2°$ grid, where there are totally 360 crustal types. The seven crust layers are listed from the Earth's surface to the Moho boundary as: ice, water, soft sediments, hard sediments, upper crust, middle crust, and lower crust. Each $2°\times2°$ cell (one $2°\times2°$ grid layer) is assigned to one kind of crustal type where the compressional and shear wave velocity (V_P, V_S), density ρ and the upper and lower boundaries are given explicitly for each individual layer.

The determination of the geometry of the shallow layer is discussed as follows. First, we focus on the upper surface of the shallow layer, namely the topographic surface. A digital terrain/elevation model (DTM/DEM) with a specific grid resolution can be used to represent the topographic surface. This representation depends on a discretization due to the fact that DTM/DEM is usually given at scattered locations or on geographical grids. For the numerical evaluation, the global digital topographic model DTM2006.0 mentioned before can be used: this is a model created to supplement EGM2008, and it can provide elevation on land areas and bathymetry on ocean areas for an arbitrary point. However, this is inconsistent with our case. What we need is the topographic surface on both continents and ocean surface. This inconsistency can be simply eliminated by setting DTM2006.0 heights on ocean areas to zero. In the ocean surfaces, a better choice is to use the Danish National Space Center data set DNSC08 mean sea surface (MSS), established from an integration of satellite altimetry data with a time span from 1993 to 2004 (Andersen and Knudsen, 2009; Andersen et al., 2010). Hence, a new upper surface of the shallow layer is established by combining DTM2006.0 on land areas and DNSC08 MSS on ocean surfaces.

Second, we have to choose the lower surface of the shallow layer, namely the surface $\partial\Gamma$ (Cf. Figure 1). Theoretically, $\partial\Gamma$ can be a closed surface in a quite arbitrary shape that lies inside the geoid (Shen, 2006). Since the geoid undulations vary within the range of ±120 m, it is easy to determine the approximate position of the surface $\partial\Gamma$. In order to simplify the description and calculations, we choose the EGM2008 geoid as an initial reference surface. Now, the shallow layer model (including upper and lower surfaces, and density distribution) has been established. Then, a new surface that extends from the reference surface downward to a depth of 15 meters is constructed, which is referred to as the lower surface and denoted as $\partial\Gamma$ (Cf. Figure 3). In this case, it is guaranteed that the real geoid locates in the domain outside the surface $\partial\Gamma$. Now both the upper and lower surfaces of the shallow layer have been determined.

Then, we apply the CMM described in section 2.2 to calculate the gravitational potential generated by the shallow layer.

4. Evaluation of global geoid model

As an example, in this section we apply the new method (Shen, 2006) to the global geoid determination, and provide a 30'×30' global geoid model, which is evaluated by globally available GPS benchmarks (GPSBMs). The main contents are referred to Shen and Han (2012b).

4.1. A global geoid

Based on the new method (see section 2), taking the value W_0=62636856.0 m²/s² and solving Eq.(4), we obtain a 30'×30' global geoid as shown in Figure 4. For convenience, hereinafter, the 30'×30' global geoid determined by the new method (Shen, 2006) is referred to as the calculated global geoid, while the 30'×30' global geoid computed based on EGM2008 is referred to as the EGM2008 global geoid.

4.2. Comparisons with the EGM2008 global geoid

The differences between the calculated global geoid and the EGM2008 global geoid are shown in Figure 5, and the statistical results are listed in Table 1.

According to Figure 5 and Table 1, significant differences can be found between these two geoid models in plateau, mountainous regions, the Antarctic and Greenland ice sheet, while in other regions, namely in plain areas and the ocean area (STD=1.5cm), the two geoid models show good agreement with each other. Specifically, good agreements occur in Australia, Arctic region, Europe, the USA and Africa. The STDs of the differences are 4.6cm, 7.3cm, 7.5cm, 14.2cm and 14.8cm, respectively. Large deviations can be found in South America, Asia and Antarctica, and the STDs are 19.0cm, 25.8cm and 28.4cm, respectively. The deviations of the two models are very large in China (STD=31.0cm), especially in the Tibetan region, Western China (extreme value ±2m). However, the STD drops to 15.6cm in the eastern China, where the lands are relatively flat. The STD of the entire land area obtained is 8.84 cm, and the STD over the globe is 2.9cm (cf. Table 1).

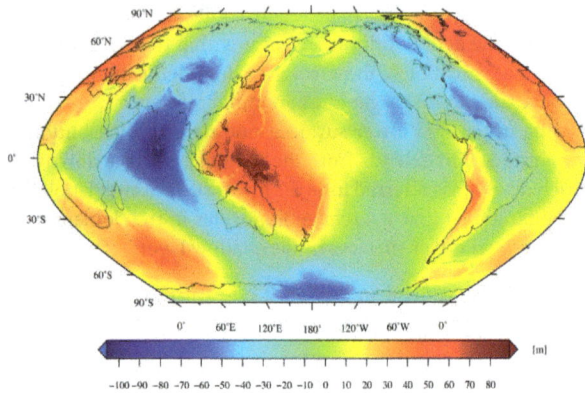

Figure 4. global geoid based on the new method (Shen and Han, 2012b)

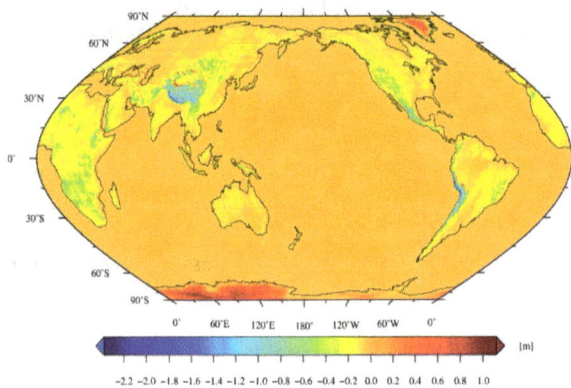

Figure 5. The differences between calculated global geoid and the EGM2008 global geoid (Shen and Han, 2012b)

Region	Max	Min	Mean	STD
Globe	1.138	-2.389	-0.026	0.029
Ocean area	0.369	-0.576	-0.001	0.015
Australia	0.094	-0.306	-0.060	0.046
Arctic region	0.632	-0.576	-0.031	0.073
Europe	0.325	-0.702	-0.073	0.075
USA	0.171	-1.171	-0.138	0.142
Africa	0.270	-1.188	-0.202	0.148
South America	0.942	-2.263	-0.107	0.190
Asia (including China)	1.010	-2.389	-0.153	0.258
Antarctica	1.138	-0.406	0.223	0.284
China	1.010	-2.389	-0.238	0.310
Eastern China	0.336	-1.132	-0.146	0.156

Table 1. Statistics of the differences between calculated global geoid and the EGM2008 global geoid (unit: m) (Shen and Han, 2012b)

4.3. Comparisons with GPS/leveling data

Two GPS/leveling datasets, the GPSBMs09 released by NGS (National Geodetic Survey, http://www.ngs.noaa.gov/GEOID/GPSonBM09/) and the Australian GPS/leveling data (http://www.ga.gov.au/ausgeoid/nvalcomp.jsp, Hu, 2011), were served for testing the calculated global geoid and the EGM2008 global geoid. GPSBMs09 dataset includes 20446 GPS/leveling benchmarks in the conterminous US (except Alaska and Hawaii), and 1474 points were taken away based on the rejection code given by NGS. Australian GPS/leveling data set includes 2614 GPS/leveling benchmarks. The distributions of all GPS/leveling benchmarks are shown in Figure 6.

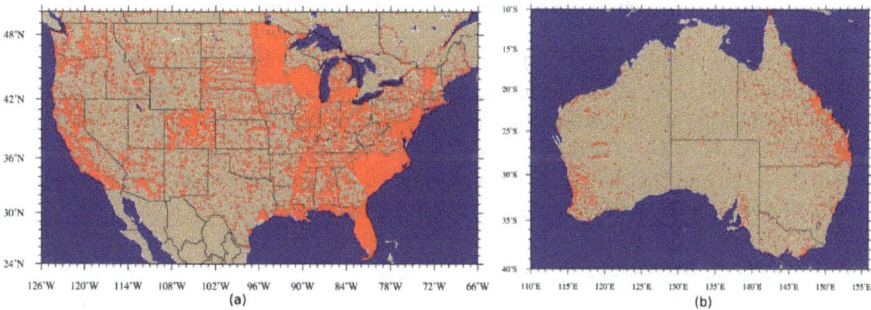

Figure 6. a) Distribution of the GPSBMs in USA; (b) Distribution of the GPSBMs in Australia (Shen and Han, 2012b)

The differences between the US GPSBMs09, Australian GPSBM dataset and the calculated global geoid are shown in Figure 7. And the differences between the GPS/leveling datasets and the EGM2008 global geoid, the GGM03S geoid, the GOCO03S geoid are similar to those with respect to the calculated geoid, so they are not shown here. Table 2 lists the statistics of the differences among the US GPSBMs09 as well as Australian GPSBM dataset and the calculated global geoid, the EGM2008 geoid, GGM03S geoid (Grace-only, Tapley et al., 2007) and the GOCO03S geoid (Grace and GOCE combined, Mayer-Gürret et al., 2012).

Comparisons and validations show that: the 30'×30' calculated global geoid is identical to the 30'×30' EGM2008 global geoid, and an overall standard deviation of the differences between the two models is at centimeter level. The calculated geoid and the EGM2008 geoid (both degree/order 360) perform better than the satellite-only geoids (degree/order 180 for GGM03S geoid and degree/order 250 for GOCO03S geoid), see Table 2. The accuracy of the 30'×30' calculated global geoid in the USA is 28cm while in Australia it is 14cm.

	Selected case		Max	Min	Mean	STD
USA	GPS/leveling geoid –	EGM2008 global geoid	0.184	-1.087	-0.473	0.2820
		Calculated global geoid	0.184	-0.874	-0.468	0.2807
		GGM03S global geoid	0.258	-1.134	-0.475	0.2825
		GOCO03S global geoid	0.244	-1.110	-0.472	0.2838
Australia	GPS/leveling geoid –	EGM2008 global geoid	0.700	-0.086	0.477	0.1361
		Calculated global geoid	0.700	-0.086	0.478	0.1362
		GGM03S global geoid	0.819	-0.037	0.478	0.1467
		GOCO03S global geoid	0.798	-0.029	0.484	0.1478

Table 2. The validation results of the 30'×30' geoid (unit: m) (Shen and Han, 2012b)

Due to the relatively low degree (360), the 30'×30' calculated global geoid is almost identical to the 30'×30' EGM2008 global geoid and there is no noticeable improvement. However, another study of the authors shows that there is a significant improvement in the calculated geoid with respect to the EGM2008 geoid if we determine a geoid with a higher resolution (e.g., 5'×5') (Shen and Han, 2012a). The reason is that the lateral and radial density variations have been taken into account by the new method and the short-wavelength part of the geoid has been refined. A detailed study is needed to consider the error sources (e.g., the errors existed in CRUST2.0 model) in the geoid modeling using the new method in further investigations. Moreover, an updated version of CRUST2.0, CRUST1.0, will be released in this year (http://igppweb.ucsd.edu/~gabi/crust2.html) and this significant update will greatly improve the accuracy of the geoid determined based on the new method.

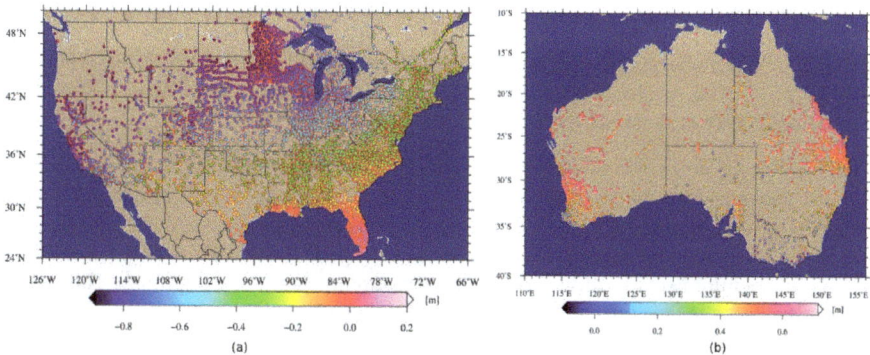

Figure 7. a) Differences between the GPS/leveling geoid and the calculated geoid in USA; (b) Differences between the GPS/leveling geoid and the calculated geoid in Australia. (Shen and Han, 2012b)

5. Discussions and conclusions

By conventional methods, e.g., Stokes method, for the purpose of the mass adjustment, one needs the orthometric height (above the geoid), which is determined by leveling and gravimetry. Since the measurement errors increase with the propagation distance of the spirit leveling, it can not be guaranteed that the orthometric height could achieve the centimeter-level accuracy globally, especially in the mountain areas. This is one of the disadvantages in Stokes' method. This disadvantage might be avoided by introducing the new method (Shen, 2006).

As an application example of this new method, a global 30'×30' geoid was provided (see section 4; see also Shen and Han, 2012b), which takes full advantage of the most recently published models and data sets, namely, EGM2008, DTM2006.0, CRUST2.0 and DNSC08. A model of the shallow layer with 3D density distribution has also been established to implement the new method (see section 2.2). To calculate the gravitational contribution of the shallow layer, a combination modeling method is used in the computations, and the iterative spherical harmonic analysis and synthesis procedures have been presented to determine the gravitational potential $V_0^*(P)$ in the domain $\bar{\Gamma}$. Validations show that the calculated geoid fits GPSBMs as good as the EGM2008 geoid within one centimeter level. Moreover, if more GPSBMs are available, the validation may reveal more details about the calculated geoid.

The accuracy of the calculated geoid depends on those of the models involved in the computations, as well as the methodology itself. The errors in EGM2008, DTM2006.0 and CRUST2.0 dominate the errors in the calculated geoid. EGM2008 is currently the best and most reliable global geopotential model, with about 10cm-level precision in average globally. The errors in elevations from DTM2006.0, which is a supplement to EGM2008, may introduce large errors in geoid height determination (e.g., Merry, 2003; Kiamehr and Sjöberg, 2005). Compared to the high-resolution geopotential model and DEM, CRUST2.0 provides density and stratification information in a relatively poor resolution (2°×2°). In order to maintain the same resolution in each computational step, one has to interpolate CRUST2.0 to finer resolutions. Uncertainties of the CRUST2.0 model and the interpolation process may yield unacceptable errors (Han and Shen, 2012b). Optimistically, better results could be achieved after a new updated crust density model (CRUST1.0; see Laske, 2011) is released.

The global geoid determined by the new method may offer complementary information to map the geological structures (Shen and Han, 2012c). The new method is also applicable to determining a regional geoid (especially in mountainous areas; see Shen and Han, 2012a; Han and Shen, 2012) using the publicly available datasets (e.g., EGM2008, DTM2006, CRUST2.0), without the requirements of additional gravity measurements and spirit leveling. This is one of the advantages of the new method.

Author details

WenBin Shen* and Jiancheng Han

Wuhan University, China

References

[1] Andersen, O. B., and P. Knudsen, 2009: DNSC08 mean sea surface and mean dynamic topography models. J. Geophys. Res., 2009, 114, C11001, doi:10.1029/2008JC005179.

[2] Andersen, O. B., P. Knudsen, and P. Berry, 2010: The DNSC08GRA global marine gravity field from double retracked satellite altimetry. *J Geod.*, 84(X.3), DOI: 10.1007/s00190-009-0355-9.

[3] Anderson, E. G., 1976: The effect of topography on solutions of Stokes' problem. Unisurv S-14, Rep, School of Surveying, University of New South Wales, Kensington.

[4] Bassin, C., G. Laske, and G. Masters, 2000: The current limits of resolution for surface wave tomography in North America. *EOS Trans AGU*, 81, F897.

[5] Burša, M., S. Kenyon, J. Kouba, Z. Šíma, V. Vatrt, V. Vítek, and M. Vojtíšková, 2007: The geopotential value W_0 for specifying the relativistic atomic time scale and a global vertical reference system. *J Geod.*, 81, 103-110.

[6] Dziewonski, A. M., and D. L. Anderson, 1981: Preliminary reference Earth model. *Phys. Earth Planet. Inter.*, 25: 297-356.

[7] Farr, T. G., P. A. Rosen, E. Caro, R. Crippen, R. Duren, S. Hensley, M. Kobrick, M. Paller, E. Rodriguez, L. Roth, D. Seal, S. Shaffer, J. Shimada, J. Umland, M. Werner, M. Oskin, D. Burbank, and D. Alsdorf, 2007: The Shuttle Radar Topography Mission. *Rev. Geophys.*, 45, RG2004.

[8]   Grafarend, E. W., 1994: What is a geoid? In: Geoid and its Geophysical Interpretations, Vanicek P, Christou N (eds), CRC Press, Boca Raton, 3-32.

[9]   Grombein, T., K. Seitz, and B. Heck, 2010: Modelling topographic effects in GOCE gravity gradients. BMBF Geotechnologien Statusseminar "Erfassung des Systems Erde aus dem Weltraum III", 04. Oktober 2010, Universität Bonn.

[10] Han, J., and W. B. Shen, 2012: Geoid determination with different density hypotheses: A case study in the Xinjiang and Tibetan region. Presented at the 3rd TibXS, August 26-29, 2012, Chengdu, China.

[11] Heck, B., and K. Seitz, 2007: A comparison of the tesseroid, prism and point-mass methodes for mass reductions in gravity field modeling. *J Geod.*, 81, 121-136.

[12] Heiskanen, W. A., and H. Moritz, 1967: Physical geodesy. Freeman and Company, San Francisco

[13] Hofmann-Wellenhof, B., and H. Moritz, 2005: Physical Geodesy. Springer, Vienna and New York.

[14] Hu, G. R., 2011: GPS/leveling data sets of Australia. Personal communication, 2011.

[15] Kiamehr, R., and L. E. Sjöberg, 2005: Effect of the SRTM global DEM on the determination of a high-resolution geoid model: a case study in Iran. *J Geod.*, 79, 540-551.

[16] Kenyon, S., J. Factor, N. Pavlis, and S. Holmes, 2007: Towards the next Earth gravitational model. Society of Exploration Geophysicists 77th Annual Meeting 2007, San Antonio, Texas, USA, September 23-28.

[17] Kong, X. Y., J. M. Guo, and Z. Q. Liu, 2001: Foundation of the Geodesy(1st ed). Wuhan University Press, Wuhan. (in Chinese)

[18] Laske, G., 2011: The release date of CRUST1.0. Personal communication.

[19] Listing, J. B., 1872: Regarding our present knowledge of the figure and size of the earth. Rep Roy Soc Sci Gottingen66.

[20] Mayer-Gürr T., et al., 2012: The new combined satellite only model GOCO03s. Abstract submitted to GGHS2012, Venice (Poster).

[21] Merry, C. L., 2003: DEM-induced errors in developing a quasi-geoid model for Africa. *J Geod.*, 77, 537-542.

[22] Mohr, P. J., B. N. Taylor, and D. B. Newell, 2008: CODATA recommended values of the fundamental physical constants: 2006. *Rev. Mod. Phys.*, 80, 633-730.

[23] Molodensky, M. S., V. F. Eremeev and M. I. Yurkina, 1962: Methods for study of the external gravitational field and figure of the Earth. Israeli Programme for the Translation of Scientific Publications, Jerusalem.

[24] Mooney, W. D., G. Laske, and G. T. Masters, 1998: CRUST5.1: A global crustal model at 5°×5°. *J. Geophys. Res.*, 103, 727-747.

[25] Nagy, D., G. Papp, and J. Benedek, 2000: The gravitational potential and its derivatives for the prism. *J Geod.*,74, 552-560.

[26] Nagy, D., G. Papp, and J. Benedek, 2002: Corrections to "The gravitational potential and its derivatives for the prism". *J Geod.*,76, 475.

[27] NGS, 2009: GPS on bench marks (GPSBM) used to make GEOID09. http:// www.ngs.noaa. gov/GEOID/GPSonBM09/

[28] Pavlis, N. K., J. K. Factor, and S. A. Holmes, 2007: Terrain-related gravimetric quantities computed for the next EGM. In: Proceedings of the 1st International Symposium of the International Gravity Field Service Vol. 18. Harita Dergisi, Istanbul, pp 318-323.

[29] Pavlis, N .K., S. A. Holmes, S. C. Kenyon, and J. K. Factor, 2008: An Earth gravitational model to degree 2160: EGM2008. Presented at the 2008 General Assembly of the European Geosciences Union, Vienna, 13-18 April 2008.

[30] Pavlis, N. K., S. A. Holmes, S. C. Kenyon, and J. K. Factor, 2012: The development and evaluation of the Earth Gravitational Model 2008 (EGM2008), J. Geophys. Res., 117, B04406, doi:10.1029/2011JB008916

[31] Shen, W. B., 2006: An approach for determining the precise global geoid. Presented at 1st International Symposium of the IGFS, Aug. 30 - Sept. 2, 2006, Istanbul.

[32] Shen, W.B., 2007: An approach for determining the precise global geoid. Proceedings of the 1st International Symposium of the International Gravity Field Service Vol. 18. Harita Dergisi, Istanbul, pp. 318-323

[33] Shen, W. B., and J. Han, 2012a: Improved geoid determination based on the shallow layer method: A case study using EGM2008 and CRUST2.0 in the Xinjiang and Tibetan regions. *Terrestrial, Atmospheric and Oceanic Sciences.* (accepted)

[34] Shen, W. B., and J. Han, 2012b. The 30'×30' global geoid model determined based on a new method and its validation. Geomatics and Information Science of Wuhan University, 37:1135-1139. (in Chinese)

[35] Shen, W. B., and J. Han, 2012c: An idea for refining the crust density model based on gravimetric information and GPS benchmarks. Presented at the 3rd TibXS, August 26-29, 2012, Chengdu, China.

[36] Stokes, G. G., 1849: On the variation of gravity at the surface of the Earth. *Trans Cambridge Phil. Soc.*, 672.

[37] Tapley B., J. Ries, S. Bettadpur, D. Chambers, M. Cheng, F. Condi, S. Poole, 2007: The GGM03 Mean Earth Gravity Model from GRACE, Eos Trans. AGU, 88(52), Fall Meet. Suppl., Abstract G42A-03.

[38] Tsoulis, D., 2004: Spherical harmonic analysis of the CRUST 2.0 global crustal model. *J Geod.*, 78: 7-11.

[39] Tsoulis, D., P. Novák, and M. Kadlec, 2009: Evaluation of precise terrain effects using high-resolution digital elevation models. *J. Geophys. Res.*, 114, 02404.

Permissions

The contributors of this book come from diverse backgrounds, making this book a truly international effort. This book will bring forth new frontiers with its revolutionizing research information and detailed analysis of the nascent developments around the world.

We would like to thank Shuanggen Jin, for lending his expertise to make the book truly unique. He has played a crucial role in the development of this book. Without his invaluable contribution this book wouldn't have been possible. He has made vital efforts to compile up to date information on the varied aspects of this subject to make this book a valuable addition to the collection of many professionals and students.

This book was conceptualized with the vision of imparting up-to-date information and advanced data in this field. To ensure the same, a matchless editorial board was set up. Every individual on the board went through rigorous rounds of assessment to prove their worth. After which they invested a large part of their time researching and compiling the most relevant data for our readers. Conferences and sessions were held from time to time between the editorial board and the contributing authors to present the data in the most comprehensible form. The editorial team has worked tirelessly to provide valuable and valid information to help people across the globe.

Every chapter published in this book has been scrutinized by our experts. Their significance has been extensively debated. The topics covered herein carry significant findings which will fuel the growth of the discipline. They may even be implemented as practical applications or may be referred to as a beginning point for another development. Chapters in this book were first published by InTech; hereby published with permission under the Creative Commons Attribution License or equivalent.

The editorial board has been involved in producing this book since its inception. They have spent rigorous hours researching and exploring the diverse topics which have resulted in the successful publishing of this book. They have passed on their knowledge of decades through this book. To expedite this challenging task, the publisher supported the team at every step. A small team of assistant editors was also appointed to further simplify the editing procedure and attain best results for the readers.

Our editorial team has been hand-picked from every corner of the world. Their multi-ethnicity adds dynamic inputs to the discussions which result in innovative outcomes. These outcomes are then further discussed with the researchers and contributors who give their valuable feedback and opinion regarding the same. The feedback is then collaborated with the researches and they are edited in a comprehensive manner to aid the understanding of the subject.

Apart from the editorial board, the designing team has also invested a significant amount of their time in understanding the subject and creating the most relevant covers. They scrutinized every image to scout for the most suitable representation of the subject and create an appropriate cover for the book.

The publishing team has been involved in this book since its early stages. They were actively engaged in every process, be it collecting the data, connecting with the contributors or procuring relevant information. The team has been an ardent support to the editorial, designing and production team. Their endless efforts to recruit the best for this project, has resulted in the accomplishment of this book. They are a veteran in the field of academics and their pool of knowledge is as vast as their experience in printing. Their expertise and guidance has proved useful at every step. Their uncompromising quality standards have made this book an exceptional effort. Their encouragement from time to time has been an inspiration for everyone.

The publisher and the editorial board hope that this book will prove to be a valuable piece of knowledge for researchers, students, practitioners and scholars across the globe.

List of Contributors

Robert Heinkelmann
Deutsches Geodätisches Forschungsinstitut (DGFI), Bayerische Akademie der Wissenschaften, Munich, Germany

Kewen Sun
Hefei University of Technology, China

Marios Smyrnaios and Steffen Schön
Institut für Erdmessung, Leibniz-Universität Hannover, Hannover, Germany

Marcos Liso Nicolás
Institut für Nachrichtentechnik Technische Universität Braunschweig, Braunschweig, Germany

Shuanggen Jin
Shanghai Astronomical Observatory, Chinese Academy of Sciences, China

Vyacheslav Kunitsyn, Elena Andreeva, Ivan Nesterov and Artem Padokhin
M. Lomonosov Moscow State University, Faculty of Physics, Moscow, Russia

Vladislav V. Demyanov
Irkutsk State Railway University, Russia

Yury V. Yasyukevich
Institute of Solar-Terrestrial Physics, the Russian Academy of Sciences, the Siberian Branch, Irkutsk, Russia

Shuanggen Jin
Shanghai Astronomical Observatory, Chinese Academy of Sciences, Shanghai, China

Jiuhou Lei and Xiankang Dou
CAS Key Laboratory of Geospace Environment, University of Science and Technology of China, Hefei, Anhui, China

Guangming Chen and Jiyao Xu
State Key Laboratory for Space Weather, Center for Space Sciences and Applied Research, Chinese Academy of Sciences, Beijing, China

Sung-Ho Na
Korea Astronomy and Space Science Institute, Korea

WenBin Shen and Jiancheng Han
Wuhan University, China